About IFPRI

The International Food Policy Research Institute (IFPRI), established in 1975, provides research-based policy solutions to sustainably reduce poverty and end hunger and malnutrition. The Institute conducts research, communicates results, optimizes partnerships, and builds capacity to ensure sustainable food production, promote healthy food systems, improve markets and trade, transform agriculture, build resilience, and strengthen institutions and governance. Gender is considered in all of the Institute's work. IFPRI collaborates with partners around the world, including development implementers, public institutions, the private sector, and farmers' organizations. IFPRI is a member of the CGIAR Consortium.

About IFPRI's Peer Review Process

IFPRI books are policy-relevant publications based on original and innovative research conducted at IFPRI. All manuscripts submitted for publication as IFPRI books undergo an extensive review procedure that is managed by IFPRI's Publications Review Committee (PRC). Upon submission to the PRC, the manuscript is reviewed by a PRC member. Once the manuscript is considered ready for external review, the PRC submits it to at least two external reviewers who are chosen for their familiarity with the subject matter and the country setting. Upon receipt of these blind external peer reviews, the PRC provides the author with an editorial decision and, when necessary, instructions for revision based on the external reviews. The PRC reassesses the revised manuscript and makes a recommendation regarding publication to the director general of IFPRI. With the director general's approval, the manuscript enters the editorial and production phase to become an IFPRI book.

Southern African Agriculture and Climate Change
A Comprehensive Analysis

Edited by Sepo Hachigonta, Gerald C. Nelson,
Timothy S. Thomas, and Lindiwe Majele Sibanda

With the financial support of

A peer-reviewed publication

International Food Policy Research Institute
Washington, DC

Copyright © 2013 International Food Policy Research Institute.

All rights reserved. Sections of this material may be reproduced for personal and not-for-profit use without the express written permission of but with acknowledgment to IFPRI. To reproduce material contained herein for profit or commercial use requires express written permission. To obtain permission, contact the Communications Division at ifpri-copyright@cgiar.org.

The opinions expressed in this book are those of the authors and do not necessarily reflect the policies of their host institutions.

International Food Policy Research Institute
2033 K Street, NW
Washington, DC 20006-1002, USA
Telephone: +1-202-862-5600
www.ifpri.org

DOI: http://dx.doi.org/10.2499/9780896292086

Library of Congress Cataloging-in-Publication Data

Southern African agriculture and climate change : a comprehensive analysis / edited by Sepo Hachigonta . . . [et al]. — 1st ed.
   p.   cm. — (Climate change in Africa ; 3)
Includes bibliographical references and index.
ISBN 978-0-89629-208-6 (alk. paper)
  1. Agriculture—Africa, Southern. 2. Climatic changes—Africa, Southern. 3. Food crops—Yields—Africa, Southern. 4. Africa, Southern—Population. I. Hachigonta, Sepo. II. Series: Climate change in Africa ; 3.
S472.A356S68   2013
635.0968—dc23   2013021765

Cover design: Carolyn Hallowell
Book layout: Princeton Editorial Associates Inc., Scottsdale, Arizona

Contents

	Figures	vii
	Tables	xix
	Foreword	xxiii
	Acknowledgments	xxv
	Abbreviations and Acronyms	xxvii
Chapter 1	**Overview** Sepo Hachigonta, Gerald C. Nelson, Timothy S. Thomas, and Lindiwe Majele Sibanda	1
Chapter 2	**Methodology** Gerald C. Nelson, Amanda Palazzo, Daniel Mason-d'Croz, Richard Robertson, and Timothy S. Thomas	25
Chapter 3	**Botswana** Peter P. Zhou, Tichakunda Simbini, Gorata Ramokgotlwane, Timothy S. Thomas, Sepo Hachigonta, and Lindiwe Majele Sibanda	41
Chapter 4	**Lesotho** Patrick Gwimbi, Timothy S. Thomas, Sepo Hachigonta, and Lindiwe Majele Sibanda	71
Chapter 5	**Malawi** John D. K. Saka, Pickford Sibale, Timothy S. Thomas, Sepo Hachigonta, and Lindiwe Majele Sibanda	111

| Chapter 6 | **Mozambique** | 147 |

Genito A. Maure, Timothy S. Thomas, Sepo Hachigonta, and Lindiwe Majele Sibanda

| Chapter 7 | **South Africa** | 175 |

Peter Johnston, Timothy S. Thomas, Sepo Hachigonta, and Lindiwe Majele Sibanda

| Chapter 8 | **Swaziland** | 213 |

Absalom M. Manyatsi, Timothy S. Thomas, Michael T. Masarirambi, Sepo Hachigonta, and Lindiwe Majele Sibanda

| Chapter 9 | **Zambia** | 255 |

Joseph Kanyanga, Timothy S. Thomas, Sepo Hachigonta, and Lindiwe Majele Sibanda

| Chapter 10 | **Zimbabwe** | 289 |

Francis T. Mugabe, Timothy S. Thomas, Sepo Hachigonta, and Lindiwe Majele Sibanda

| | **Contributors** | 325 |
| | **Index** | 327 |

Figures

1.1	Annual average precipitation in southern Africa, 1950–2000 (millimeters per year)	3
1.2	Land cover and land use in southern Africa, 2000	4
1.3	Yields for the main rainfed crops in southern Africa, 2000 (metric tons per hectare)	7
1.4	Travel time to cities of 500,000 or more people in southern Africa, circa 2000	10
1.5	Changes in mean annual precipitation in southern Africa, 2000–2050, A1B scenario (millimeters)	13
1.6	Change in monthly mean maximum temperature in southern Africa for the warmest month, 2000–2050, A1B scenario	14
2.1	The International Model for Policy Analysis of Agricultural Commodities and Trade (IMPACT) modeling framework	26
2.2	International Model for Policy Analysis of Agricultural Commodities and Trade (IMPACT) unit of analysis, the food production unit (FPU)	31
2.3	Sample box-and-whisker graph	38
3.1	Population trends in Botswana: Total population, rural population, and percent urban, 1960–2008	45
3.2	Poverty rates in Botswana by region and urban/rural, 2003	47
3.3	Land cover and land use in Botswana, 2000	49
3.4	Protected areas in Botswana, 2009	50
3.5	Travel time to urban areas of various sizes in Botswana, circa 2000	51

3.6	Population projections for Botswana, 2010–2050	55
3.7	Gross domestic product (GDP) per capita in Botswana, future scenarios, 2010–2050	55
3.8	Changes in mean annual precipitation in Botswana, 2000–2050, A1B scenario (millimeters)	57
3.9	Change in monthly mean maximum daily temperature in Botswana for the warmest month, 2000–2050, A1B scenario (°C)	58
3.10	Yield change under climate change: Rainfed maize in Botswana, 2000–2050, A1B scenario	60
3.11	Yield change under climate change: Rainfed sorghum in Botswana, 2000–2050, A1B scenario	61
3.12	Impact of changes in GDP and population on maize in Botswana, 2010–2050	63
3.13	Impact of changes in GDP and population on sorghum in Botswana, 2010–2050	64
3.14	Number of malnourished children under five years of age in Botswana in multiple income and climate scenarios, 2010–2050	65
3.15	Share of malnourished children under five years of age in Botswana in multiple income and climate scenarios, 2010–2050	65
3.16	Kilocalories per capita in Botswana in multiple income and climate scenarios, 2010–2050	66
4.1	Population trends in Lesotho: Total population, rural population, and percent urban, 1960–2008	75
4.2	Population distribution in Lesotho, 2000 (persons per square kilometer)	76
4.3	Per capita GDP in Lesotho (constant 2000 US$) and share of GDP from agriculture (percent), 1960–2008	77
4.4	Well-being indicators in Lesotho, 1960–2008	78
4.5	Land cover and land use in Lesotho, 2000	81
4.6	Protected areas in Lesotho, 2009	82
4.7	Travel time to urban areas of various sizes in Lesotho, circa 2000	83
4.8	Population projections for Lesotho, 2010–2050	88
4.9	Gross domestic product (GDP) per capita in Lesotho, future scenarios, 2010–2050	89

4.10	Changes in mean annual precipitation in Lesotho, 2000–2050, A1B scenario (millimeters)	90
4.11	Change in monthly mean maximum daily temperature in Lesotho for the warmest month, 2000–2050, A1B scenario (°C)	92
4.12	Yield change under climate change: Rainfed maize in Lesotho, 2000–2050, A1B scenario	94
4.13	Yield change under climate change: Rainfed sorghum in Lesotho, 2000–2050, A1B scenario	95
4.14	Yield change under climate change: Rainfed wheat in Lesotho, 2000–2050, A1B scenario	96
4.15	Number of malnourished children under five years of age in Lesotho in multiple income and climate scenarios, 2010–2050	98
4.16	Share of malnourished children under five years of age in Lesotho in multiple income and climate scenarios, 2010–2050	99
4.17	Kilocalories per capita in Lesotho in multiple income and climate scenarios, 2010–2050	99
4.18	Impact of changes in GDP and population on maize in Lesotho, 2010–2050	101
4.19	Impact of changes in GDP and population on sorghum in Lesotho, 2010–2050	102
4.20	Impact of changes in GDP and population on wheat in Lesotho, 2010–2050	103
5.1	Population trends in Malawi: Total population, rural population, and percent urban, 1960–2008	114
5.2	Population distribution in Malawi, 2000 (persons per square kilometer)	116
5.3	Per capita GDP in Malawi (constant 2000 US$) and share of GDP from agriculture (percent), 1960–2008	117
5.4	Well-being indicators in Malawi, 1960–2008	118
5.5	Poverty in Malawi, circa 2005 (percentage of population below US$2 per day)	118
5.6	Land cover and land use in Malawi, 2000	119
5.7	Protected areas in Malawi, 2009	121
5.8	Travel time to urban areas of various sizes in Malawi, circa 2000	122

5.9	Yield (metric tons per hectare) and harvest area density (hectares) for rainfed maize in Malawi, 2000	125
5.10	Yield (metric tons per hectare) and harvest area density (hectares) for rainfed cassava in Malawi, 2000	126
5.11	Yield (metric tons per hectare) and harvest area density (hectares) for rainfed cotton in Malawi, 2000	126
5.12	Yield (metric tons per hectare) and harvest area density (hectares) for rainfed groundnuts in Malawi, 2000	127
5.13	Yield (metric tons per hectare) and harvest area density (hectares) for rainfed beans in Malawi, 2000	127
5.14	Population projections for Malawi, 2010–2050	128
5.15	Gross domestic product (GDP) per capita in Malawi, future scenarios, 2010–2050	129
5.16	Changes in mean annual precipitation in Malawi, 2000–2050, A1B scenario (millimeters)	130
5.17	Change in monthly mean maximum daily temperature in Malawi for the warmest month, 2000–2050, A1B scenario (°C)	132
5.18	Yield change under climate change: Rainfed maize in Malawi, 2000–2050, A1B scenario	133
5.19	Number of malnourished children under five years of age in Malawi in multiple income and climate scenarios, 2010–2050	134
5.20	Share of malnourished children under five years of age in Malawi in multiple income and climate scenarios, 2010–2050	135
5.21	Kilocalories per capita in Malawi in multiple income and climate scenarios, 2010–2050	135
5.22	Impact of changes in GDP and population on maize in Malawi, 2010–2050	137
5.23	Impact of changes in GDP and population on cassava in Malawi, 2010–2050	138
5.24	Impact of changes in GDP and population on cotton in Malawi, 2010–2050	139
6.1	Population trends in Mozambique: Total population, rural population, and percent urban, 1960–2008	149
6.2	Population distribution in Mozambique, 2000 (persons per square kilometer)	150

6.3	Per capita GDP in Mozambique (constant 2000 US$) and share of GDP from agriculture (percent), 1960–2008	151
6.4	Well-being indicators in Mozambique, 1960–2008	152
6.5	Poverty in Mozambique, circa 2005 (percentage of population below US$2 per day)	153
6.6	Land cover and land use in Mozambique, 2000	154
6.7	Protected areas in Mozambique, 2009	155
6.8	Travel time to urban areas of various sizes in Mozambique, circa 2000	156
6.9	Yield (metric tons per hectare) and harvest area density (hectares) for rainfed maize in Mozambique, 2000	159
6.10	Yield (metric tons per hectare) and harvest area density (hectares) for rainfed cassava in Mozambique, 2000	160
6.11	Population projections for Mozambique, 2010–2050	161
6.12	Gross domestic product (GDP) per capita in Mozambique, future scenarios, 2010–2050	161
6.13	Change in mean annual precipitation in Mozambique, 2000–2050, A1B scenario (millimeters)	163
6.14	Change in monthly mean maximum daily temperature in Mozambique for the warmest month, 2000–2050, A1B scenario (°C)	164
6.15	Yield change under climate change: Rainfed maize in Mozambique, 2000–2050, A1B scenario	165
6.16	Number of malnourished children under five years of age in Mozambique in multiple income and climate scenarios, 2010–2050	167
6.17	Share of malnourished children under five years of age in Mozambique in multiple income and climate scenarios, 2010–2050	167
6.18	Kilocalories per capita in Mozambique in multiple income and climate scenarios, 2010–2050	168
6.19	Impact of changes in GDP and population on rainfed maize in Mozambique, 2010–2050	169
6.20	Impact of changes in GDP and population on rainfed cassava in Mozambique, 2010–2050	170

7.1	Population trends in South Africa: Total population, rural population, and percent urban, 1960–2008	176
7.2	Population distribution in South Africa, 2000 (persons per square kilometer)	177
7.3	Per capita GDP in South Africa (constant 2000 US$) and share of GDP from agriculture (percent), 1960–2008	177
7.4	Well-being indicators in South Africa, 1960–2008: Life expectancy and under-five mortality rate	179
7.5	Well-being indicators in South Africa: Prevalence of HIV infection, 1990–2008	179
7.6	Poverty in South Africa, circa 2005 (percentage of population below US$2 per day)	180
7.7	Land cover and land use in South Africa, 2000	181
7.7	Protected areas in South Africa, 2009	182
7.9	Travel time to urban areas of various sizes in South Africa, circa 2000	183
7.10	Value of production of agricultural commodities by type in South Africa, 2007/2008–2011/2012	187
7.11	Gross value of food consumption expenditure in South Africa, 2007/2008–2011/2012	187
7.12	Yield (metric tons per hectare) and harvest area density (hectares) for maize in South Africa, 2005	188
7.13	Area planted with maize, South Africa, 1993/1994–2010/2011 (thousands of hectares)	190
7.14	Maize production, South Africa, 1993/1994–2010/2011 (thousands of metric tons)	190
7.15	Yield (metric tons per hectare) and harvest area density (hectares) for wheat in South Africa, 2005	191
7.16	Area planted, production, and yields for wheat in South Africa, 1990/1991–2010/2011	192
7.17	Yield (kilograms per hectare) and harvest area density (hectares) for sugarcane in South Africa, 2005	193
7.18	Population projections for South Africa, 2010–2050	195
7.19	Gross domestic product (GDP) per capita in South Africa, future scenarios, 2010–2050	195

7.20	Change in monthly mean maximum daily temperature in South Africa for the warmest month, 2000–2050, A1B scenario (°C)	197
7.21	Yield change under climate change: Rainfed maize in South Africa, 2000–2050, A1B scenario	199
7.22	Yield change under climate change: Rainfed wheat in South Africa (excluding Western Cape), 2000–2050, A1B scenario	200
7.23	Impact of changes in GDP and population on maize in South Africa, 2010–2050	202
7.24	Impact of changes in GDP and population on wheat in South Africa, 2010–2050	203
7.25	Trends in wheat production and area planted in South Africa, 1970/1971–2011/2012	204
7.26	Impact of changes in GDP and population on sugarcane in South Africa, 2010–2050	205
7.27	Number of malnourished children under five years of age in South Africa in multiple income and climate scenarios, 2010–2050	206
7.28	Share of malnourished children under five years of age in South Africa in multiple income and climate scenarios, 2010–2050	206
7.29	Kilocalories per capita in South Africa in multiple income and climate scenarios, 2010–2050	207
8.1	Population trends in Swaziland: Total population, rural population, and percent urban, 1960–2008	214
8.2	Population distribution in Swaziland, 2000 (persons per square kilometer)	215
8.3	Per capita GDP in Swaziland (constant 2000 US$) and share of GDP from agriculture (percent), 1960–2008	217
8.4	Maize production in Swaziland, 2004/2005–2009/2010	218
8.5	Well-being indicators in Swaziland, 1960–2008	219
8.6	Land cover and land use in Swaziland, 2000	221
8.7	Travel time to urban areas of various sizes in Swaziland, circa 2000	224
8.8	Yield (metric tons per hectare) and harvest area density (hectares) for rainfed maize in Swaziland, 2000	227

8.9	Yield (metric tons per hectare) and harvest area density (hectares) for irrigated sugarcane in Swaziland, 2000	228
8.10	Population projections for Swaziland, 2010–2050	229
8.11	Gross domestic product (GDP) per capita in Swaziland, future scenarios, 2010–2050	230
8.12	Change in mean annual precipitation in Swaziland, 2000–2050, A1B scenario (millimeters)	231
8.13	Change in monthly mean maximum daily temperature in Swaziland for the warmest month, 2000–2050, A1B scenario (°C)	233
8.14	Yield change under climate change: Rainfed maize in Swaziland, 2000–2050, A1B scenario	234
8.15	Impact of changes in GDP and population on maize in Swaziland, 2010–2050	236
8.16	Impact of changes in GDP and population on sugarcane in Swaziland, 2010–2050	237
8.17	Impact of changes in GDP and population on cotton in Swaziland, 2010–2050	238
8.18	Number of malnourished children under five years of age in Swaziland in multiple income and climate scenarios, 2010–2050	240
8.19	Share of malnourished children under five years of age in Swaziland in multiple income and climate scenarios, 2010–2050	240
8.20	Kilocalories per capita in Swaziland in multiple income and climate scenarios, 2010–2050	241
9.1	Population trends in Zambia: Total population, rural population, and percent urban, 1960–2008	256
9.2	Population distribution in Zambia, 2000 (persons per square kilometer)	257
9.3	Per capita GDP in Zambia (constant 2000 US$) and share of GDP from agriculture (percent), 1960–2008	258
9.4	Well-being indicators in Zambia, 1960–2008	260
9.5	Poverty in Zambia, circa 2005 (percentage of population below US$2 per day)	261
9.6	Land cover and land use in Zambia, 2000	262
9.7	Protected areas in Zambia, 2009	264

9.8	Travel time to urban areas of various sizes in Zambia, circa 2000	265
9.9	Yield (metric tons per hectare) and harvest area density (hectares) for rainfed maize in Zambia, 2000	268
9.10	Yield (metric tons per hectare) and harvest area density (hectares) for rainfed cassava in Zambia, 2000	269
9.11	Yield (metric tons per hectare) and harvest area density (hectares) for rainfed cotton in Zambia, 2000	269
9.12	Yield (metric tons per hectare) and harvest area density (hectares) for rainfed groundnuts in Zambia, 2000	270
9.13	Population projections for Zambia, 2010–2050	271
9.14	Gross domestic product (GDP) per capita in Zambia, future scenarios, 2010–2050	271
9.15	Changes in mean annual precipitation in Zambia, 2000–2050, A1B scenario (millimeters)	273
9.16	Changes in monthly mean maximum daily temperature in Zambia for the warmest month, 2000–2050, A1B scenario (°C)	274
9.17	Yield change under climate change: Rainfed maize in Zambia, 2000–2050, A1B scenario	276
9.18	Number of malnourished children under five years of age in Zambia in multiple income and climate scenarios, 2010–2050	277
9.19	Share of malnourished children under five years of age in Zambia in multiple income and climate scenarios, 2010–2050	277
9.20	Kilocalories per capita in Zambia in multiple income and climate scenarios, 2010–2050	278
9.21	Impact of changes in GDP and population on maize in Zambia, 2010–2050	279
9.22	Impact of changes in GDP and population on cassava in Zambia, 2010–2050	280
9.23	Impact of changes in GDP and population on cotton in Zambia, 2010–2050	281
10.1	Population trends in Zimbabwe: Total population, rural population, and percent urban, 1960–2008	291
10.2	Population distribution in Zimbabwe, 2000 (persons per square kilometer)	291

10.3	Per capita GDP in Zimbabwe (constant 2000 US$) and share of GDP from agriculture, 1960–2008 (percent)	293
10.4	Well-being indicators in Zimbabwe, 1960–2008	294
10.5	Land cover and land use in Zimbabwe, 2000	296
10.6	Protected areas in Zimbabwe, 2009	297
10.7	Travel time to urban areas of various sizes in Zimbabwe, circa 2000	298
10.8	Yield (metric tons per hectare) and harvest area density (hectares) for rainfed maize in Zimbabwe, 2000	300
10.9	Yield (metric tons per hectare) and harvest area density (hectares) for rainfed cotton in Zimbabwe, 2000	300
10.10	Yield (metric tons per hectare) and harvest area density (hectares) for rainfed sorghum in Zimbabwe, 2000	301
10.11	Yield (metric tons per hectare) and harvest area density (hectares) for rainfed millet in Zimbabwe, 2000	301
10.12	Yield (metric tons per hectare) and harvest area density (hectares) for rainfed groundnuts in Zimbabwe, 2000	302
10.13	Population projections for Zimbabwe, 2010–2050	303
10.14	Gross domestic product (GDP) per capita in Zimbabwe, future scenarios, 2010–2050	303
10.15	Changes in mean annual precipitation in Zimbabwe, 2000–2050, A1B scenario (millimeters)	305
10.16	Changes in monthly mean maximum daily temperature in Zimbabwe for the warmest month, 2000–2050, A1B scenario (°C)	306
10.17	Yield change under climate change: Rainfed maize in Zimbabwe, 2000–2050, A1B scenario	307
10.18	Yield change under climate change: Rainfed sorghum in Zimbabwe, 2000–2050, A1B scenario	308
10.19	Impact of changes in GDP and population on maize in Zimbabwe, 2010–2050	310
10.20	Impact of changes in GDP and population on cotton in Zimbabwe, 2010–2050	311
10.21	Impact of changes in GDP and population on sorghum in Zimbabwe, 2010–2050	312

10.22	Impact of changes in GDP and population on millet in Zimbabwe, 2010–2050	313
10.23	Number of malnourished children under five years of age in Zimbabwe in multiple income and climate scenarios, 2010–2050	314
10.24	Share of malnourished children under five years of age in Zimbabwe in multiple income and climate scenarios, 2010–2050	314
10.25	Kilocalories per capita in Zimbabwe in multiple income and climate scenarios, 2010–2050	315

Tables

1.1	Average harvest area of leading agricultural commodities in southern Africa, 2006–2008 (hectares)	5
1.2	Population of southern Africa, annualized growth rate, and urban growth rate, 1988 and 2008	8
1.3	Income of southern Africans (GDP per capita and share of GDP from agriculture), 1988 and 2008	9
1.4	Under-five mortality and life expectancy at birth in southern Africa, 1988 and 2008	10
1.5	Projected population in southern Africa, 2010–2050 (millions)	11
1.6	GDP per capita scenarios for southern Africa, 2010–2050 (constant 2000 US$)	12
1.7	Maize projections for southern Africa showing yield, area, and production, 2010–2050	16
1.8	Millet projections for southern Africa showing yield, area, and production, 2010–2050	17
1.9	Sorghum projections for southern Africa showing yield, area, and production, 2010–2050	18
2.1	GCM and SRES scenario global average changes, 2000–2050	28
2.2	Gross domestic product (GDP) and population choices for the three overall scenarios	33
2.3	Global average scenario per capita gross domestic product growth rate, 1990–2000 and 2010–2050 (percent per year)	33
2.4	Summary statistics for population and per capita gross domestic project, 2010 and 2050	34

2.5	Noncaloric determinants of global child malnutrition, 2010 and 2050	36
2.6	Mean price elasticities used for southern African countries in IMPACT, 2010 and 2050	37
3.1	Population growth rates in Botswana, 1960–2008 (percent)	46
3.2	Income and poverty indicators for Botswana, 1984/1985, 1993/1994, and 2002/2003	47
3.3	Agricultural and national gross domestic product (GDP) for Botswana, 2000–2007	52
3.4	Estimation of requirements, production, and deficits of maize and sorghum/millet in Botswana, 2001/2002–2006/2007 (metric tons)	53
3.5	Beef exports from Botswana to the EU, 2001–2007 (metric tons)	53
4.1	Population growth rates in Lesotho, 1960–2008 (percent)	74
4.2	Education and labor statistics for Lesotho, 1980s, 1990s, and 2000s	79
4.3	Harvest area of leading agricultural commodities in Lesotho, 2006–2008 (thousands of hectares)	85
4.4	Consumption of leading food commodities in Lesotho, 2003–2005 (thousands of metric tons)	85
5.1	Population growth rates in Malawi, 1960–2008 (percent)	114
5.2	Education and labor statistics for Malawi, 2000s	117
5.3	Harvest area of leading agricultural commodities in Malawi, 2006–2008 (thousands of hectares)	123
5.4	Value of production of leading agricultural commodities in Malawi, 2005–2007 (millions of US$)	124
5.5	Consumption of leading food commodities in Malawi, 2003–2005 (thousands of metric tons)	124
5.6	Climate change issues and recommendations from stakeholder consultations in Malawi	140
6.1	Population growth rates in Mozambique, 1960–2008 (percent)	149
6.2	Education and nutrition statistics for Mozambique, 2000s	151
6.3	Harvest area of leading agricultural commodities in Mozambique, 2006–2008 (thousands of hectares per year)	157

6.4	Value of production of leading agricultural commodities in Mozambique, 2005–2007 (millions of US$)	158
6.5	Consumption of leading food commodities in Mozambique, 2003–2005 (thousands of metric tons)	158
7.1	Population growth rates in South Africa, 1960–2008 (percent)	176
7.2	Education and labor statistics for South Africa, 2007	178
7.3	Value of production of leading agricultural commodities in South Africa, 2000 and 2008 (millions of US$)	186
7.4	Distribution of irrigated area in South Africa, by province, 1999 (hectares)	194
8.1	Population growth rates in Swaziland, 1960–2008 (percent)	214
8.2	Education and labor statistics for Swaziland, 2000s	219
8.3	Land use in each ecological zone of Swaziland, 1994	222
8.4	Area and production of different crops in Swaziland, 2006–2008	226
8.5	Consumption of leading food commodities in Swaziland, 2003–2005 (thousands of metric tons)	226
8.6	Livelihood strategies and coping mechanisms practiced by rural communities in Swaziland, 2010	243
8.7	Legislation in Swaziland relevant to climate change	246
9.1	Population growth rates in Zambia, 1960–2008 (percent)	256
9.2	Education and labor statistics for Zambia, 2000s	259
9.3	Harvest area of leading agricultural commodities in Zambia, 2006–2008 (thousands of hectares)	266
9.4	Consumption of leading food commodities in Zambia, 2003–2005 (thousands of metric tons)	267
10.1	Population growth rates in Zimbabwe, 1960–2008 (percent)	292
10.2	Education and labor statistics for Zimbabwe, 1900s and 2000s	294
10.3	Harvest area of leading agricultural commodities in Zimbabwe, 2006–2008 (thousands of hectares)	299
10.4	Consumption of leading food commodities in Zimbabwe, 2003–2005 (thousands of metric tons)	299
10.5	Civil society–driven projects and initiatives in Zimbabwe for strengthening the adaptive capacity of farmers and areas of implementation, 2000s and ongoing	316

Foreword

The world's population is projected to grow from 7 billion in 2012 to around 9 billion by 2050. In Africa south of the Sahara, the population is likely to surge from around 850 million today to around 1.7 billion in 2050. Southern Africa alone will make up almost 14 percent of the population of Africa south of the Sahara and almost 3 percent of the world's population in 2050. Most of the people comprising this population increase are expected to live in urban areas and to have higher incomes than currently is the case, which will result in increased demand for food. The challenge of meeting this food demand in a sustainable manner will be enormous. When one takes into account the effects of climate change (higher temperatures, shifting seasons, more frequent and extreme weather events, flooding, and drought) on food production, that challenge grows even more daunting. The global food price spikes of 2008, 2010, and 2012 are harbingers of a troubled future for global food security.

At the end of 2010, IFPRI published *Food Security, Farming, and Climate Change to 2050: Scenarios, Results, Policy Options,* a research monograph by Gerald Nelson and a team of IFPRI researchers that quantitatively assessed the additional challenges to sustainable food security that climate change would bring, focusing on global outcomes but also including national and subnational results. Three years later, Nelson and a group of leading agriculturists and climate change researchers have written this monograph, which draws out those national results based on a detailed global model and enhances them with country-specific analysis and insights for southern Africa.

This is one of three publications (covering West, East, and southern Africa) that make up IFPRI's *Climate Change in Africa* series. It provides the most comprehensive analysis to date of the scope of climate change as it relates to food security in southern Africa, including who will be most affected and what policymakers can do to facilitate adaptation. Augmenting the text are dozens of detailed maps that provide graphical representations of the range of food security challenges and the special threats from climate change.

Using a comprehensive integrated empirical analysis, the authors generated information to better guide national development agendas on climate change and have suggested that policymakers should (1) incorporate climate change adaptation strategies in short- and long-term national development planning; (2) develop national capacity in the skills and tools needed for technical assessments, planning, and policy development in the context of climate change; (3) promote sustainable agriculture initiatives that target vulnerable communities; and (4) enhance investments in relevant economic sectors, in particular the agricultural sector. *Southern African Agriculture and Climate Change* will be indispensable to a wide range of readers, including the policymakers, development workers, and researchers who tackle these inextricably linked issues.

Shenggen Fan
Director General, International Food Policy Research Institute

Acknowledgments

The editors of this monograph and the authors of the individual chapters thank the following organizations for their financial support: the EU through its support for the Climate Change, Agriculture, and Food Security Research Program of the CGIAR (CCAFS); the Federal Ministry for Economic Cooperation and Development, Germany; and the Bill and Melinda Gates Foundation and their respective home institutions. We also thank the Food, Agriculture, and Natural Resources Policy Analysis Network (FANRPAN), IFPRI, the Rockefeller Foundation, the International Development Research Centre (IDRC), and CCAFS for the encouragement they have provided. We give special recognition to Sepo Hachigonta of FANRPAN. He identified counterpart national scientists to undertake the national reports and provided invaluable intellectual leadership in managing the challenging process of coordinating and supporting many different authors while leading the development of the regional overview chapter. We thank Gerald C. Nelson and Lindiwe Majele Sibanda for having the vision and courage to commit to this initiative. Any errors or omissions remain the responsibility of the authors.

Abbreviations and Acronyms

A1B	greenhouse gas emissions scenario that assumes fast economic growth, a population that peaks midcentury, and the development of new and efficient technologies, along with a balanced use of energy sources
A2	greenhouse gas emissions scenario that assumes a very heterogeneous world with continuously increasing global population and regionally oriented economic growth that is more fragmented and slower than in other storylines
AGOA	African Growth and Opportunity Act
AIACC	Assessment of Impacts and Adaptation to Climate Change
AR4	Fourth Assessment Report of the Intergovernmental Panel on Climate Change
B1	greenhouse gas emissions scenario that assumes a population that peaks midcentury (like A1B) but with rapid changes toward a service and information economy and the introduction of clean and resource-efficient technologies
CAADP	Comprehensive African Agricultural Development Program
CANGO	Coordinating Assembly of Non Governmental Organizations
CCAA	Climate Change Adaptation in Africa
CCAFS	The Climate Change, Agriculture, and Food Security Research Program of the CGIAR

CNR	National Meteorological Research Center, France
CNRM-CM3	National Meteorological Research Center–Climate Model 3
CSIRO	Commonwealth Scientific and Industrial Research Organisation, Australia
CSIRO MARK 3	Climate model developed at the Australia Commonwealth Scientific and Industrial Research Organization
DFID	Department for International Development
DM	Disaster Management
DSSAT	Decision Support Software for Agrotechnology Transfer
ECHAM 5	fifth-generation climate model developed at the Max Planck Institute for Meteorology (Hamburg)
FANRPAN	Food, Agriculture, and Natural Resources Policy Analysis Network
FAO	Food and Agriculture Organization of the United Nations
FMD	foot and mouth disease
FPU	food production unit
GCM	general circulation model
GDP	gross domestic product
IAM	integrated assessment model
IDRC	International Development Research Centre
IFPRI	International Food Policy Research Institute
IFRC	International Federation of Red Cross and Red Crescent Societies
IMPACT	International Model for Policy Analysis of Agricultural Commodities and Trade
IPCC	Intergovernmental Panel on Climate Change
IUCN	International Union for Conservation of Nature
KDDP	Komati Downstream Development Project
LHWP	Lesotho Highlands Water Project
LIMID	Livestock Management and Infrastructure Development Program

LUSIP	Lower Usuthu Smallholder Irrigation Project
MDG	Millennium Development Goal
MIROC 3.2	Model for Interdisciplinary Research on Climate, developed at the University of Tokyo Center for Climate System Research
MOA	Ministry of Agriculture
NAMBOARD	National Agricultural Marketing Board
NAMPAADD	National Master Plan for Arable Agriculture and Dairy Development
NAPA	National Adaptation Programme of Action [on Climate Change]
NDP	National Development Plan
NGO	nongovernmental organization
NMC	National Maize Corporation
OVCS	orphaned and vulnerable children
R	South African rand
SACU	South African Customs Union
SADC	Southern Africa Development Community
SADP	Swaziland Agricultural Development Program
SARDC	Southern African Research and Documentation Centre
SCCP	Swaziland Climate Change Programme
SCF	seasonal climate forecast
SNL	Swazi Nation land
SPAM	Spatial Production Allocation Model
SRES	Special Report on Emissions Scenarios, a report by the Intergovernmental Panel on Climate Change that was published in 2000
SSA	Swaziland Sugar Association
SWADE	Swaziland Water and Agricultural Development Enterprise
TDL	title deed land
UN	United Nations

UNDP	United Nations Development Programme
UNFCCC	United Nations Framework Convention on Climate Change
UNPOP	United Nations Department of Economic and Social Affairs–Population Division
US$	US dollars
UZ	University of Zimbabwe

Chapter 1

OVERVIEW

Sepo Hachigonta, Gerald C. Nelson, Timothy S. Thomas, and Lindiwe Majele Sibanda

According to the Fourth Assessment Report of the Intergovernmental Panel on Climate Change (IPCC) (Le Treut et al. 2007, 96), climate is defined as average weather over a period of time, ranging from months to millions of years. Climate is usually described in terms of the mean and variability of temperature, precipitation, and wind. The IPCC defines an extreme climate event as an event that is rare within its statistical reference distribution at a particular place.

Extreme weather events (such as droughts, floods, and changes in the frequency and intensity of dry spells) already negatively affect agriculture in most parts of Africa. Higher temperatures tend to reduce yields of crops by reducing soil moisture content and the length of the growing season, and in most places they tend to encourage weed and pest proliferation. Greater variations in precipitation patterns increase the likelihood of crop failures and long-run production declines (Nelson et al. 2009).

There is a growing concern that climate change will intensify existing agricultural problems in developing countries, where communities are directly dependent on the natural environment and are underresourced to adequately adapt to extreme changes in climate (Meinke et al. 2006; Ziervogel et al. 2008). This is particularly true for those communities that rely on rainfed agriculture for their livelihoods. Some of the projected climate change impacts are

- direct—on crops and livestock productivity,
- indirect—on availability and prices of food, both domestic and in international markets, and
- indirect—on income generated from agricultural production at farm and national levels.

This monograph builds on previous research that focused on regional and global effects of climate change, including Nelson et al. (2009, 2010) and ADB and IFPRI (2009). This chapter provides a regional overview for southern Africa. Eight chapters look at the effects of climate change on eight countries in southern Africa: Botswana, Lesotho, Malawi, Mozambique, South Africa, Swaziland, Zambia, and Zimbabwe.

The main objective of this study is to analyze the range of plausible impacts of climate change by the year 2050, focusing almost entirely on crops. We use both crop models and global partial equilibrium models, informed by four different climate models and three socioeconomic scenarios. In some of the chapters on countries in which the livestock sector is important, we highlight some key aspects of this sector using secondary literature. We discuss agricultural adaptation options throughout this monograph.

Our goal is to provide policymakers and others concerned with climate change, agriculture, and food policy with guidance on the range of the impacts of climate change and some information as to how climate change might affect various regions differently. We provide some suggestions for policies that could most help each country prepare for the future impacts of climate change.

There is, inherently, significant uncertainty about how to model the way these effects play out over the surface of the earth. It is thus important to review the tools used to generate outputs in this monograph.

- Global circulation models (GCMs) depict the physics and chemistry of the atmosphere and its interactions with oceans and the land surface. Several GCMs have been developed independently around the world, and four have been used for this analysis: CNRM-CM3, CSIRO Mark 3, ECHAM 5, and MIROC 3.2 (see details on these GCMs in the methodology chapter).[1]

- Integrated assessment models (IAMs) simulate the interactions between humans and their surroundings, including industrial activities, transportation, and agriculture and other land uses, to estimate the emissions of various greenhouse gases (including carbon dioxide, methane, and nitrous oxide, the most influential). Several independent IAMs exist as well.

- The emissions simulation results of the IAMs are made available to the GCMs as inputs that alter atmospheric chemistry. The end result is a set of estimates of precipitation and temperature values around the globe, usually at two-degree intervals (about 200 kilometers at the equator).

---

[1] CNRM-CM3 is National Meteorological Research Center–Climate Model 3. CSIRO Mark 3 is a climate model developed at the Australia Commonwealth Scientific and Industrial Research Organisation. ECHAM 5 is a fifth-generation climate model developed at the Max Planck Institute for Meteorology in Hamburg. MIROC 3.2 is the Model for Interdisciplinary Research on Climate, developed at the University of Tokyo Center for Climate System Research.

Review of Current Trends

Climate and Agriculture of Southern Africa

Figure 1.1 shows the annual precipitation distribution for southern Africa in the early part of the 21st century. The humid (or semihumid) areas in the northern parts of Angola, Zambia, Mozambique, and Malawi receive the highest rainfall during the summer season, which lasts from October to March. The high precipitation over these areas is partly associated with the southward migration of rainfall systems from the north, such as the Intertropical Convergence Zone. A reduced amount of precipitation is observed south of the region, especially over the southwest arid regions of southern Africa. The distribution of rainfall, both within the season (intraseasonal) and over years, is subject to high variability, with some regions often having their peak season extended or reduced (Reason, Hachigonta, and Phaladi 2005).

The region has a very diverse vegetation cover. The arid western parts (referred to as "sparse" and "open" grasslands in Figure 1.2) have little crop farming activity. The largest croplands are observed in the Free State province of South Africa. Southern Zambia, Zimbabwe, and Limpopo Province of South Africa have occasional patches of cropland. Some countries in the region have relatively small crop sectors due to unfavorable climatic

FIGURE 1.1 Annual average precipitation in southern Africa, 1950–2000 (millimeters per year)

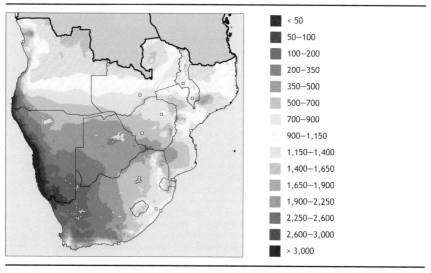

Source: WorldClim version 1.4 (Hijmans et al. 2005).

FIGURE 1.2 Land cover and land use in southern Africa, 2000

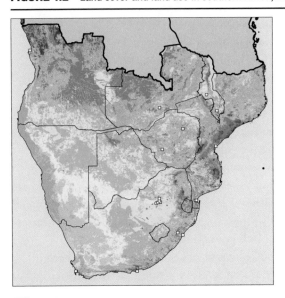

- ■ Tree cover, broadleaved, evergreen
- ■ Tree cover, broadleaved, deciduous, closed
- ▢ Tree cover, broadleaved, open
- ■ Tree cover, broadleaved, needle-leaved, evergreen
- ▢ Tree cover, broadleaved, needle-leaved, deciduous
- ▢ Tree cover, broadleaved, mixed leaf type
- ▢ Tree cover, broadleaved, regularly flooded, fresh water
- ■ Tree cover, broadleaved, regularly flooded, saline water
- ▢ Mosaic of tree cover/other natural vegetation
- ■ Tree cover, burnt
- ▢ Shrub cover, closed-open, evergreen
- ▢ Shrub cover, closed-open, deciduous
- ▢ Herbaceous cover, closed-open
- ▢ Sparse herbaceous or sparse shrub cover
- ▢ Regularly flooded shrub or herbaceous cover
- ▢ Cultivated and managed areas
- ▢ Mosaic of cropland/tree cover/other natural vegetation
- ▢ Mosaic of cropland/shrub/grass cover
- ▢ Bare areas
- ▢ Water bodies
- ▢ Snow and ice
- ■ Artificial surfaces and associated areas
- ▢ No data

Source: GLC2000 (Bartholome and Belward 2005).

conditions, as is the case with Botswana and Namibia, though the livestock sector is very important in both.

Agriculture remains the primary source of employment and income for most of the rural population of the Southern Africa Development Community (SADC). In Malawi, about 39 percent of the gross domestic product (GDP) is from agriculture (FAO 2005). In Zimbabwe, an estimated 80 percent of the population directly depends on agriculture; of this over 60 percent are small-scale farmers (NOAA 2002; Raes et al. 2004). In Zambia, agriculture contributes about 18 percent of GDP; small-scale farmers contribute about 60 percent of farming outputs, and a large share of production is maize (ODI 2002). South Africa has a dual agricultural system, including both well-developed commercial farming and more subsistence-based production in the deep rural areas; agriculture contributes about 3 percent to South Africa's GDP (South Africa, GCIS 2009).

The region as a whole has more than 50 percent of its agricultural land allocated to cereals, with maize (the main staple crop) accounting for more than 40 percent of the total harvested area (Table 1.1). South Africa is the largest maize producer in the region, mainly due to the contribution of irrigated farmlands, followed by Zimbabwe. Namibia, Swaziland, and Botswana have the smallest areas in maize.

TABLE 1.1 Average harvest area of leading agricultural commodities in southern Africa, 2006–2008 (hectares)

Country	Maize	Millet	Rice (paddy)	Sorghum	Wheat
Total	9,199,950	988,690	250,235	848,518	811,903
Angola	1,113,333	358,333	13,333	0	2,467
Botswana	52,333	1,033	0	25,833	383
Lesotho	166,990	0	0	36,739	25,519
Malawi	1,525,050	43,452	57,749	73,115	1,713
Mozambique	1,471,333	59,000	163,333	342,000	1,967
Namibia	21,635	245,267	0	20,000	2,189
South Africa	2,461,082	21,000	1,400	65,317	716,500
Swaziland	47,264	0	50	1,000	213
Zambia	711,330	44,688	14,119	24,950	18,549
Zimbabwe	1,629,600	215,917	250	259,564	42,403

Source: FAOSTAT (FAO 2010).

Millet and sorghum are also important crops, especially in the drier areas. Some countries, such as Botswana and Lesotho, cannot meet their national demand for maize and sorghum through domestic production; the large deficits are traditionally met by imports from South Africa. Zimbabwe and Angola have the highest yield of millet per hectare; sorghum is mostly grown in Mozambique. Most maize production over the region is rainfed, except in South Africa, which has high yields from irrigated areas in the central region of the country (Figure 1.3). Wheat is produced mainly under irrigation in South Africa and Zimbabwe (UNEP 2009). South Africa has by far the highest yield per hectare of wheat in the region, averaging more than four tons (see Figure 1.3).

Demographic and Economic Indicators

Population

Rapid population growth in southern Africa will likely increase its vulnerability to the effects of climate change. In addition, other demographic trends—such as urbanization in coastal areas and encroachment of populations into ecologically marginal areas—can exacerbate climate risks (Jiang and Hardee 2009).

In 2008, southern Africa had an estimated population of 135 million people (Table 1.2). South Africa alone represents over 35 percent of southern Africa's population and 73 percent of the GDP of the region; Swaziland represents less than 1 percent of the SADC population and less than 1 percent of regional GDP (UNEP 2009). As can be seen in Table 1.2, Angola has had the greatest increase in population in southern Africa during the past two decades, followed by Zambia. The smallest amount of growth was in Zimbabwe: after a sharp increase in population (between 3.3 and 3.7 percent) from 1980 to 2000, its growth declined to 1.9 percent in the 1990s and 0.0 between 2000 and 2008 (World Bank 2009). The absence of population growth in Zimbabwe during the past decade can be attributed to the country's weakened economy and the resulting emigration to neighboring countries and overseas; approximately 3–4 million Zimbabweans migrated during the past decade, mainly to South Africa, Botswana, and Britain (Ploch 2010).

The urban growth rate in southern Africa has exceeded that of the rural population. The greatest decadal urban population increase was in Angola and Mozambique, where it increased by about 10 percent between 1988 and 2008. The highest rural growth rate was in Zambia, at 4.2 percent from

FIGURE 1.3 Yields for the main rainfed crops in southern Africa, 2000 (metric tons per hectare)

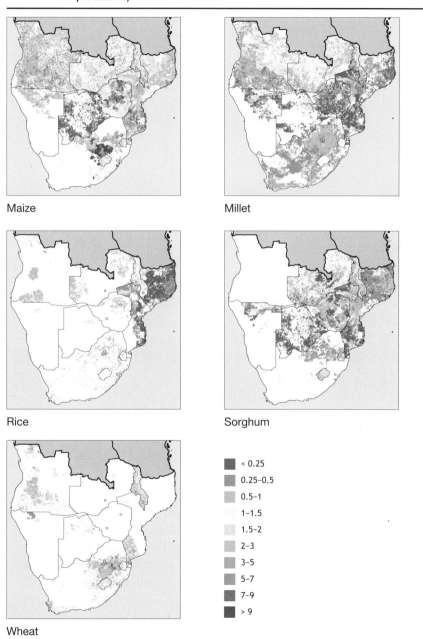

Source: SPAM (Spatial Production Allocation Model) (You and Wood 2006; You, Wood, and Wood-Sichra 2006, 2009).

TABLE 1.2 Population of southern Africa, annualized growth rate, and urban growth rate, 1988 and 2008

Country	Total population (millions) 1988	Total population (millions) 2008	Annualized growth rate, 1988–2008 (%)	Urban growth rate, 1988–2008 (%)
Total	88.03	135.05	2.7	5.1
Angola	10.09	18.02	3.9	10.0
Botswana	1.29	1.90	2.4	7.3
Lesotho	1.55	2.02	1.5	7.5
Malawi	8.65	14.28	3.3	9.3
Mozambique	13.33	21.78	3.2	10.9
Namibia	1.30	2.11	3.1	6.1
South Africa	33.73	48.69	2.2	3.6
Swaziland	0.80	1.17	2.3	3.3
Zambia	7.45	12.62	3.5	2.4
Zimbabwe	9.84	12.46	1.3	3.4

Source: *World Development Indicators* (World Bank 2009).

1988 to 2008. During the postindependence era, countries in southern Africa have rapidly urbanized. Rural–urban migration is a response to differences in economic opportunities and incomes between the towns and the rural areas.

Income

With the exception of Zambia (which shows a decrease) and Zimbabwe (for which no data were available for 2008), the southern African countries included here had increases in per capita GDP from 1988 to 2008 (Table 1.3). During this period, Mozambique had the highest rate of growth in GDP as a result of its recovery from civil war. Botswana had the second-highest rate of growth in GDP per capita in SADC, which is also reflected in its reduced poverty level, which declined from 47 percent in 1993/1994 to 30 percent in 2002/2003. However, the gains in GDP per capita and poverty reduction for most SADC countries have fallen short of the targets of the United Nations (UN) Millennium Development Goals—especially the requirement to halve both extreme poverty and hunger by 2015.

Zambia's GDP per capita started to decline rapidly between 1975 and 1999, reaching its lowest level in the mid-1990s (World Bank 2009). This decline can be attributed to several factors: nationalization (which brought much of the economic base under Zambian government management),

TABLE 1.3 Income of southern Africans (GDP per capita and share of GDP from agriculture), 1988 and 2008

Country	GDP per capita (constant 2000 US$) 1988	GDP per capita (constant 2000 US$) 2008	Rate of GDP growth, 1988–2008 (%)	Share of GDP from agriculture (%) 1988	Share of GDP from agriculture (%) 2008
Angola	838	1,357	3.1	16	10
Botswana	2,183	4,440	5.2	6	2
Lesotho	308	525	3.5	25	7
Malawi	134	165	1.1	50	34
Mozambique	174	365	5.5	43	28
Namibia	1,912	2,692	2.0	12	8
South Africa	3,223	3,764	2.5	6	3
Swaziland	1,039	1,559	0.8	16	8
Zambia	413	387	–0.3	17	21
Zimbabwe	608	n.a.	n.a.	16	n.a.

Source: *World Development Indicators* (World Bank 2009).
Note: GDP = gross domestic product; n.a. = not available; US$ = US dollars.

fluctuations in copper prices, and the mass closure of most mines and industries due to the economic liberalization policy of the mid-1990s. Since 2000, Zambia has seen a steady increase in GDP per capita, reaching almost $400 in 2008 (see Table 1.3).

Table 1.3 shows the decreasing contribution of agriculture to GDP, with Lesotho showing the highest percentage reduction over the past two decades. The agricultural sector in Lesotho accounted for less than 10 percent of GDP in 2008, even though agriculture is the primary source of income for more than half of the population (LVAC 2008).

Well-being indicators

Generally, the past two decades show a declining regional under-five mortality rate (Table 1.4). This might be attributed to improved health services and access to health services in most countries, especially in the rural areas, with their special challenges in addressing medical needs. Although the under-five mortality rate shows a positive change over the past two decades, the rates remain far worse than those of most other countries in the world. Life expectancy at birth, in fact, declined significantly over the past two decades in most countries in the region; one factor in this decrease is the high incidence of HIV/AIDS.

The availability of transport infrastructure influences access to markets and vulnerability to food scarcity (Paavola 2003). The region's transport

TABLE 1.4 Under-five mortality and life expectancy at birth in southern Africa, 1988 and 2008

	Under-five mortality (deaths per 1,000)		Life expectancy at birth (years)	
Country	1988	2008	1988	2008
Angola	257	158	42	47
Botswana	57	40	63	51
Lesotho	102	84	59	43
Malawi	209	110	49	48
Mozambique	200	168	44	42
Namibia	87	67	62	53
South Africa	63	59	62	51
Swaziland	96	91	60	46
Zambia	163	170	51	45
Zimbabwe	95	90	60	44

Source: *World Development Indicators* (World Bank 2009).

infrastructure is generally good in urban areas, as reflected in the short travel times between cities of more than 500,000 people. However, most of the rural districts are poorly accessible (Figure 1.4). South Africa, Swaziland, and Zimbabwe have the shortest travel times, whereas Namibia has the longest.

FIGURE 1.4 Travel time to cities of 500,000 or more people in southern Africa, circa 2000

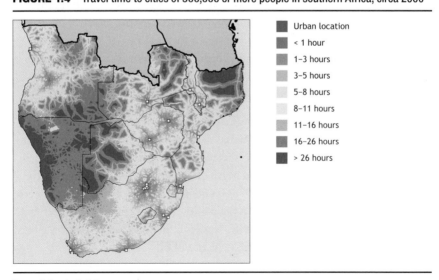

Source: Authors' calculations.

Regional Future Scenarios

Population

Continuing population growth in developing countries will worsen food insecurity over the region (IPCC 2007). Table 1.5 shows the outcomes for southern Africa by 2050 for the three population scenarios mentioned earlier. The low variant reflects an increase in HIV/AIDS-related deaths as well as a reduction in the birth rate. The median (baseline) variant is most likely, showing small annual increases over the next 40 years; the overall result, however, is an increase of 70 percent. The projections assume that the standard of living, life expectancy, and nutritional status of the population will be improved.

The population projections for southern Africa show varying growth rates for different countries. Overall, the regional population is shown increasing by about 70 percent (from 142 million to about 241 million) by 2050 under the median-variant scenario. Based on UN estimates, the highest rate of population growth by 2050 is shown for Angola, Malawi, Mozambique, and Zambia.

Income

Table 1.6 shows the three GDP per capita scenarios used for this study. These are the result of combining three sets of GDP projections with the three population projections from the UN Population Division. The optimistic scenario

TABLE 1.5 Projected population in southern Africa, 2010–2050 (millions)

Country	2010	2050		
		Low variant	Median variant	High variant
Angola	18,993	37,224	42,267	47,675
Botswana	1,978	2,335	2,758	3,220
Lesotho	2,084	2,056	2,491	2,970
Malawi	15,692	32,019	36,575	41,456
Mozambique	23,406	38,268	44,148	50,480
Namibia	2,212	3,076	3,588	4,141
South Africa	50,492	47,536	56,802	67,051
Swaziland	1,202	1,463	1,749	2,061
Zambia	13,257	25,302	28,957	32,870
Zimbabwe	12,644	18,930	22,178	25,731

Source: UNPOP (2009).

TABLE 1.6 GDP per capita scenarios for southern Africa, 2010–2050 (constant 2000 US$)

Country	2010	2050 Pessimistic	2050 Baseline	2050 Optimistic
Angola	877	3,548	4,002	8,378
Botswana	5,353	3,686	25,628	48,646
Lesotho	645	1,850	3,166	6,279
Malawi	166	656	744	2,488
Mozambique	327	1,186	1,812	2,885
Namibia	2,754	4,082	14,239	26,654
South Africa	4,987	5,409	15,473	29,941
Swaziland	1,717	2,709	8,026	15,455
Zambia	502	1,791	2,454	4,254
Zimbabwe	523	1,326	1,539	5,296

Source: Computed from GDP data from the World Bank Economic Adaptation to Climate Change project (World Bank 2010), from the Millennium Ecosystem Assessment (2005) reports, and from population data from the United Nations (UNPOP 2009).
Notes: Entries in the 2010 column are for the baseline scenario. GDP = gross domestic product; US$ = US dollars.

combines high GDP growth with low population growth. The baseline scenario combines medium GDP growth with medium population growth. Finally, the pessimistic scenario combines low GDP growth with high population growth. The GDP per capita projections show an increase in all three scenarios by 2050.

Climate Change Scenarios

Although the general consequences of climate change are becoming increasingly well known, great uncertainty remains about how the effects of climate change will play out in specific locations. Plants and animals that can neither quickly adapt to new conditions nor relocate to new areas will become extinct as a result of climate change (CGIAR 2009). This will destabilize vital ecosystems and erode the genetic base for future crops and livestock. Figure 1.5 shows changes in average precipitation over southern Africa between 2000 and 2050 as modeled by four GCMs, each using the A1B scenario.[2] Figure 1.6 shows the changes in average maximum temperatures. In

---

[2] The A1B scenario describes a world of very rapid economic growth, low population growth, and rapid introduction of new and more efficient technologies, with moderate resource use and a balanced use of technologies. It represents the "best-case" scenario.

FIGURE 1.5 Changes in mean annual precipitation in southern Africa, 2000–2050, A1B scenario (millimeters)

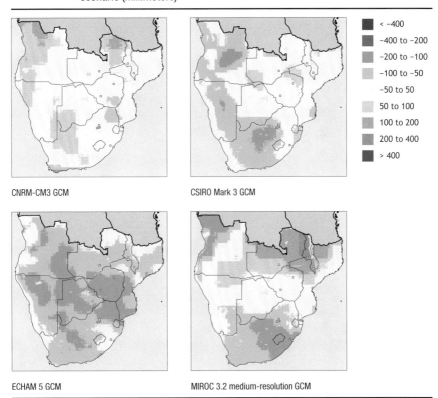

Source: Authors' calculations based on Jones, Thornton, and Heinke (2009).
Notes: A1B = greenhouse gas emissions scenario that assumes fast economic growth, a population that peaks midcentury, and the development of new and efficient technologies, along with a balanced use of energy sources; CNRM-CM3 = National Meteorological Research Center–Climate Model 3; CSIRO Mark 3 = climate model developed at the Australia Commonwealth Scientific and Industrial Research Organisation; ECHAM 5 = fifth-generation climate model developed at the Max Planck Institute for Meteorology in Hamburg; GCM = general circulation model; MIROC 3.2 = Model for Interdisciplinary Research on Climate, developed by the University of Tokyo Center for Climate System Research.

each set of figures, the legend colors are identical; a specific color represents the same change in temperature or precipitation across the models.

A quick glance at these figures shows that differences exist. For example, Figure 1.5 shows that the ECHAM 5 model projects that all of southern Africa will become drier, while the other models show a mixture, with areas that are wetter, other areas that are drier, and still other areas that will remain unchanged. The CNRM-CM3 model results show the central part of South Africa getting marginally wetter, while the other model results show the same area getting drier. Other studies, such as one conducted at the University of

FIGURE 1.6 Change in monthly mean maximum temperature in southern Africa for the warmest month, 2000–2050, A1B scenario

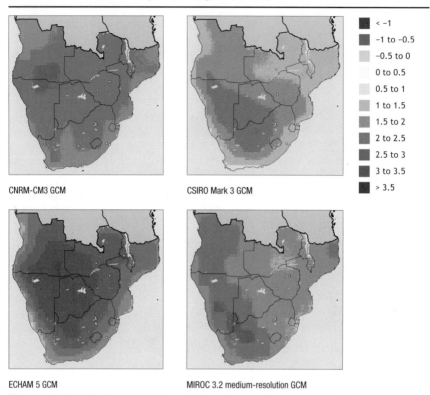

Source: Authors' calculations based on Jones, Thornton, and Heinke (2009).
Notes: A1B = greenhouse gas emissions scenario that assumes fast economic growth, a population that peaks midcentury, and the development of new and efficient technologies, along with a balanced use of energy sources; CNRM-CM3 = National Meteorological Research Center–Climate Model 3; CSIRO Mark 3 = climate model developed at the Australia Commonwealth Scientific and Industrial Research Organisation; ECHAM 5 = fifth-generation climate model developed at the Max Planck Institute for Meteorology in Hamburg; GCM = general circulation model; MIROC 3.2 = Model for Interdisciplinary Research on Climate, developed by the University of Tokyo Center for Climate System Research.

Cape Town using downscaled climate models, show the eastern parts of South Africa becoming wetter by 2050. In Figure 1.6, although all the models show temperature increases for most of southern Africa, they differ on the amount of increase, with the ECHAM 5 model showing a much hotter future.

The precipitation figures shown here illustrate the range of potential climate outcomes using current modeling capabilities and indicate the uncertainty inherent in estimating the impacts of climate change. Policymakers therefore need to develop plans that will allow for flexibility in responsiveness to climate change.

Scenarios Showing Changes in Crops

Tables 1.7–1.9 show simulations from the International Model for Policy Analysis of Agricultural Commodities and Trade (IMPACT) associated with maize, millet, and sorghum, respectively, for 2010 and 2050, based on the GDP per capita and population growth scenarios and climate change scenarios discussed earlier. Each featured crop has three categories, representing yield, area, and production. The production of maize over the region increases in all scenarios by 2050. However, in some countries the increased production will not meet local demand, resulting in an increase in net maize imports.

IMPACT shows the world price of maize increasing from $120 per ton to over $200 per ton by 2050, which would have a serious implication for affordability for the majority of the population living below the poverty line.[3] The average yield of maize is shown to increase by 2050, but the harvested area for maize is shown to remain about the same except in South Africa, where it is projected to decline by a large amount. Similar outcomes are shown for the yield and production of sorghum and millet; the area under these crops are projected to increase, however, and this would partially compensate for the reduction in area harvested for maize.

Conclusions and Recommendations

Smallholder farmers account for the majority of agriculture production in the region. This makes the agricultural sector in the region very vulnerable to the impacts of climate change, especially given the poor travel times and already low yields there. Rapidly growing populations and higher rates of evapotranspiration (resulting from increased temperatures) can be expected to put pressure on the region's water resources, increasing the rate of rural–urban migration.

Most governments in southern Africa are making efforts to highlight the impacts of climate on agriculture. For instance, in Botswana, National Vision 2016 strives to promote irrigation, drought-resistant crops, and selective breeding of drought-tolerant livestock while diversifying the agricultural base and conserving scarce agricultural land resources. This is especially encouraging because Botswana is likely to remain a net importer of grains such as maize and sorghum. In addition, deployment of better farming technologies is envisaged to increase yields for both crops—depending on the successful uptake of these technologies by farmers through effective subsidies.

---

3 All dollar figures in this monograph are constant 2000 US dollars, and all tons are metric tons.

TABLE 1.7 Maize projections for southern Africa showing yield, area, and production, 2010–2050

	2010			2050						
				Yield (MT/ha)		Area (thousands of ha)		Production (thousands of MT)		
Country	Yield (MT/ha)	Area (thousands of ha)	Production (thousands of MT)	Min	Max	Min	Max	Min	Max
Angola	0.55	1,140	631	1.25	1.39	1,133	1,191	1,417	1,660
Botswana	0.20	56	11	0.29	0.41	43	55	12	22
Lesotho	0.67	139	94	1.80	2.34	142	155	273	333
Malawi	0.88	1,399	1,232	0.98	1.06	1,320	1,437	1,298	1,522
Mozambique	1.02	1,290	1,314	1.93	2.06	1,250	1,360	2,409	2,796
Namibia	1.32	25	33	1.36	1.73	22	23	30	40
South Africa	3.86	2,845	10,979	5.72	6.56	1,981	2,175	11,332	13,777
Swaziland	0.81	42	34	1.35	1.69	25	28	34	45
Zambia	2.00	546	1,090	3.06	3.19	541	589	1,657	1,877
Zimbabwe	0.57	1,353	770	1.14	1.31	1,265	1,377	1,445	1,807

Source: Based on analysis conducted for Nelson et al. (2010).

Notes: Min (minimum) represents the smallest projected number from the simulations based on the CSIRO A1B, CSIRO B1, MIROC A1B, and MIROC B1 climate model/scenarios combined with the pessimistic, baseline, and optimistic scenarios. Max (maximum) represents the largest of the twelve simulated values. A1B = greenhouse gas emissions scenario that assumes fast economic growth, a population that peaks midcentury, and the development of new and efficient technologies, along with a balanced use of energy sources; B1 = greenhouse gas emissions scenario that assumes a population that peaks midcentury (like the A1B), but with rapid changes toward a service and information economy and the introduction of clean and resource efficient technologies; CSIRO = climate model developed at the Australia Commonwealth Scientific and Industrial Research Organisation; MIROC = Model for Interdisciplinary Research on Climate, developed at the University of Tokyo Center for Climate System Research; ha = hectares; MT = metric tons.

TABLE 1.8 Millet projections for southern Africa showing yield, area, and production, 2010–2050

Country	2010			2050						
	Yield (MT/ha)	Area (thousands of ha)	Production (thousands of MT)	Yield (MT/ha)		Area (thousands of ha)		Production (thousands of MT)		
				Min	Max	Min	Max	Min	Max	
Angola	0.27	366	97	1.00	1.06	495	534	507	566	
Botswana	0.36	4	2	1.23	1.54	5	6	7	9	
Lesotho	0	0	0	0	0	0	0	0	0	
Malawi	0.40	40	16	1.37	1.44	45	48	62	69	
Mozambique	0.25	110	27	0.38	0.40	148	158	57	63	
Namibia	0.21	266	56	0.64	0.83	335	355	222	290	
South Africa	0.53	21	11	1.35	1.45	19	20	26	28	
Swaziland	0	0	0	0	0	0	0	0	0	
Zambia	0.42	63	26	1.43	1.50	87	92	124	137	
Zimbabwe	0.33	128	42	1.31	1.44	168	179	225	256	

Source: Based on analysis conducted for Nelson et al. (2010).

Notes: Millet is not grown in Lesotho or Swaziland. Min (minimum) represents the smallest projected number from the simulations based on the CSIRO A1B, CSIRO B1, MIROC A1B, and MIROC B1 climate model/scenarios combined with the pessimistic, baseline, and optimistic scenarios. Max (maximum) represents the largest of the twelve simulated values. A1B = greenhouse gas emissions scenario that assumes fast economic growth, a population that peaks midcentury, and the development of new and efficient technologies, along with a balanced use of energy sources; B1 = greenhouse gas emissions scenario that assumes a population that peaks midcentury (like the A1B), but with rapid changes toward a service and information economy and the introduction of clean and resource efficient technologies; CSIRO = climate model developed at the Australia Commonwealth Scientific and Industrial Research Organisation; MIROC = Model for Interdisciplinary Research on Climate, developed at the University of Tokyo Center for Climate System Research; ha = hectares; MT = metric tons.

TABLE 1.9 Sorghum projections for southern Africa showing yield, area, and production, 2010–2050

Country	2010			2050						
	Yield (MT/ha)	Area (thousands of ha)	Production (thousands of MT)	Yield (MT/ha)		Area (thousands of ha)		Production (thousands of MT)		
				Min	Max	Min	Max	Min	Max	
Angola	0	0	0	0	0	0	0	0	0	
Botswana	0.34	58	19	0.81	1.08	69	83	57	87	
Lesotho	0.60	35	21	2.48	2.80	51	53	129	147	
Malawi	0.64	73	46	1.44	1.49	106	108	152	160	
Mozambique	0.48	582	282	1.12	1.15	841	862	945	992	
Namibia	0.35	21	7	1.29	1.52	27	28	36	42	
South Africa	2.79	94	263	4.49	4.91	92	95	414	461	
Swaziland	0.61	1	1	1.92	1.95	2	3	5	5	
Zambia	0.74	32	23	1.61	1.64	35	36	57	59	
Zimbabwe	0.61	140	85	1.55	1.66	171	175	265	288	

Source: Based on analysis conducted for Nelson et al. (2010).

Notes: Sorghum is not grown in Angola. Min (minimum) represents the smallest projected number from the simulations based on the CSIRO A1B, CSIRO B1, MIROC A1B, and MIROC B1 climate model/scenarios combined with the pessimistic, baseline, and optimistic scenarios. Max (maximum) represents the largest of the twelve simulated values. A1B = greenhouse gas emissions scenario that assumes fast economic growth, a population that peaks midcentury, and the development of new and efficient technologies, along with a balanced use of energy sources; B1 = greenhouse gas emissions scenario that assumes a population that peaks midcentury (like the A1B), but with rapid changes toward a service and information economy and the introduction of clean and resource efficient technologies; CSIRO = climate model developed at the Australia Commonwealth Scientific and Industrial Research Organisation; MIROC = Model for Interdisciplinary Research on Climate, developed at the University of Tokyo Center for Climate System Research; ha = hectares; MT = metric tons.

Because agriculture employs 60–70 percent of the labor force in Lesotho, climate scenarios showing adverse effects on food production and water supply pose a particular threat to a great proportion of the labor force and their extended families. However, despite the existing gaps in mainstreaming climate change issues, Lesotho is taking important steps toward increasing awareness through outreach programs, the creation of institutional frameworks for action, and the adaption of drought-resistant crop varieties. In addition to the recommendations presented below, the Lesotho government could also pay attention to capacity building in terms of climate change knowledge and promoting needed skills development across all the sectors of development as well as in learning institutions across different disciplines. Effective communications technologies need to be made available to poor and inaccessible communities to allow broad access to seasonal weather forecasts and early warning systems.

Scenarios for Swaziland paint a rather uncomfortable picture, with temperatures shown increasing by as much as 2.5°C by 2050 and precipitation decreasing by as much as 100 millimeters in the highveld and the lowveld—the ecological zones where most of the crops are produced. In developing Swaziland's climate change policy, the National Committee (which is made up of government officers, to the exclusion of public or private organizations and representatives of nongovernmental organizations [NGOs]) needs to engage all relevant stakeholders to represent a broad range of interests on this topic. There is a need to improve the understanding of local communities regarding climate change. At the same time, continued dialogue between researchers and policymakers would provide mutual learning opportunities and ensure that the knowledge produced by researchers is both useful and used.

There will be no one-size-fits-all solution, because each country in the region faces different challenges. For instance, as in most countries in the region, agriculture in Zambia is mainly dependent on rainfall and thus vulnerable to climate variability and change. The development of an integrated approach, tailored to address local challenges, will be essential to develop effective climate change adaptation and mitigation strategies. There is a strong drive by organizations in civil society and the private sector to implement climate change adaptation strategies for agriculture, especially in rural areas.

In Zimbabwe, models show a shift in agroecological zones as a result of climate change. Some of the maize-producing areas would need to adopt drought-tolerant varieties or crops like sorghum and millet. With improved availability of agriculture inputs, yields and production are projected to increase for maize, while the harvest area will increase for sorghum and millet. About 70 percent of the population is resident in semiarid areas where the area under crop production would decrease as a result of climate variability and change.

Scenarios for Mozambique show that the country faces risks from a changing climate, with loss of cultivated and managed lands projected for most of southern Mozambique. The modeled results show the production and yield of cassava increasing slowly until 2025 and then declining from 2025 to 2050. This potential outcome reinforces the need to implement the Mozambican National Adaptation Programme of Action for agriculture. It would help if the government enhanced the road network in the north of Mozambique so that the expected increase in cassava and maize production can be efficiently distributed throughout the country when the need arises in the near future.

Agricultural production in South Africa is mainly commercial, and though adapting to climate change is not less expensive for commercial farmers, in most cases it repays the extra investment. Commercial agriculture provides food for the majority of South Africans (and significant amounts for export to neighboring countries); shifts in climate patterns, the suitability of ecological regions, and market prices have a major influence on plantings and yields and therefore on production and returns. Our main recommendation for the country in this chapter is to educate decisionmakers, policymakers, and water users regarding the implications of climate change on agriculture, identifying the various vulnerabilities. Adaptation frameworks and actions will be needed to build resilience and to guarantee food security for the country and the region. Because vulnerability to climate change is highly variable throughout South Africa, policymakers should tailor policies to local conditions (Gbetibouo and Ringler 2009).

Some adaptation recommendations identified in the country chapters include the following:

- Incorporation of climate change adaptation in long-term planning and developmental programs, including budgetary allocations for climate change adaptation strategies.
- Smallholder irrigation development targeting vulnerable communities.
- Capacity building in skills and tools for technical assessments, planning, and policy development in the context of climate change.
- Awareness raising regarding climate change issues to gather support for action on the part of governments, NGOs, the private sector, and the public.

- Development and promotion of drought-tolerant and heat-tolerant crop varieties and hardy livestock and development of linkages between research institutions and extension institutions for more rapid transmission of information to farmers.
- Increased availability of weather and climate information.

Despite government efforts, gaps and disconnects still exist between climate change science and adaptation action. Important components are still needed: plans and policies that incorporate climate change issues, education on climate change among communities, and information on outreach programs as they are developed. Critical interventions must be based on developing institutional frameworks for action, drought-resistant crop varieties, smallholder water harvesting and supply strategies, and strategies for alleviating poverty aimed at making people more resilient to climate variability.

The next chapter, on methodology, presents some of the technical information needed for the reader to understand at a deeper level many of the results in this chapter and each of the country chapters. The rest of the chapters focus on impacts of climate on agriculture in individual countries.

References

ADB and IFPRI (Asian Development Bank and International Food Policy Research Institute). 2009. *Building Climate Resilience in the Agriculture Sector in Asia and the Pacific.* Mandaluyong City, Philippines: Asian Development Bank. www.adb.org/publications/building-climate-resilience-agriculture-sector-asia-and-pacific.

Bartholome, E., and A. S. Belward. 2005. "GLC2000: A New Approach to Global Land Cover Mapping from Earth Observation Data." *International Journal of Remote Sensing* 26 (9): 1959–1977.

CGIAR. 2009. "Climate, Agriculture, and Food Security: A Strategy for Change." Accessed June 7, 2011. http://ccafs.cgiar.org/sites/default/files/pdf/CC_for_COP15_Final_LR_2.pdf.

FAO (Food and Agriculture Organisation of the United Nations). 2005. "FAO/WFP Crop and Food Supply Assessment Mission to Malawi." Accessed July 8, 2011. www.fao.org/docrep/008/j5509e/j5509e00.htm.

———. 2010. FAOSTAT. Rome. http://faostat.fao.org.

Gbetibouo, G. A., and C. Ringler. 2009. *Mapping South African Farming Sector Vulnerability to Climate Change and Variability: A Sub-national Assessment.* IFPRI Discussion Paper 00885. Washington, DC: International Food Policy Research Institute.

Hijmans, R. J., S. E. Cameron, J. L. Parra, P. G. Jones, and A. Jarvis. 2005. "Very High Resolution Interpolated Climate Surfaces for Global Land Areas." *International Journal of Climatology* 25: 1965–1978. www.worldclim.org.

IPCC (Intergovernmental Panel on Climate Change). 2007. "Summary for Policymakers." In *Climate Change 2007: Impacts, Adaptation, and Vulnerability,* edited by M. L. Parry, O. F. Canziani, J. P. Palutikof, P. J. van der Linden, and C. E. Hanson. Contribution of Working Group II to the Fourth Assessment Report of the Intergovernmental Panel on Climate Change. Cambridge: Cambridge University Press.

Jiang, L., and K. Hardee. 2009. *How Do Recent Population Trends Matter to Climate Change?* Working paper. Washington, DC: Population Action International.

Jones, P. G., P. K. Thornton, and J. Heinke. 2009. *Generating Characteristic Daily Weather Data Using Downscaled Climate Model Data from the IPCC's Fourth Assessment.* Project report. Nairobi, Kenya: International Livestock Research Institute.

Le Treut, H., R. Somerville, U. Cubasch, Y. Ding, C. Mauritzen, A. Mokssit, T. Peterson, and M. Prather. 2007. "Historical Overview of Climate Change." In *Climate Change 2007: The Physical Science Basis,* edited by S. Solomon, D. Qin, M. Manning, Z. Chen, M. Marquis, K. B. Averyt, M. Tignor, and H. L. Miller. Contribution of Working Group I to the Fourth Assessment Report of the Intergovernmental Panel on Climate Change. Cambridge: Cambridge University Press.

LVAC (Lesotho Vulnerability Assessment Commission). 2008. *Lesotho Food Security and Vulnerability Monitoring Report.* Maseru.

Meinke, H., R. Nelson, P. Kokic, R. Stone, R. Selvaraju, and W. Baethgen. 2006. "Actionable Climate Knowledge: From Analysis to Synthesis. *Climate Research* 33: 101–110.

Millennium Ecosystem Assessment. 2005. *Ecosystems and Human Well-being: Synthesis.* Washington, DC: Island Press. www.maweb.org/en/Global.aspx.

Nelson, G., M. Rosegrant, J. Koo, R. Robertson, T. Sulser, T. Zhu, C. Ringler, S. Msangi, A. Palazzo, M. Batka, M. Magalhaes, R. Valmonte-Santos, M. Ewing, and D. Lee. 2009. *Climate Change: Impacts on Agriculture and Costs of Adaptation.* Food Policy Report. Washington, DC: International Food Policy Research Institute.

Nelson, G. C., M. W. Rosegrant, A. Palazzo, I. Gray, C. Ingersoll, R. Robertson, S. Tokgoz, et al. 2010. *Food Security, Farming, and Climate Change to 2050: Scenarios, Results, Policy Options.* Washington, DC: International Food Policy Research Institute.

NOAA (National Oceanic and Atmospheric Administration). 2002. "Climate and the Republic of Zimbabwe: Can Today's Climate Science Help over Tomorrow's Catastrophe?" Office of Global Programs for Africa, Silver Spring, MD, US. www.cip.ogp.noaa.gov.

ODI (Overseas Development Institute). 2002. "Trends in the Zambian Agriculture Sector." Technical Report, Department for International Development.

Paavola, J. 2003. "Vulnerability to Climate Change in Tanzania: Sources, Substance, and Solutions." Paper presented at the inaugural workshop of Southern Africa Vulnerability Initiative (SAVI), June 19–21, in Maputo, Mozambique.

Ploch, L. 2010. "Zimbabwe: The Transitional Government and Implications for U.S. Policy." Accessed February 7, 2011. www.fas.org/sgp/crs/row/RL34509.pdf.

Raes, D., A. Sithole, A. Makarau, and J. Milford. 2004. "Valuation of First Planting Dates Recommended by Criteria Currently Used in Zimbabwe." *Agricultural and Forest Meteorology* 125: 177–185.

Reason, C. J. C., S. Hachigonta, and R. F. Phaladi. 2005. "Interannual Variability in Rainy Season Characteristics over the Limpopo Region of Southern Africa." *International Journal of Climatology* 25: 1835–1853.

South Africa, GCIS (Government Communication and Information System). 2009. "South Africa Yearbook 2008/09." Pretoria. www.gcis.gov.za/resource.

UNEP (United Nations Environment Programme). 2009. "Climate Change and Variability in Southern Africa: Impacts and Adaptation in the Agricultural Sector." Accessed February 7, 2011. www.unep.org/themes/freshwater/documents/climate_change_and_variability_in_the_southern_africa.pdf.

UNPOP (United Nations Department of Economic and Social Affairs–Population Division). 2009. *World Population Prospects: The 2008 Revision.* New York. http://esa.un.org/unpd/wpp/.

World Bank. 2009. *World Development Indicators.* Washington, DC.

———. 2010. *The Costs of Agricultural Adaptation to Climate Change.* Washington, DC.

You, L., and S. Wood. 2006. "An Entropy Approach to Spatial Disaggregation of Agricultural Production." *Agricultural Systems* 90 (1–3): 329–347.

You, L., S. Wood, and U. Wood-Sichra. 2006. "Generating Global Crop Distribution Maps: From Census to Grid." Paper presented at the International Association of Agricultural Economists Conference in Brisbane, Australia, August 11–18.

———. 2009. "Generating Plausible Crop Distribution and Performance Maps for Sub-Saharan Africa Using a Spatially Disaggregated Data Fusion and Optimization Approach." *Agricultural Systems* 99 (2–3): 126–140.

Ziervogel, G., A. Cartwright, A. Tas, J. Adejuwon, F. Zermoglio, M. Shale, and B. Smith. 2008. *Climate Change and Adaptation in African Agriculture.* SEI Report. New York: Rockefeller Foundation.

Chapter 2

METHODOLOGY

Gerald C. Nelson, Amanda Palazzo, Daniel Mason-d'Croz, Richard Robertson, and Timothy S. Thomas

Modeling the impacts of climate change presents a complex challenge arising from the wide-ranging processes underlying the working of markets, ecosystems, and human behavior. The analytical framework used in this monograph integrates modeling components that range from the macro to the micro to model a range of processes, from those driven by economics to those that are essentially biological in nature. This chapter brings together in one place the technical details associated with models used in this monograph along with other technical information that is common to most or all of the chapters. Figure 2.1 provides a diagram of the links among the three models used: the International Model for Policy Analysis of Agricultural Commodities and Trade (IMPACT) of the International Food Policy Research Institute (IFPRI) (Rosegrant et al. 2008), a partial equilibrium agriculture model that emphasizes policy simulations; a hydrology model incorporated into IMPACT; and the Decision Support Software for Agrotechnology Transfer (DSSAT) crop model suite (Jones et al. 2003), which is used to estimate yields of crops under varying management systems and climate change scenarios.

General Circulation Models (GCMs) and Climate Scenarios

GCMs model the physics and chemistry of the atmosphere and its interactions with oceans and the land surface. Several GCMs have been developed independently around the world. For the Fourth Assessment Report (AR4) of the Intergovernmental Panel on Climate Change (IPCC), 23 GCMs made some model results publicly available. Results from four are used in this monograph.

This chapter draws heavily on Nelson et al. (2010).

FIGURE 2.1 The International Model for Policy Analysis of Agricultural Commodities and Trade (IMPACT) modeling framework

[Figure: IMPACT modeling framework diagram showing Model inputs and scenario definitions (Urban growth & changes in food habits (demand elasticities); Income growth projections; Population growth projections; Supply, demand, and trade data from FAOSTAT, IFPRI, UN, World Bank, and others; Area elasticities with respect to crop prices; Yield elasticities with respect to crop, labor, and capital prices; Area and yield annual growth rates) feeding into Model calculations (food) (Domestic prices f(world price, trade wedge, marketing margin); Supply projection; Demand projection; Net trade exports — imports; Malnutrition results; World trade balance imports = exports; Adjust world price; Iteration for world market clearing) and Water simulation (Water supply — Renewable H₂O, Effective H₂O for irrigated and rainfed crops; Water demand — Irrigation, Livestock, Domestic, Industry, Environment; Climate scenarios Rainfall, runoff, potential reference evapotranspiration; Update inputs Go to next year).]

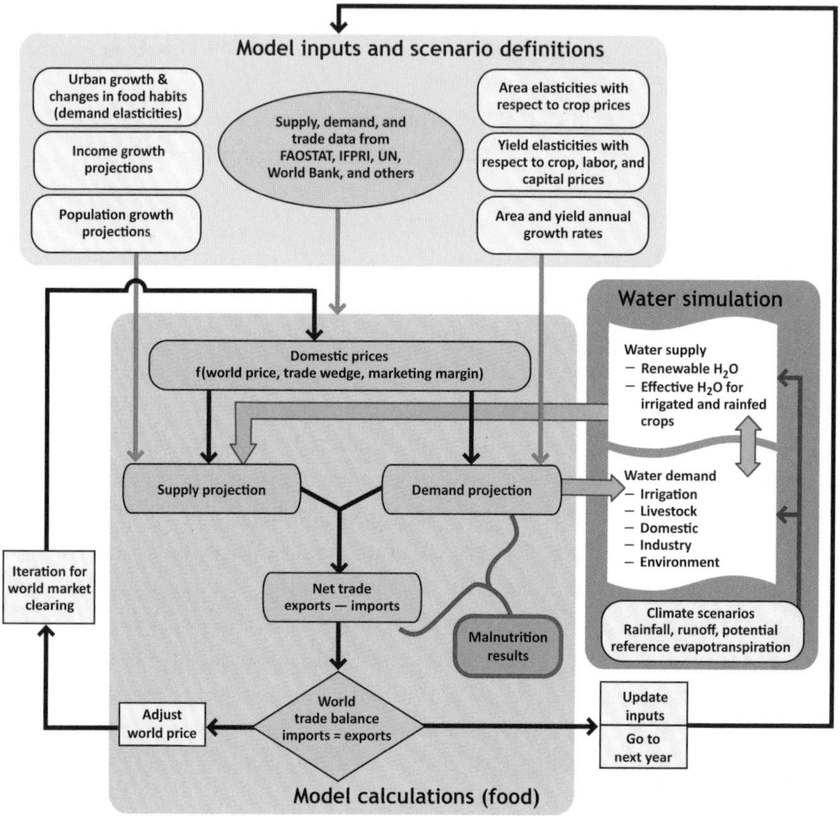

Source: Nelson et al. (2010).
Note: FAOSTAT = *FAOSTAT Database on Agriculture* (FAO 2010); IFPRI = International Food Policy Research Institute; UN = United Nations.

The GCMs create estimates of precipitation and temperature values around the globe, often at something close to 2-degree intervals (about 200 kilometers at the equator) for most models. This is very coarse and may hide important differences on a more local scale. To have finer resolution, it is common to "downscale" the data. Data downscaled by Jones, Thornton, and Heinke (2009) provide precipitation and temperature data at a 5-arc-minute resolution (around 9 kilometers at the equator, smaller away from it, which is called 10-kilometer resolution in this monograph).

Greenhouse gas emissions alter atmospheric chemistry, ultimately increasing temperatures and altering precipitation patterns. AR4 had three scenarios of greenhouse gas emissions pathways: B1, A1B, and A2.[1] Scenario B1 was a low-emissions scenario, which by 2013 is not looking very realistic. Scenarios A2 and A1B are higher-emission scenarios, with similar trajectories through 2050 but different ones after 2050. Because this monograph is primarily concerned with changes to 2050, we elected to focus on scenario A1B when presenting the biophysical effects of climate change on crop yields, but we used both B1 and A1B in the IMPACT model to provide a wider range of scenario outcomes.

To illustrate the range of potential effects on crops, we used results from four GCMs, CNRM-CM3, CSIRO Mark 3, ECHAM 5, and MIROC 3.2 medium resolution.[2] For inputs into the IMPACT model, results from only two GCMs were used, the CSIRO Mark 3 and the MIROC 3.2 medium-resolution models. The rationale for doing that can be seen more clearly in Table 2.1, in which we see that the lowest levels of precipitation change and lowest levels of temperature change are given by the CSIRO GCM and the highest levels of precipitation and temperature change are given by the MIROC GCM.

In the country analyses in the other chapters of this monograph, we display two kinds of maps that show spatially differentiated predictions of the GCMs. One shows changes in annual rainfall, and the other shows changes in the mean daily maximum temperature for the warmest month. The changes in the latter are determined by finding the month in 2000 with the highest mean daily maximum temperature and subtracting that mean value from the mean value for the month with the highest mean daily maximum temperature for 2050.

The Spatial Production Allocation Model (SPAM)

SPAM is a set of raster datasets showing harvest area, production, and yield for 20 crops or aggregates of crops and for three management systems (irrigated, high-input rainfed, and low-input rainfed, with the latter two combined

---

1 B1 is a greenhouse gas emissions scenario that assumes a population that peaks midcentury (like A1B), but with rapid changes toward a service and information economy, and the introduction of clean and resource-efficient technologies. A1B is a greenhouse gas emissions scenario that assumes fast economic growth, a population that peaks midcentury, and the development of new and efficient technologies, along with a balanced use of energy sources. A2 is a greenhouse gas emissions scenario that assumes a very heterogeneous world with continuously increasing global population and regionally oriented economic growth that is more fragmented and slower than in other storylines (Nakicenovic et al. 2000).

2 CNRM-CM3 is National Meteorological Research Center–Climate Model 3. CSIRO Mark 3 is a climate model developed at the Australia Commonwealth Scientific and Industrial Research Organisation. ECHAM 5 is a fifth-generation climate model developed at the Max Planck Institute for Meteorology in Hamburg. MIROC 3.2 is the Model for Interdisciplinary Research on Climate, developed at the University of Tokyo Center for Climate System Research.

TABLE 2.1 GCM and SRES scenario global average changes, 2000–2050

		Change between 2000 and 2050 in the annual averages			
GCM	SRES scenario	Precipitation (percent)	Precipitation (millimeters)	Minimum temperature (°C)	Maximum temperature (°C)
CSIRO	**B1**	**0.0**	**0.1**	**1.2**	**1.0**
CSIRO	**A1B**	**0.7**	**4.8**	**1.6**	**1.4**
CSIRO	A2	0.9	6.5	1.9	1.8
ECHAM 5	B1	1.6	11.6	2.1	1.9
CNRM-CM3	B1	1.9	14.0	1.9	1.7
ECHAM 5	A2	2.1	15.0	2.4	2.2
CNRM-CM3	A2	2.7	19.5	2.5	2.2
ECHAM 5	A1B	3.2	23.4	2.7	2.5
MIROC	A2	3.2	23.4	2.8	2.6
CNRM-CM3	A1B	3.3	23.8	2.6	2.3
MIROC	**B1**	**3.6**	**25.7**	**2.4**	**2.3**
MIROC	**A1B**	**4.7**	**33.8**	**3.0**	**2.8**

Source: Nelson et al. (2010).

Notes: In this table and elsewhere in the text, a reference to a particular year for a climate realization, such as 2000 or 2050, in fact refers to mean values around that year. For example, the data described as 2000 in this table are representative of the period 1950–2000. The data described as 2050 are representative of the period 2041–2060. GCM scenario combinations in bold are the ones used in the climate scenario analysis. A1B = greenhouse gas emissions scenario that assumes fast economic growth, a population that peaks midcentury, and the development of new and efficient technologies, along with a balanced use of energy sources; B1 = greenhouse gas emissions scenario that assumes a population that peaks midcentury (like A1B), but with rapid changes toward a service and information economy, and the introduction of clean and resource-efficient technologies; CNRM-CM3 = climate model developed by the National Meteorological Research Center; CSIRO = climate model developed at the Australia Commonwealth Scientific and Industrial Research Organisation; ECHAM 5 = fifth-generation climate model developed at the Max Planck Institute for Meteorology (Hamburg); GCM = general circulation model; MIROC = Model for Interdisciplinary Research on Climate, developed by the University of Tokyo Center for Climate System Research; SRES = *Special Report on Emissions Scenarios*, a report by the Intergovernmental Panel on Climate Change that was published in 2000.

in this monograph to produce a rainfed total). The model employs a cross-entropy approach to manage inputs with different levels of likelihood in indicating the specific locations of agricultural production (You and Wood 2006; You, Wood, and Wood-Sichra 2006, 2009).

SPAM spatially allocates crop production from large reporting units (administrative units such as those at the province or district level) to a raster grid at a spatial resolution of 5 arc-minutes. The allocation model works by inferring likely production locations from multiple indicators that, in addition to subnational crop production statistics, also include satellite data on land cover, maps of irrigated areas, biophysical crop suitability assessments, population density, and secondary data on irrigation and rainfed production.

In some of the maps presented in this monograph, SPAM areas are reported in units of hectares per raster cell. A 5 arc-minute grid cell is just over 8,500 hectares at the equator, which is a reasonable value to use when gauging how great a proportion of the cell might be used by the crop shown in the map.

SPAM areas are used in the regional overview (Chapter 1) to provide weights for calculating provincial and national yield changes due to climate change. Cells with greater current levels of that crop are weighted higher when aggregating the crop model results. This was also the approach used for aggregating crop model results to the national level for use in the IMPACT model.

DSSAT

DSSAT is a software package used for modeling crop production (Jones et al. 2003). The software "grows" the crop in daily time increments, and therefore daily weather data are required. With climate models, we have only monthly statistics on the weather. DSSAT, however, overcomes this limitation by including a weather simulator that can convert monthly statistics into simulated daily weather. In this analysis, the weather is simulated many different times and the outcome averaged over several growing seasons. The result is more of a long-term yield perspective that will not be unduly influenced by any individual stochastic extreme in the simulation.

The soil data used were adapted by John Dimes and Jawoo Koo from the Harmonized World Soil Database (HWSD ver. 1.1) by Batjes et al. (2009). They are simplified to 27 types of soil, each with high, medium, or low levels of soil organic carbon; deep, medium, or shallow rooting depth; and major components of sand, loam, or clay. Some grid cells had more than one soil type represented, and when that was the case, the dominant type was used.

DSSAT has parameters to model different varieties of each crop. For our work, we picked what seemed an appropriate variety and used it in all locations and time periods investigated. DSSAT is somewhat limiting for the purposes of studying the effects of climate change on agriculture because it includes only 26 different crops, excluding many that we would have an interest in studying, including sweet potatoes and yams and most perennials, cash crops, and tree crops.

DSSAT requires the user to input the planting date of a crop. For rainfed crops, it is assumed that a crop is planted in the first month of a four-month period in which the monthly average maximum temperature does not exceed 37°C (about 99°F), the monthly average minimum temperature does not drop below 5°C (about 41°F), and the monthly total precipitation is not less than 60 millimeters. In the tropics, the planting month begins with the rainy season. The particular mechanism for determining the start of the rainy season at any location is to look for the block of four months that gets the most rainfall. The month before that block is called the beginning of the rainy season. For irrigated crops, the first choice is the rainfed planting month.

DSSAT has an option to include CO_2 fertilization effects at different levels of CO_2 atmospheric concentration. For this study, all results use a setting of 369 parts per million, which essentially assumes no yield gains from CO_2 fertilization. A short summary of the reasons for and against including CO_2 fertilization is found in Nelson et al. (2010, 14, text and footnote):

> Plants produce more vegetative matter as atmospheric concentrations of CO_2 increase. The effect depends on the nature of the photosynthetic process used by the plant species. So-called C3 plants use CO_2 less efficiently than C4 plants, so C3 plants are more sensitive to higher concentrations of CO_2. It remains an open question whether these laboratory results translate to actual field conditions. A recent report on field experiments on CO_2 fertilization (Long et al. 2006) finds that the effects in the field are approximately 50 percent less than in experiments in enclosed containers. Another report (Zavala et al. 2008) finds that higher levels of atmospheric CO_2 increase the susceptibility of soybean plants to the Japanese beetle and of maize to the western corn rootworm. Finally, a recent study (Bloom et al. 2010) finds that higher CO_2 concentrations inhibit the assimilation of nitrate into organic nitrogen compounds. So the actual field benefits of CO_2 fertilization remain uncertain.

Some use of nitrogen fertilizer is assumed in all our crop models. For almost all countries in Africa, the level of use is 20 kilograms of nitrogen per hectare (regardless of crop). For Madagascar and parts of South Africa, the level is 100 kilograms of nitrogen per hectare (regardless of crop). Levels were set appropriately for the rest of the world in the global modeling.

DSSAT is used in two ways in this monograph. It is used directly for each country and for the region to compute changes in yields from the climate of 2000 to the climate of 2050. DSSAT is also used to provide results for each country of the world so that IMPACT can control for climate effects. The global work and the regional work were very similar though not perfectly identical because they were produced by two different teams. One example of the differences is the spatial resolution, which was 15 arc-minutes (30 kilometers) for the global team and 5 arc-minutes (10 kilometers) for the regional team.

IMPACT

IMPACT was initially developed at IFPRI to project global food supply, food demand, and food security to the year 2020 and beyond (Rosegrant et al. 2008). It is a partial equilibrium agricultural model that includes 32 crop and livestock commodities, including cereals, soybeans, roots and tubers, meats, milk, eggs, oilseeds, oilcakes and meals, sugar, and fruits and vegetables. IMPACT has 115 regions, which are usually countries (though in a few cases several countries are aggregated together, and in rare cases, a country may be subdivided), with specified supply, demand, and prices for agricultural commodities.

Large regions are further divided into major river basins. The result, portrayed in Figure 2.2, is 281 spatial units called food production units (FPUs). The model links the various countries and regions through international trade, using a series of linear and nonlinear equations to approximate the underlying production and demand relationships. World agricultural commodity prices are determined annually at levels that clear international markets. Growth in

FIGURE 2.2 International Model for Policy Analysis of Agricultural Commodities and Trade (IMPACT) unit of analysis, the food production unit (FPU)

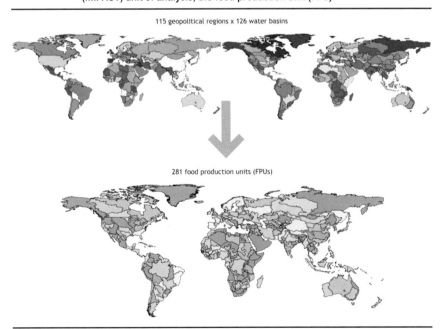

Source: Nelson et al. (2010).

crop production in each country is determined by crop and input prices, exogenous rates of productivity growth and area expansion, investment in irrigation, and water availability. Demand is a function of prices, income, and population growth. We distinguish four categories of commodity demand: food, feed, biofuels feedstock, and other uses.

From DSSAT to IMPACT

For input into the IMPACT model, DSSAT is run for five crops—rice, wheat, maize, soybeans, and groundnuts—at 15-arc-minute intervals for the locations where the SPAM dataset shows that the crop is currently grown. The results from this analysis are then aggregated to the IMPACT FPU level.

In extending these results to other crops in IMPACT, the primary assumption is that plants with the same photosynthetic metabolic pathway will react similarly to any given climate change effect in a particular geographic region. Millet, sorghum, sugarcane, and maize all use the C4 pathway. Millet and sugarcane are assumed to have the same productivity effects from climate change as maize in the same geographic regions. Sorghum effects for the Africa region were modeled explicitly, but for the rest of the world the maize productivity effects were assumed to apply to sorghum as well. The remainder of the crops use the C3 pathway. The climate effects for the C3 crops not directly modeled in DSSAT follow the average for wheat, rice, soy, and groundnuts from the same geographic region, with the following two exceptions. The IMPACT commodities of "other grains" and dryland legumes are directly mapped to the DSSAT results for wheat and groundnuts, respectively.

Income and Population Drivers

Differences in gross domestic product (GDP) and population growth define the overall scenarios, with all other driver values remaining the same across the three scenarios. Table 2.2 documents the GDP and population growth choices for the three overall scenarios.

The GDP and population growth rates combine to generate the three scenarios of per capita GDP growth. The results by region are shown in Table 2.3, and statistics for population and per capita GDP are shown in Table 2.4. The baseline scenario (see Table 2.4) has just over 9 billion people in 2050; the optimistic scenario results in a substantially smaller number, 7.9 billion; the pessimistic scenario results in 10.4 billion people. For developed countries, the differences among the three scenarios are relatively small, with little overall population growth: the population ranges from just over 1 billion to 1.3 billion in 2050 compared to 1 billion in 2010. For the developing countries as a

TABLE 2.2 Gross domestic product (GDP) and population choices for the three overall scenarios

Category	Pessimistic	Baseline	Optimistic
GDP (constant 2000 US$)	Lowest of the four GDP growth rate scenarios from the Millennium Ecosystem Assessment GDP scenarios (Millennium Ecosystem Assessment 2005) and the rate used in the baseline (next column)	Based on rates from a World Bank Economics of Adaptation to Climate Change study (World Bank 2010), updated for Africa south of the Sahara and South Asian countries	Highest of the four GDP growth rates from the Millennium Ecosystem Assessment GDP scenarios (Millennium Ecosystem Assessment 2005) and the rate used in the baseline (previous column)
Population	UN High variant, 2008 revision	UN medium variant, 2008 revision	UN low variant, 2008 revision

Source: Nelson et al. (2010).
Note: UN = United Nations; US$ = US dollars.

TABLE 2.3 Global average scenario per capita gross domestic product growth rate, 1990–2000 and 2010–2050 (percent per year)

		2010–2050		
Category	1990–2000	Pessimistic	Baseline	Optimistic
Developed countries	2.7	0.74	2.17	2.56
Developing countries	3.9	2.09	3.86	5.00
Low-income developing countries	4.7	2.60	3.60	4.94
Middle-income developing countries	3.8	2.21	4.01	5.11
World	2.9	0.86	2.49	3.22

Sources: *World Development Indicators* for 1990–2000 (World Bank 2009) and Nelson et al. (2010) calculations for 2010–2050.

group, the total 2010 population of 5.8 billion becomes 6.9–9 billion in 2050, depending on scenario.

As Table 2.4 shows, the average world per capita income, beginning at $6,600 in 2010, ranges from $8,800 to $23,800 in 2050, depending on scenario. The gap between average per capita income in developed and developing countries was large in 2010: developing countries' per capita income was only 5.6 percent of that in developed countries. Regardless of scenario, the relative difference is reduced over time: the developing-country income increases to between 8.6 percent and 14.0 percent of developed-country income in 2050, depending on overall scenario. Middle- and low-income developing

TABLE 2.4 Summary statistics for population and per capita gross domestic project, 2010 and 2050

Category	2010	2050 Optimistic	2050 Baseline	2050 Pessimistic
Population (millions)				
World	6,870	7,913	9,096	10,399
Developed countries	1,022	1,035	1,169	1,315
Developing countries	5,848	6,877	7,927	9,083
Middle-income developing countries	4,869	5,283	6,103	7,009
Low-income developing countries	980	1,594	1,825	2,074
East Africa	361	879	777	682
Southern Africa	142	276	240	207
West Africa	300	697	618	545
Income per capita (2000 US$)				
World	6,629	23,760	17,723	8,779
Developed countries	33,700	93,975	79,427	43,531
Developing countries	1,897	13,190	8,624	3,747
Middle-income developing countries	2,194	15,821	10,577	4,531
Low-income developing countries	420	4,474	2,094	1,101
East Africa	204	565	1,161	1,778
Southern Africa	1,961	2,725	5,892	11,499
West Africa	363	816	1,695	3,185

Source: Nelson et al. (2010).
Notes: 2010 income per capita is for the baseline scenario. US$ = US dollars.

countries' 2010 per capita income values are 6.5 percent and 2.6 percent, respectively, of the developed-country income. By 2050 the share increases to between 10.4 percent and 16.8 percent for middle-income developing countries, depending on overall scenario. For the low-income developing countries, however, the 2050 ratios remain low—between 2.5 percent and 4.8 percent.

The reader should be somewhat cautious in interpreting the results based on the three different scenarios. The optimistic scenario is optimistic not just for one country but for the entire world. This means that we cannot look at the impact from assuming that a single country is able to reduce its population growth rate or increase its GDP while the rest of the world continues at the same GDP and population growth rate. Rather, we only have the case in which the entire world has a higher GDP and lower population growth rate as well. This means that changing scenarios change supply and demand for the whole world, not just for one country.

Metrics for Human Well-being

Physical human well-being has many determinants. Calorie availability is a key element in low-income countries, where malnutrition and poverty are serious problems. Distribution, access, and supporting resources can enhance or reduce an individual's calorie availability. Similarly, child malnutrition has many determinants, including calorie intake (Rosegrant et al. 2008). The relationship used to estimate the number of malnourished children is based on a cross-country regression relationship estimated by Smith and Haddad (2000) that takes into account female access to secondary education, the quality of maternal and child care, and health and sanitation.[3] The IMPACT model provides data on per capita calorie availability by country; the other determinants are assumed to remain the same across the overall scenarios. Table 2.5 shows the 2010 and 2050 values for the noncaloric determinants of child malnutrition, aggregated to low- and middle-income countries. The small decline in female relative life expectancy in 2050 for the middle-income countries is primarily caused by a decline in China, where it is expected that male life expectancy will gradually move up rather than female life expectancy moving down.

Agricultural Vulnerability to Climate Change

There are many dimensions of agricultural vulnerability to climate change: vulnerability of agricultural systems, communities, households, and individuals to climate change. Vulnerability is influenced by the degree of exposure and

---

3 Because it is a partial equilibrium model, IMPACT has no feedback mechanisms from the effects of climate change on productivity to income. This means that it cannot directly estimate the poverty effects of agricultural productivity declines from climate change. However, the reduced-form function that relates child malnutrition to calorie availability and other determinants implicitly includes the effects of real income change on child malnutrition. Hertel, Burke, and Lobell (2010) used a general equilibrium model to explicitly estimate the effects of climate change on poverty. They found that the poverty impacts to 2030 "depend as much on where impoverished households earn their income as on the agricultural impacts themselves, with poverty rates in some non-agricultural household groups rising by 20–50% in parts of Africa and Asia under these price changes, and falling by equal amounts for agriculture-specialized households elsewhere in Asia and Latin America" (577).

TABLE 2.5 Noncaloric determinants of global child malnutrition, 2010 and 2050

	Clean water access (percent)[a]		Female schooling (percent)[b]		Female relative life expectancy[c]	
Country category	2010	2050	2010	2050	2010	2050
Middle-income countries	86.8	98.4	71.6	81.7	1.066	1.060
Low-income countries	69.0	85.8	54.9	61.6	1.044	1.048

Sources: Population-weighted aggregations in Nelson et al. (2010) based on data from 2000 with expert extrapolations to 2050. Original data sources include the World Health Organization's Global Database on Child Growth and Malnutrition; the United Nations Administrative Committee on Coordination–Subcommittee on Nutrition; the World Bank's *World Development Indicators* (World Bank 2009); FAOSTAT (FAO 2010); and the United Nations Educational, Scientific, and Cultural Organization's UNESCOSTAT database. Aggregations are weighted by population shares and are based on the baseline population growth scenario.
[a]Share of population with access to safe water.
[b]Total female enrollment in secondary education (any age group) as a percentage of the female age group corresponding to national regulations for secondary education.
[c]Ratio of female to male life expectancy at birth.

sensitivity to that exposure. Household-level vulnerability is most often associated with threats to livelihoods. Livelihoods can be inadequate because of resource constraints and low productivity (e.g., farmers with too little land and no access to fertilizer) or because farmer operate in a risky environment (e.g., droughts that cause harvest failure).

Potential impacts of climate change on vulnerability to food insecurity include both direct nutritional effects (changes in consumption quantities and composition) and livelihood effects (changes in employment opportunities and the cost of acquiring adequate nutrition). Climate change can affect each of these dimensions. This monograph focuses on the productivity effects of climate change that translate into changes in calorie availability and on the effects on child malnutrition. At this point the methodology and data to provide quantitative estimates of livelihood vulnerability are not available.

For the calorie estimates used in this monograph, we used price and income elasticities from an earlier global study (Nelson et al. 2010). Table 2.6 presents a weighted average of these elasticities for important crops in the southern African countries included in this monograph. The own-price elasticities are large (in absolute terms) relative to those reported in other studies. Other studies of starchy staple own-price elasticities report values that range from −0.05 to −0.3, whereas the elasticities in Table 2.6 are in the range of −0.45 to −0.72. The global modeling results in large price increases, which offset the effects of income increases in the pessimistic scenario and in some cases in the optimistic scenario as well.

TABLE 2.6 Mean price elasticities used for southern African countries in IMPACT, 2010 and 2050

	2010		2050	
Food	Income	Own price	Income	Own price
Beef	0.596	−0.602	0.502	−0.566
Poultry	0.534	−0.513	0.431	−0.453
Maize	0.110	−0.606	0.081	−0.612
Wheat	0.458	−0.615	0.439	−0.628
Cassava	0.154	−0.717	0.036	−0.662
Sugar	0.304	−0.562	0.284	−0.582

Source: Authors' calculations based on data from FAOSTAT (FAO 2010).
Notes: The numbers are weighted averages based on national consumption of each food item in 2000. IMPACT = International Model for Policy Analysis of Agricultural Commodities and Trade.

Although it is not possible to recalibrate elasticities and generate all new results for this monograph in a timely manner, the IMPACT model is being revised continuously, and future results will be based on improved elasticities.

Travel Time Maps

We developed databases that show simulated travel times to towns and cities of various sizes. The analysis begins with information on how long it would take someone to travel through a small region, roughly 10 kilometers on a side. This information is developed by overlaying various global information system datasets, including ones for roads, rivers and other water bodies, urban areas, and international boundaries. Each feature has a particular speed associated with it, and there is a default speed for areas without detailed information.

Once the time to travel across the regions is developed, the only other data required are the locations of the towns and cities of interest. We used cities and towns from two sources: CIESIN et al. (2004) and the *World Gazetteer* (Helders 2005). ArcView 3.2 was used to calculate the shortest travel time to any point in the specified cities and towns dataset.

Box-and-Whisker Graphs

A box-and-whisker graph summarizes a variety of information for a variable in a relatively straightforward diagram. A sample box-and-whisker graph is shown in Figure 2.3. The horizontal lines at the top and bottom of the

FIGURE 2.3 Sample box-and-whisker graph

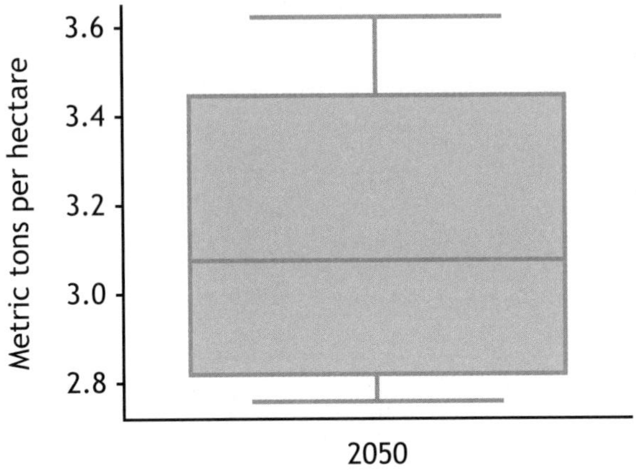

[Box-and-whisker plot with y-axis labeled "Metric tons per hectare" ranging from 2.8 to 3.6, and x-axis labeled "2050"]

Sources: Authors, using StataCorp (2009) and Tukey (1977).

diagram are the "whiskers" and show the minimum and maximum values of the variable. The top and bottom edges of the rectangle, the "box," show the 75th and 25th percentile of the variable under consideration. The horizontal divider line inside the box is the median value of the data.

These graphs were generated using Stata (StataCorp 2009) with Tukey's (1977) formula for setting the upper and lower whisker values, which Stata calls "adjacent values."

Now that we have given a general overview of the models and some of the data reviewed in this monograph, we are ready to see the results of the models applied to each of the countries studied in the chapters that follow.

References

Batjes, N., K. Dijkshoorn, V. van Engelen, G. Fischer, A. Jones, L. Montanarella, M. Petri, et al. 2009. *Harmonized World Soil Database.* Laxenburg, Austria: International Institute for Applied Systems Analysis (IIASA).

Bloom, A. J., M. Burger, R. Assensio, J. Salvador, and A. B. Cousins. 2010. "Carbon Dioxide Enrichment Inhibits Nitrate Assimilation in Wheat and Arabidopsis." *Science* 328: 899–902.

CIESIN (Center for International Earth Science Information Network), Columbia University, IFPRI (International Food Policy Research Institute), World Bank, and CIAT (Centro Internacional de Agricultura Tropical). 2004. *Global Rural–Urban Mapping Project (GRUMP), Alpha Version: Population Density Grids.* Palisades, NY, US: Socioeconomic Data and Applications Center (SEDAC), Columbia University. http://sedac.ciesin.columbia.edu/gpw.

FAO (Food and Agriculture Organisation of the United Nations). 2010. FAOSTAT. Accessed December 22, 2010. http://faostat.fao.org.

Helders, S. 2005. *World Gazetteer Database.* Accessed June 7, 2007. http://world-gazetteer.com/.

Hertel, T. M., M. B. Burke, and D. B. Lobell. 2010. "The Poverty Implications of Climate-Induced Crop Yield Changes by 2030." *Global Environmental Change* 20 (4): 577–585.

Jones, J. W., G. Hoogenboom, C. H. Porter, K. J. Boote, W. D. Batchelor, L. A. Hunt, P. W. Wilkens, U. Singh, A. J. Gijsman, and J. T. Ritchie. 2003. "The DSSAT Cropping System Model." *European Journal of Agronomy* 18 (3–4): 235–265.

Jones, P. G., P. K. Thornton, and J. Heinke. 2009. *Generating Characteristic Daily Weather Data Using Downscaled Climate Model Data from the IPCC's Fourth Assessment.* Project report. Nairobi, Kenya: International Livestock Research Institute.

Long, S. P., E. A. Ainsworth, A. D. B. Leakey, J. Nosberger, and D. R. Ort. 2006. "Food for Thought: Lower-than-Expected Crop Yield Stimulation with Rising CO_2 Concentrations." *Science* 312 (5782): 1918–1921. doi:10.1126/science.1114722.

Millennium Ecosystem Assessment. 2005. *Ecosystems and Human Well-being: Synthesis.* Washington, DC: Island Press. http://www.maweb.org/en/Global.aspx.

Nakicenovic, N., et al. 2000. *Special Report on Emissions Scenarios: A Special Report of Working Group III of the Intergovernmental Panel on Climate Change.* Cambridge: Cambridge University Press. www.grida.no/climate/ipcc/emission/index.htm.

Nelson, G. C., M. W. Rosegrant, A. Palazzo, I. Gray, C. Ingersoll, R. Robertson, S. Tokgoz, et al. 2010. *Food Security, Farming, and Climate Change to 2050: Scenarios, Results, Policy Options.* Washington, DC: International Food Policy Research Institute.

Rosegrant, M. W., S. Msangi, C. Ringler, T. B. Sulser, T. Zhu, and S. A. Cline. 2008. *International Model for Policy Analysis of Agricultural Commodities and Trade (IMPACT): Model Description.* Washington, DC: International Food Policy Research Institute.

Smith, L., and L. Haddad. 2000. *Explaining Child Malnutrition in Developing Countries: A Cross-Country Analysis.* Washington, DC: International Food Policy Research Institute.

StataCorp. 2009. Stata: Release 11. Statistical Software. College Station, TX, US: StataCorp.

Tukey, J. W. 1977. *Exploratory Data Analysis.* Reading, MA, US: Addison–Wesley.

World Bank. 2009. *World Development Indicators.* Washington, DC.

———. 2010. *Economics of Adaptation to Climate Change: Synthesis Report.* Washington, DC. http://climatechange.worldbank.org/content/economics-adaptation-climate-change-study-homepage.

You, L., and S. Wood. 2006. "An Entropy Approach to Spatial Disaggregation of Agricultural Production." *Agricultural Systems* 90 (1–3): 329–347.

You, L., S. Wood, and U. Wood-Sichra. 2006. "Generating Global Crop Distribution Maps: From Census to Grid." Paper presented at the International Association of Agricultural Economists Conference in Brisbane, Australia, August 11–18.

———. 2009. "Generating Plausible Crop Distribution and Performance Maps for Sub-Saharan Africa Using a Spatially Disaggregated Data Fusion and Optimization Approach." *Agricultural Systems* 99 (2–3): 126–140.

Zavala, J. A., C. L. Casteel, E. H. DeLucia, and M. R. Berenbaum. 2008. "Anthropogenic Increase in Carbon Dioxide Compromises Plant Defense against Invasive Insects." *Proceedings of the National Academy of Sciences, USA* 105 (13): 5129–5133. doi:10.1073/pnas.0800568105.

Chapter 3

BOTSWANA

Peter P. Zhou, Tichakunda Simbini, Gorata Ramokgotlwane, Timothy S. Thomas, Sepo Hachigonta, and Lindiwe Majele Sibanda

In this chapter we assess the vulnerability of Botswana's agriculture to climate change, with special emphasis on impacts on the poor. The agriculture sector is inherently vulnerable to climate change, and Botswana's semiarid climate already severely limits agricultural production. Following a broad overview of current economic and demographic indicators, land use, and agricultural performance, we summarize the current state of institutional policy, programs, and strategies relating to agriculture and climate change. The section "Scenarios for the Future" summarizes projected economic and demographic trends, as well as the results of the modeled biophysical scenarios relating to climate and crop production as they apply to Botswana. The analysis of these scenarios addresses specific areas of agricultural vulnerability to climate change, taking into account projected trends in global agricultural commodity prices in relation to climate change.

The project methodology included a literature review as well as stakeholder consultations. The literature review, to evaluate agriculture production, policies, strategies, and programs, was complemented by stakeholder consultation meetings with representatives of the Ministry of Agriculture (MoA), the Department of Meteorological Services, and the Department of Environmental Affairs to gain insights into the performance, constraints, and activities of the agricultural sector. Complete statistical information is not yet available, because most of the departments are still compiling statistical databases of the country's agriculture production. For example, the Crop Production Department is working on a database for the past 30 years that will indicate changes in crop yields for each district and crop.

The Policy Context of Agriculture Production

Botswana's *Vision 2016* (Botswana Vision 2016 Council 1997), developed in 1997, has guided national and sectoral development planning since then. It states that agriculture in Botswana will be productive, profitable, and sustainable and will make a full contribution to economic development, poverty alleviation, food security, quality of life, and sustainable use of natural resources. *Vision 2016* does not have targets for assessing or alleviating the vulnerability of agriculture to climate change, perhaps because it was formulated before climate change was recognized by Botswana's government as an issue.

National Development Plan (*NDP*) 9 (Botswana, Ministry of Finance and Development Planning 2003), for the period 2003–2009, stressed the need to be less dependent on diamond mining and to diversify the agricultural production base. Under that plan, the government adopted the following strategies for irrigation and water development:

- Develop a gender-sensitive irrigation policy.
- Establish two irrigation schemes using treated effluent, at Lobatse and Francistown.
- Establish an irrigation systems testing center.
- Construct new dams, and assist farmers in rehabilitating and upgrading existing dams.
- Continue the well rehabilitation program.
- Explore rainwater harvesting technologies in settlements to promote backyard gardening.
- Contract private companies to complement government efforts in developing irrigation and water resources.

Under NDP 9, 48 wells and 37 dams were rehabilitated for livestock purposes, and this effort is continuing under NDP 10. The irrigation schemes that were developed are listed in the *National Master Plan for Arable Agriculture and Dairy Development* (*NAMPAADD*) (Botswana, Ministry of Agriculture 2010c).

The Integrated Support Program for Arable Agriculture Development (Botswana, Ministry of Agriculture 2010a) was introduced in 2008 to replace the Arable Land Development Program, and it was meant to address the challenges facing arable farmers as well as the inherently low productivity of the arable subsector. The primary objectives of the program are to

- increase grain production,
- promote food security at the household and national levels,

- commercialize agriculture through mechanization,
- facilitate access to farm inputs and credit, and
- improve extension outreach.

The performance of the arable subsector is highly dependent on the availability of water. It is therefore essential to improve the availability of water to cluster areas. The program aims to drill new boreholes and purchase and rehabilitate existing boreholes. The program has a component for the provision of farming inputs such as draft power, fertilizer, and seeds. Fertilizer and draft power are provided free of charge to all farmers with 5 hectares of land or less (which includes virtually all poor farming households); free seed is provided to all farmers with 15 hectares or less. For farms that exceed these size thresholds, inputs are provided at subsidized rates.

The Livestock Management and Infrastructure Development Program (LIMID), Phase 1 (Botswana, Ministry of Agriculture 2010b), is a government program to improve food security and eliminate poverty, aimed specifically at the livestock sector. The program focuses on resource-poor households, though without specific attention to gender, and provides three packages. The program objectives are to

- promote food security through the improved productivity of cattle, small stock, and Tswana chickens;
- improve livestock management;
- improve range resource use and conservation;
- eradicate poverty; and
- provide infrastructure for the safe and hygienic processing of poultry (or meat).

The results of the LIMID evaluation demonstrated that more women than men participated in LIMID, especially in the resource-poor component. The small stock livestock population (sheep and goats) increased from 9,007 to 11,405 in the districts where the program was piloted, representing an increase of 25.4 percent achieved through the program.

The NAMPAADD program, also designed to address food security and poverty alleviation, focuses on upgrading communal farms (crops and dairy) to commercial farms. It aims to

- improve food security at the household and national levels, with an emphasis on household food security;
- diversify the agricultural production base;
- increase agricultural output and productivity;
- increase employment opportunities for the rapidly growing labor force;
- provide a secure agricultural and productive environment for agricultural producers; and
- conserve scarce agricultural and land resources for future generations.

An evaluation of the performance of the rainfed agriculture sector in the *NAMPAADD* (Botswana, Ministry of Agriculture 2010c) shows that the arable farming sector is severely hampered by unfavorable agroclimatic conditions, including endemic drought and high summer temperatures. Although the NAMPAADD program is aware of the impact of climatic conditions on small-scale farmers, it does not articulate a clear action plan for addressing the vulnerability of agriculture to climate change.

Review of Current Trends

Botswana is a landlocked country in the middle of southern Africa, bordered by Namibia to the west, Zambia and Zimbabwe to the northeast, and South Africa to the south. Predominantly flat tableland, the country covers 581,730 square kilometers at a mean altitude of 1,000 meters. The country is divided into 10 districts: Central, Chobe, Ghanzi, Kgalagadi, Kgatleng, Kweneng, North-East, Ngamiland, South-East, and Southern.

Botswana's climate is semiarid, characterized by warm winters, hot summers, low rainfall, and high levels of evapotranspiration. In the center of the country is the Kalahari Desert sandveld, covering two-thirds of the country and suitable mainly for livestock and wildlife. The eastern part of the country consists mostly of loamy clay soils that are more fertile and better suited to crop production. In the northeastern part of the country is the wet sandveld, characterized by green wetlands and an abundance of wildlife. The remainder of the country is transition sandveld.

Rainfall is erratic, averaging 450 millimeters per year and ranging from 650 millimeters in the northeast (Chobe) to just 250 millimeters in the southwest (Gemsbok National Park). The rainfall season varies from year to year and is punctuated by periods of severe drought. Summer temperatures can soar to 44°C between October and March, averaging between 35°C and 40°C during midday. The hottest months are December and January. Night temperatures seldom fall below 26°C; winters are mostly dry, with daytime temperatures of about 27°C, dropping to around 7°C at night. Occasionally it drops below freezing, with July being the coldest month.

Botswana's climate greatly limits the country's food production capacity. Only about 5 percent of the country's land area is suitable for cultivation, and less than 1 percent is currently cultivated, mainly in the eastern part of the country where conditions are favorable for crop production. According to the Ministry of Agriculture (2010c), the frequency of droughts has been increasing over the past few years, from once every four years to once every two years; thus, the years 2001–2003, 2005–2006, and 2006–2007 were drought years.

Economic and Demographic Indicators

Population

The population of Botswana is currently estimated at just over 1.8 million. According to the 2001 population census (Botswana, Central Statistics Office 2001), 55 percent of the population resides in urban areas and the remaining 45 percent in rural areas. Figure 3.1 shows the country's total and rural

FIGURE 3.1 Population trends in Botswana: Total population, rural population, and percent urban, 1960–2008

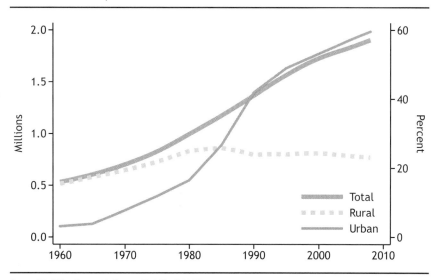

Source: *World Development Indicators* (World Bank 2009).

TABLE 3.1 Population growth rates in Botswana, 1960–2008 (percent)

Decade	Total growth rate	Rural growth rate	Urban growth rate
1960–1969	2.7	2.3	11.9
1970–1979	3.5	2.5	11.1
1980–1989	3.2	−0.2	12.6
1990–1999	2.4	0.2	4.9
2000–2008	1.2	−0.6	2.6

Source: Authors' calculations based on *World Development Indicators* (World Bank 2009).

population (left axis) and the share of urban population (right axis). The urban population is estimated to have increased from 41.9 percent in 1990 to 57.4 percent in 2005 and is projected to reach 64.6 percent in 2015 (*National Development Plan 9*) (Botswana, Ministry of Finance and Development Planning 2003). This urbanization trend is driven not only by migration from rural areas but also by the reclassification of settlements from rural to "urban villages" when their population reaches 5,000.

Additional information concerning rates of population growth is shown in Table 3.1, indicating a negative population growth rate in rural areas for the years 1980–1989 and 2000–2008. This has had a negative impact on the agriculture sector, depriving the rural sector of much-needed labor. The MoA reports a rising proportion of aged individuals among communal farmers.

Income

The *Population and Housing Census* of 2001 (Botswana, Central Statistics Office 2001) and the subsequent *Household Income and Expenditure Survey* of 2002/2003 (Botswana, Central Statistics Office 2004) found 394,272 households in Botswana, of which 41.4 percent were rural. The report highlighted income disparities among urban, urban village, and rural households, with the rural households earning the least and urban households earning two to three times as much as the rural households.

The report also indicated that the poverty levels (people living on less than US$1 per day) were declining, from 46.7 percent in 1993/1994 to 30.3 percent by 2002/2003. However, the gains in poverty reduction may still result in Botswana's falling short of its *Vision 2016* target, halving the poverty level to 23 percent by 2006 and reducing it to zero by 2016. Figure 3.2 shows that, in spite of significant gains made in reducing poverty in towns and cities, the poverty level in rural areas still remains significantly high. While the poverty

FIGURE 3.2 Poverty rates in Botswana by region and urban/rural, 2003

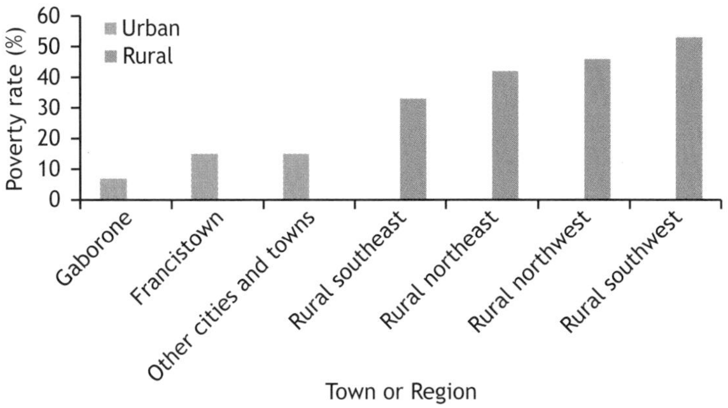

Source: Botswana, Central Statistics Office (2004).

levels in towns are less than 15 percent, in rural areas they range from 33 percent to as high as 53 percent (South-West District).

Table 3.2 indicates that income distribution worsened between 1993/1994 and 2002/2003, as reflected by an increase in the Gini coefficient from 0.537 to 0.573. Significantly, the proportion of households without livestock (cattle) rose from 50.2 to 62.5 percent; the decline in cattle raising may reflect urban migration, and it may also indicate a cause of increasing rural poverty, because people sold their cattle during the droughts. As the poverty rate has gone down, the gap between poor and rich has widened. So either way, despite the development gains and lower poverty rates, this development has taken place very unequally.

TABLE 3.2 Income and poverty indicators for Botswana, 1984/1985, 1993/1994, and 2002/2003

Variable	1984/1985	1993/1994	2002/2003
Poverty rate (percent of population)		46.7	30.3
Income distribution (Gini coefficient, disposable income)	0.556	0.537	0.573
Livestock ownership (percent of households)			
Households without cattle	50.2	54.6	62.5
Households without goats	46.9		63.0
Households without sheep			92.3
Households without chickens			59.1

Source: Botswana, Central Statistics Office (2004).
Note: Blank cells indicate that no values were reported for those years.

Land Use Overview

Figure 3.3 shows land cover and land use in Botswana as of 2000. Over 80 percent of Botswana's vegetation is classified as shrub cover under the classification of the Food and Agriculture Organization of the United Nations (FAO). Gazetted forests (government property) occupy an area of 4,555 square kilometers (Botswana, Ministry of Minerals, Energy, and Water Resources 2009). The land cover is dominated by savannah (mixed tree and grass systems) of various forms. The major plant communities include shrub savannah, tree savannah, and closed tree savannah on rocky hills, semiarid shrub savannah, aquatic grassland, dry deciduous forest, and woodland. The southwestern parts of the country are characterized by shrub savannah, whereas the extreme southwest (the driest region) has sparse vegetation and rolling sand dunes. Vegetation becomes denser toward the north and east, changing to open tree savannah, then woodlands and dry forest (Botswana, Ministry of Minerals, Energy, and Water Resources 2009).

Figure 3.4 shows the locations of protected areas in Botswana, including parks and reserves. These locations provide important protection for fragile environments, which are also important for the tourism industry. The game reserves, national parks, quarantine areas, and wildlife management areas are sensitive places and are not suitable for any agricultural activity. Other land-use categories in Botswana include government initiative areas, such as the Botswana Livestock Development Corporation ranches, freehold farms, game reserves, national parks, pastoral/arable/residential areas, quarantine camps, Tribal Grazing Land Policy ranches, wildlife management areas, forest reserves, commercial farms, government ranches, and residential land-use areas.

Figure 3.5 shows the travel times to urban areas of various sizes, which are potential markets for agricultural products and potential sources for agricultural inputs and consumer goods for rural households. Most of Botswana's population is concentrated in the eastern part of the country, and travel in these areas takes far less time (at most 5–8 hours to reach a town of 10,000 people). In the central and western parts of the country, it takes at least 16 hours to reach a town of 10,000 people.

Currently only 45 percent of farmers have access to roads, 17 percent to electricity, 22 percent to telecommunication, 65 percent to water for livestock and domestic use, and 43 percent to water for irrigation. The absence or limitations of infrastructure in production zones is associated with poor agricultural performance and stagnating farm incomes.

The Government of Botswana has planned to implement an Agriculture Infrastructure Development Initiative in NDP 10 for the planning period

FIGURE 3.3 Land cover and land use in Botswana, 2000

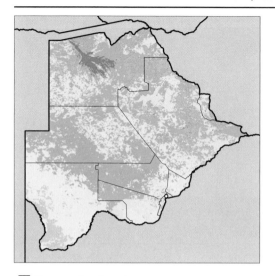

- Tree cover, broadleaved, evergreen
- Tree cover, broadleaved, deciduous, closed
- Tree cover, broadleaved, open
- Tree cover, broadleaved, needle-leaved, evergreen
- Tree cover, broadleaved, needle-leaved, deciduous
- Tree cover, broadleaved, mixed leaf type
- Tree cover, broadleaved, regularly flooded, fresh water
- Tree cover, broadleaved, regularly flooded, saline water
- Mosaic of tree cover/other natural vegetation
- Tree cover, burnt
- Shrub cover, closed-open, evergreen
- Shrub cover, closed-open, deciduous
- Herbaceous cover, closed-open
- Sparse herbaceous or sparse shrub cover
- Regularly flooded shrub or herbaceous cover
- Cultivated and managed areas
- Mosaic of cropland/tree cover/other natural vegetation
- Mosaic of cropland/shrub/grass cover
- Bare areas
- Water bodies
- Snow and ice
- Artificial surfaces and associated areas
- No data

Source: GLC2000 (Bartholome and Belward 2005).

FIGURE 3.4 Protected areas in Botswana, 2009

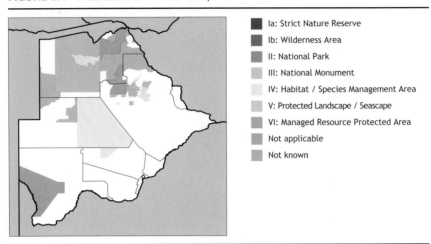

Sources: Protected areas are from the World Database on Protected Areas (UNEP and IUCN 2009). Water bodies are from the World Wildlife Fund's Global Lakes and Wetlands Database (Lehner and Döll 2004).

2010–2015/2016. The main aim of the initiative is to provide roads and electricity in priority agriculture production zones.

Agriculture Overview

Overall Performance

The contribution of agriculture to gross domestic product (GDP) in Botswana has declined drastically, from 40 percent at the country's independence in 1966 to 1.6 percent in 2007 (Botswana, Ministry of Agriculture 2010d), mainly due to rapid growth in the mining sector during the same period. More than 80 percent of the sector's GDP is from livestock production; crop production contributes slightly less than 20 percent. Both livestock and crop production are dominated by communal farms. The agriculture sector accounts for 30 percent of the country's employment.

Table 3.3 shows that over the period from 2000 to 2007, agricultural GDP grew at only 0.4 percent per year compared to the average growth rate of 5.6 percent of the national GDP for the same period. This low growth rate was mainly due to extensive droughts (indicated in Table 3.3 by a D in the column for each drought year).

In spite of significant investment by the Government of Botswana to boost agricultural production, major challenges hamper productivity, including

FIGURE 3.5 Travel time to urban areas of various sizes in Botswana, circa 2000

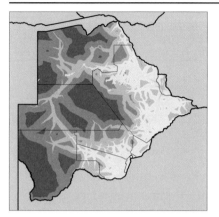

To cities of 500,000 or more people

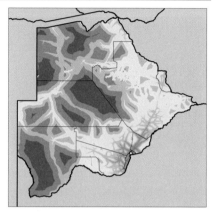

To cities of 100,000 or more people

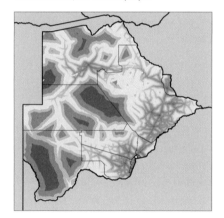

To towns and cities of 25,000 or more people

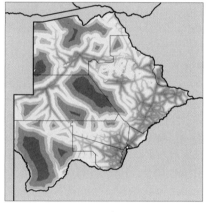

To towns and cities of 10,000 or more people

- Urban location
- < 1 hour
- 1–3 hours
- 3–5 hours
- 5–8 hours
- 8–11 hours
- 11–16 hours
- 16–26 hours
- > 26 hours

Source: Authors' calculations.

TABLE 3.3 Agricultural and national gross domestic product (GDP) for Botswana, 2000–2007

Years	2000/ 2001	2001/ 2002	2002/ 2003	2003/ 2004	2004/ 2005	2005/ 2006	2006/ 2007	Annual 2000–2007[a]
Drought years (D)		D	D		D	D	D	
Agricultural GDP	9.9	−2.6	1.8	2.8	−11.0	−3.6	6.8	0.4
Total GDP	9.0	1.6	9.7	2.8	9.7	0.5	6.1	5.6

Source: Botswana, Ministry of Agriculture (2010d).
[a] Average figures for 2000–2007.

not only adverse agroclimatic and agroecological conditions but also a risk-averse communal farm culture, low levels of skills, an aging farming population, and mistargeted and poorly formulated agricultural programs (Batisani 2010). Batisani reports a 31-year decreasing trend in land cultivation across Botswana, highly correlated to the previous year's rainfall. The increasing frequency of droughts is likely to exacerbate this trend.

Crop Production

Crop production in Botswana is dominated by grains and horticulture. There are two main farming systems, subsistence and commercial farming. Most of the subsistence farms are situated in the eastern hard-veld areas of the country (300,000 hectares), with better rainfall, better soils, and better access to water than in the western region. Subsistence farm production is characterized by low use of fertilizer and certified seeds, low mechanization, and limited irrigation. In contrast, commercial farms are mechanized and use modern inputs (fertilizers, hybrid seeds, and pesticides)

Table 3.4 shows that Botswana had a production deficit averaging 99 percent for maize and 46 percent for sorghum for the period 2001–2007. These deficits were filled through imports from South Africa, creating food dependence and vulnerability to food price inflation.

Livestock Production

Livestock production is the mainstay of the rural population, which constitutes 42.6 percent of Botswana's total population. Like crop production, animal production consists of communal and commercial production; it is dominated by beef production. About 85 percent of the national herd is grazed on communal lands and is hampered by the shortage of water; herders depend mainly on borehole water. Cattle production is the only source of agricultural exports for Botswana. Average beef production has been declining. as shown in Table 3.5.

TABLE 3.4 Estimation of requirements, production, and deficits of maize and sorghum/millet in Botswana, 2001/2002–2006/2007 (metric tons)

Crop	Year	Requirements	Production	Deficit
Maize	2001/2002	124,000	17,412	−106,588
	2002/2003	131,000	1,633	−129,367
	2003/2004	125,000	6,220	−118,780
	2004/2005	141,000	2,586	−138,414
	2005/2006	140,000	10,467	−129,533
	2006/2007	144,000	751	−143,249
Sorghum/millet	2001/2002	62,000	31,625	−30,375
	2002/2003	55,000	54,362	−638
	2003/2004	65,000	35,134	−29,866
	2004/2005	64,000	21,164	−42,836
	2005/2006	60,000	25,604	−34,396
	2006/2007	57,000	26,193	−30,807

Source: Botswana, Ministry of Agriculture, various years.

TABLE 3.5 Beef exports from Botswana to the EU, 2001–2007 (metric tons)

Export item/total exports/export quota or fulfilment	2001	2002	2003	2004	2005	2006	2007
Chilled meat	8,118	4,781	5,376	4,692	4,198	2,986	1,119
Frozen meat	8,110	5,010	3,825	5,433	3,375	3,140	5,336
Total exports	16,228	9,791	9,201	10,125	7,573	6,126	6,455
Export quota	18,916	18,916	18,916	18,916	18,916	18,916	18,916
Quota fulfillment	86%	52%	49%	54%	40%	32%	34%

Source: NDP 10–BMC Reports (Botswana, Ministry of Finance and Development Planning 2010; Botswana Meat Commission 2010).

Beef production is hampered by the lack of quality breeding stock, the lack of infrastructure in production areas, poor livestock husbandry, and diseases.

Table 3.5 shows beef exports to the EU from 2001 to 2007. The fulfillment of the beef export quota has been declining sharply, from 86 percent in 2001 to 34 percent in 2007. The MoA attributes this decline to outbreaks of disease, including foot and mouth disease (FMD) and bovine measles (affecting 10 percent of carcasses from the Botswana Meat Commission).

Dairy production is very low, currently meeting only 3 percent of the national milk requirement. Small stock production contributes negligibly to the sector's GDP but plays an important socioeconomic role in the lives of the rural poor as a source of food and income (mostly to female-headed households). The small stock population has shown a decreasing trend, mostly due to poor management, disease, and parasite infestations. The MoA reports a high mortality rate for small stock during wet years because of the high incidence of disease. This has been especially the case in the eastern part of the country due to heartwater disease, which is now being observed in areas formerly designated free of the disease.

The Department of Veterinary Services (MoA) recognizes that most animal diseases are influenced by weather conditions, and controlling them requires restricting the movement of animals. FMD, one of the most dangerous diseases that affects the economy and the livelihoods of farmers, occurs more often during dry seasons, when buffalo and cattle come in contact at the limited watering holes.

Scenarios for the Future

Socioeconomic Scenarios

Population Growth

Figure 3.6 shows population projections for Botswana for the year 2050 using three different scenarios: pessimistic, baseline, and optimistic. The Central Statistics Office indicates that HIV and AIDS will have a significant impact on slowing the population growth rate. Slow population growth would allow the programs addressing national and household food security to serve a smaller population than otherwise. However, in combination with urbanization, it may still severely reduce the agricultural labor force.

Economic Growth and Development

Figure 3.7 presents three overall scenarios for GDP per capita derived by combining three GDP scenarios with the three population scenarios of Figure 3.6 (based on United Nations population data). The optimistic scenario combines high GDP growth with low population growth for all countries, the baseline scenario combines medium GDP growth with medium population growth, and the pessimistic scenario combines low GDP growth with the high population growth. (The agricultural modeling in the next section uses these scenarios as well.)

The optimistic scenario incorporates the low population growth rate projected by the Central Statistics Office. However, it is not clear that the GDP projections can be realized, because after 2025 output in the mining sector is expected to decline, severely affecting GDP. GDP is thus projected to rise in

FIGURE 3.6 Population projections for Botswana, 2010–2050

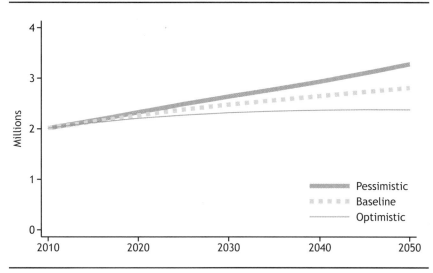

Source: UNPOP (2009).

FIGURE 3.7 Gross domestic product (GDP) per capita in Botswana, future scenarios, 2010–2050

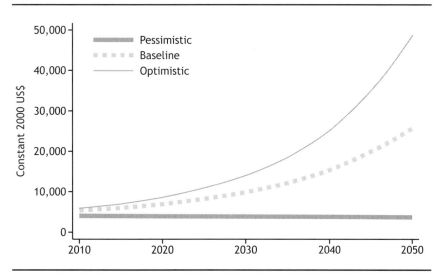

Sources: Computed from GDP data from the World Bank Economic Adaptation to Climate Change project (World Bank 2010), from the Millennium Ecosystem Assessment (2005) reports, and from population data from the United Nations (UNPOP 2009).
Note: US$ = US dollars.

line with the optimistic scenario up to 2025–2030. Without significant economic diversification, agricultural production will likely become a more significant share of GDP as mining GDP declines after 2025–2030. The projected increase in GDP per capita, coupled with population growth and continued rural–urban migration, is likely to result in increased food demand as income and quality of life improve. The increase in demand, if not matched with improvement in agriculture production, could result in increased food prices, or at least increased demand for imported foods.

Biophysical Analysis

Climate Models

Figure 3.8 shows precipitation projections for 2000–2050 based on general circulation models (GCMs) using the A1B scenario.[1] Four downscaled climate models are used: the ECHAM 5, CNRM-CM3, CSIRO Mark 3, and MIROC 3.2 GCMs.[2] CNRM-CM3, CSIRO Mark 3, and MIROC 3.2 show a minimal annual change in rainfall (–50 to 50 millimeters) in the central, northern, eastern, and western parts of the country. CSIRO Mark 3, ECHAM 5, and MIROC 3.2 all show a reduction in precipitation of between –100 and –50 millimeters in the southern and southeastern regions. CNRM-CM3 and CSIRO Mark 3 show an increase in precipitation in the southwestern or parts of the northern region. Only ECHAM 5 shows a general decrease in precipitation across most of the country, ranging between –200 and –100 millimeters.

The most pessimistic scenario thus shows a decrease in annual precipitation across Botswana of between –50 and –200 millimeters. The most probable scenario, in the sense that three out of four models considered here seem to agree, is little to no change in most of the country.

Figure 3.9 shows changes in average daily maximum temperature for the warmest month. All of the GCMs generally show an overall increase in the annual maximum temperature by margins ranging between 1.5° and 3.5°C. The ECHAM 5 GCM shows dramatic shifts of more than 3.0°C for the

---

1 The A1B scenario describes a world of very rapid economic growth, low population growth, and rapid introduction of new and more efficient technologies, with moderate resource use and a balanced use of technologies.
2 CNRM-CM3 is National Meteorological Research Center–Climate Model 3. CSIRO Mark 3 is a climate model developed at the Australia Commonwealth Scientific and Industrial Research Organisation. ECHAM 5 is a fifth-generation climate model developed at the Max Planck Institute for Meteorology in Hamburg. MIROC 3.2 is the Model for Interdisciplinary Research on Climate, developed at the University of Tokyo Center for Climate System Research.

FIGURE 3.8 Changes in mean annual precipitation in Botswana, 2000–2050, A1B scenario (millimeters)

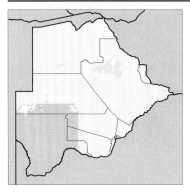

CNRM-CM3 GCM

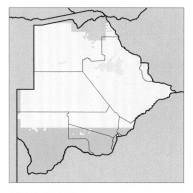

CSIRO Mark 3 GCM

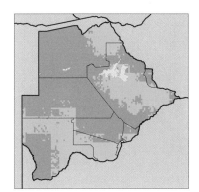

ECHAM 5 GCM

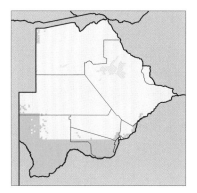

MIROC 3.2 medium-resolution GCM

- < –400
- –400 to –200
- –200 to –100
- –100 to –50
- –50 to 50
- 50 to 100
- 100 to 200
- 200 to 400
- > 400

Source: Authors' calculations based on Jones, Thornton, and Heinke (2009).
Notes: A1B = greenhouse gas emissions scenario that assumes fast economic growth, a population that peaks midcentury, and the development of new and efficient technologies, along with a balanced use of energy sources; CNRM-CM3 = National Meteorological Research Center–Climate Model 3; CSIRO = climate model developed at the Australia Commonwealth Scientific and Industrial Research Organisation; ECHAM 5 = fifth-generation climate model developed at the Max Planck Institute for Meteorology (Hamburg); GCM = general circulation model; MIROC = Model for Interdisciplinary Research on Climate, developed by the University of Tokyo Center for Climate System Research.

FIGURE 3.9 Change in monthly mean maximum daily temperature in Botswana for the warmest month, 2000–2050, A1B scenario (°C)

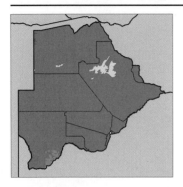
CNRM-CM3 GCM

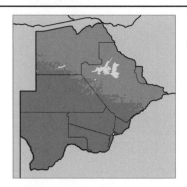
CSIRO Mark 3 GCM

ECHAM 5 GCM

MIROC 3.2 medium-resolution GCM

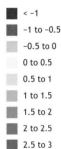

- < –1
- –1 to –0.5
- –0.5 to 0
- 0 to 0.5
- 0.5 to 1
- 1 to 1.5
- 1.5 to 2
- 2 to 2.5
- 2.5 to 3
- 3 to 3.5
- > 3.5

Source: Authors' calculations based on Jones, Thornton, and Heinke (2009).
Notes: A1B = greenhouse gas emissions scenario that assumes fast economic growth, a population that peaks midcentury, and the development of new and efficient technologies, along with a balanced use of energy sources; CNRM-CM3 = National Meteorological Research Center–Climate Model 3; CSIRO = climate model developed at the Australia Commonwealth Scientific and Industrial Research Organisation; ECHAM 5 = fifth-generation climate model developed at the Max Planck Institute for Meteorology (Hamburg); GCM = general circulation model; MIROC = Model for Interdisciplinary Research on Climate, developed by the University of Tokyo Center for Climate System Research.

entire country. This general increase in temperature would mean an increase in evapotranspiration, magnifying the effect of a decrease in precipitation and implying that the country would become drier. This effect would be felt most severely in the eastern parts of the country because this is the area with most of the agricultural production (crops and livestock) and most of the population, though the CSIRO Mark 3 and MIROC 3.2 GCMs at least indicate that the east will have smaller temperature increases than the southwest.

Crop Models

The Decision Support System for Agrotechnology Transfer (DSSAT) crop model was used to compute yields under current temperature and precipitation regimes and for the projected climates given by our four GCMs. The future yield results from DSSAT were then compared to the current or baseline yield results from DSSAT. The outputs for key crops are mapped in the following two sets of figures, which compare yields for 2050 under climate change with the yields assuming an unchanged (2000) climate.

All of the GCMs show a maize yield gain of more than 25 percent over the 2000 baseline for the North-East and Pandamatenga Districts (Figure 3.10). The CNRM-CM3 and MIROC 3.2 GCMs both show yield gains ranging from 5 percent to more than 25 percent over the baseline scenario, mostly concentrated in the Kweneng and Kgatleng Districts. For the Kgalagadi District, the GCMs show scattered effects. The Kgalagadi District could be the hardest hit by climate change; both the MIROC 3.2 and the ECHAM 5 GCMs indicate that there will be an almost total loss of land for maize.

The modeled results of the projected impacts of climate change on the production of rainfed sorghum are shown in Figure 3.11. All the GCMs show yield losses ranging from 5 to 25 percent in the Ghanzi District, the Ngamiland District, and half of the Central District. CNRM-CM3 and MIROC 3.2 both show some yield gains in parts of the Central District and in almost all of the Kweneng and Southern Districts. The GCMs differ regarding the Kgalagadi District. ECHAM 5, MIROC 3.2, and CSIRO-Mark 3 show losses of land compared to the baseline for most of the districts, whereas the CNRM-CM3 GCM projects mostly yield increases and even some addition of cultivable land. The yield losses in the western part of the country also correspond to the 31 percent yield losses projected by Chipanshi, Chanda, and Totolo (2003); however, the scenarios do not duplicate these authors' projection of yield reductions of 10 percent in the eastern parts of the country.

We note that the areas where most of the yield gains are projected are dominated by subsistence farms that are dependent on rainfed agriculture.

FIGURE 3.10 Yield change under climate change: Rainfed maize in Botswana, 2000–2050, A1B scenario

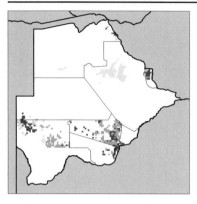
CNRM-CM3 GCM

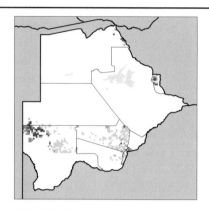
CSIRO Mark 3 GCM

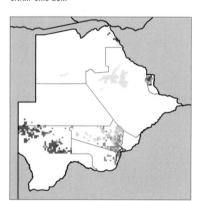
ECHAM 5 GCM

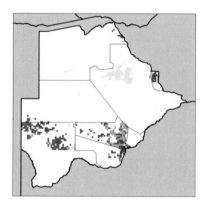
MIROC 3.2 medium-resolution GCM

- Baseline area lost
- Yield loss >25% of baseline
- Yield loss 5–25% of baseline
- Yield change within 5% of baseline
- Yield gain 5–25% of baseline
- Yield gain > 25% of baseline
- New area gained

Source: Authors' calculations.
Notes: A1B = greenhouse gas emissions scenario that assumes fast economic growth, a population that peaks midcentury, and the development of new and efficient technologies, along with a balanced use of energy sources; CNRM-CM3 = National Meteorological Research Center–Climate Model 3; CSIRO = climate model developed at the Australia Commonwealth Scientific and Industrial Research Organisation; ECHAM 5 = fifth-generation climate model developed at the Max Planck Institute for Meteorology (Hamburg); GCM = general circulation model; MIROC = Model for Interdisciplinary Research on Climate, developed by the University of Tokyo Center for Climate System Research.

FIGURE 3.11 Yield change under climate change: Rainfed sorghum in Botswana, 2000–2050, A1B scenario

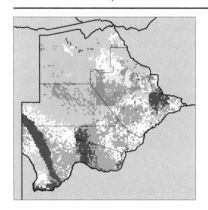

CNRM-CM3 GCM

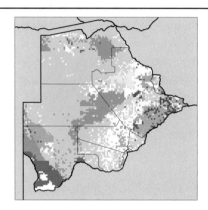

CSIRO Mark 3 GCM

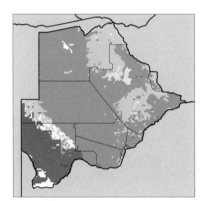

ECHAM 5 GCM

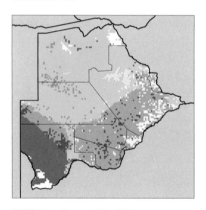

MIROC 3.2 medium-resolution GCM

- ■ Baseline area lost
- ■ Yield loss >25% of baseline
- ▫ Yield loss 5–25% of baseline
-   Yield change within 5% of baseline
- ▫ Yield gain 5–25% of baseline
- ■ Yield gain > 25% of baseline
- ■ New area gained

Source: Authors' calculations.
Notes: A1B = greenhouse gas emissions scenario that assumes fast economic growth, a population that peaks midcentury, and the development of new and efficient technologies, along with a balanced use of energy sources; CNRM-CM3 = National Meteorological Research Center–Climate Model 3; CSIRO = climate model developed at the Australia Commonwealth Scientific and Industrial Research Organisation; ECHAM 5 = fifth-generation climate model developed at the Max Planck Institute for Meteorology (Hamburg); GCM = general circulation model; MIROC = Model for Interdisciplinary Research on Climate, developed by the University of Tokyo Center for Climate System Research.

Agricultural Outcomes

Figures 3.12 and 3.13 show simulation results for future agriculture outcomes based on climate change, population growth, and GDP scenarios. Each featured crop has five graphs showing production, yield, area, net exports, and world price. The graphs show the global price of maize more than doubling by 2050, coupled with yield increases as better farming methods are implemented because, despite a reduction in the harvested area, production is expected to increase. However, the maize food security situation is not projected to improve; the country is expected to continue to import more maize than it produces, remaining vulnerable to international food price increases.

The results for sorghum are similar to those for maize. However, the price of sorghum is shown rising less sharply and remaining relatively stable after 2025. Sorghum yield, production, and area harvested are shown to be improving. Unlike in the case of maize, imports of sorghum are shown to decline due to increased production, indicating that production efficiency will improve faster than demand.

Any increase in food prices for maize and sorghum would likely have a ripple effect on livestock feed prices, thus affecting beef producer costs. This could result in reduced cereal and beef consumption and have a direct impact on the available kilocalories per capita.

Vulnerability to Climate Change

The number of malnourished children under age five and the available kilocalories per capita are used here as indirect indicators of the impact of climate change on human welfare. Figures 3.14–3.16 use "box-and-whisker" plots to illustrate these indicators under different income and climate scenarios.

Figure 3.14 shows the impact of future GDP and population scenarios on the number of malnourished children under age five. Figure 3.15 shows the share of children who are malnourished. The *number* of malnourished children is projected to continue to rise steadily until 2025 and then decline. The baseline and optimistic scenarios show this downward trend continuing through 2050, with malnutrition levels below the 2010 levels. The shorter-term increase is in line with the shorter-term decline in kilocalories per capita in Botswana (discussed below). The *share* declines more rapidly as population growth increases the total number of children.

FIGURE 3.12 Impact of changes in GDP and population on maize in Botswana, 2010–2050

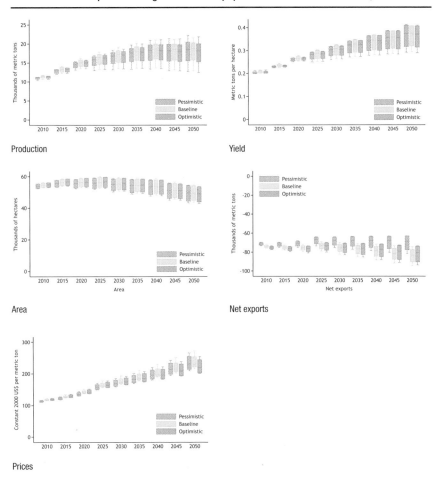

Source: Based on analysis conducted for Nelson et al. (2010).
Notes: The box and whiskers plot for each socioeconomic scenario shows the range of effects from the future climate scenarios. GDP = gross domestic product; US$ = US dollars.

Figure 3.16 shows a reduction in the availability of kilocalories per capita in Botswana until 2025 for the baseline scenario; thereafter, it is shown improving. The optimistic scenario shows a slight increase through 2025 and thereafter a much more rapid increase. The pessimistic scenario shows a steady decline through 2025 and a slower decline thereafter. The availability of kilocalories is projected to remain below the normal threshold of 2,300–2,400 kilocalories per day until 2045 in the baseline scenario.

FIGURE 3.13 Impact of changes in GDP and population on sorghum in Botswana, 2010–2050

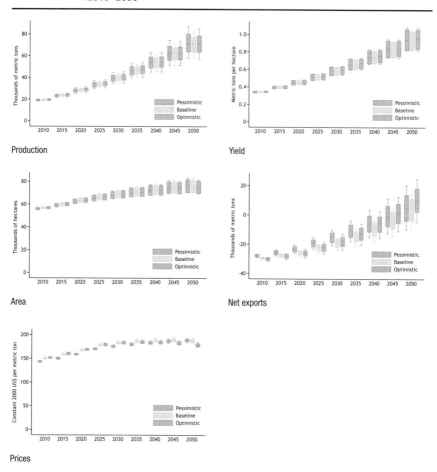

Source: Based on analysis conducted for Nelson et al. (2010).
Notes: The box and whiskers plot for each socioeconomic scenario shows the range of effects from the future climate scenarios. GDP = gross domestic product; US$ = US dollars.

Conclusions and Policy Recommendations

Botswana's semiarid climate severely limits the country's food production capacity. Of Botswana's 581,780 square kilometers, only 5 percent is suitable for cultivation. Less than 1 percent is being cultivated, mostly in the eastern parts of the country, where conditions are more favorable for crop production. In addition, the country is prone to increasingly frequent droughts, recently occurring every two years.

FIGURE 3.14 Number of malnourished children under five years of age in Botswana in multiple income and climate scenarios, 2010–2050

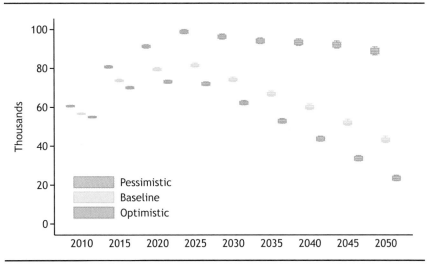

Source: Based on analysis conducted for Nelson et al. (2010).
Note: The box and whiskers plot for each socioeconomic scenario shows the range of effects from the four future climate scenarios.

FIGURE 3.15 Share of malnourished children under five years of age in Botswana in multiple income and climate scenarios, 2010–2050

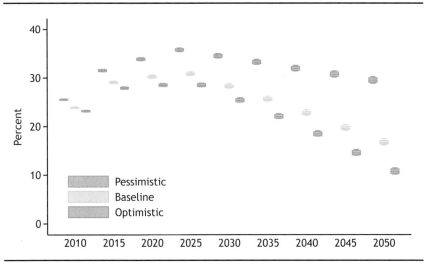

Source: Based on analysis conducted for Nelson et al. (2010).
Note: The box and whiskers plot for each socioeconomic scenario shows the range of effects from the four future climate scenarios.

FIGURE 3.16 Kilocalories per capita in Botswana in multiple income and climate scenarios, 2010–2050

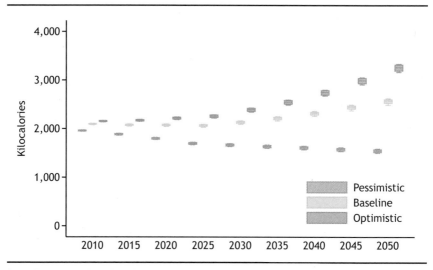

Source: Based on analysis conducted for (Nelson et al. 2010).
Note: The box and whiskers plot for each socioeconomic scenario shows the range of effects from the future climate scenarios.

Since independence, Botswana has had strong economic growth, mostly spurred by growth in the mining sector. The agriculture sector has grown more slowly, and its contribution to GDP has declined from 40 percent at independence in 1966 to just over 1 percent at present. The current government budget share for agriculture is about 5 percent. Agriculture still remains the mainstay of the rural economy, which comprises 41.4 percent of the country's households and offers employment to 30 percent of the country's employable population.

The agriculture sector in Botswana consists of two distinct sectors, subsistence and commercial farms. The communal farms—which, for those unfamiliar with Botswana agriculture, is the term used for individually owned farms, generally smallholders—cultivate 80 percent of the total planted area but produce just 38 percent of the country's total crop; they have significantly lower yields than commercial farms and are more susceptible to the effects of drought.

Both commercial and subsistence farms also raise livestock. Beef production, the only agricultural export earner in Botswana, is dominated by communal farms, which have about 85 percent of the national herd. Beef exports have been declining, however, due to the impact of such diseases as FMD.

Outbreaks of animal diseases seem to be occurring with increasing frequency, severely reducing livestock populations and production. There have been outbreaks of tick-borne diseases, heartwater disease, and other diseases in areas where they never occurred in the country's history.

Botswana had maize and sorghum production deficits (relative to national requirements) of 99 percent and 46 percent in 2001–2007. The deficits were supplied by imports from South Africa, exposing Botswana to food price inflation and adversely affecting nutrition. Studies by the FAO show that the daily average availability of kilocalories per capita in Botswana decreased, from 2,260 in 1990/1992 to 2,180 in 2001/2003, while the proportion of undernourished people in Botswana increased from 23 to 30 percent during the same period. This trend tracks the decline in agricultural output in the past two decades.

The Policy Landscape

The nation's *Vision 2016,* NDP, and related agricultural policies, plans, strategies, and programs are geared to sustaining agricultural production under the country's difficult semiarid conditions, recognizing the need to promote irrigation, drought-resistant crops, and selective breeding of drought-tolerant livestock while diversifying the agricultural base and conserving scarce agricultural land resources. The government provides subsidies to support farmers and promote improvements. The policy framework does not currently address issues related to the vulnerability of agriculture to climate change.

Most of the current MoA programs were developed to address food security at the national and household levels. The MoA is also exploring ways to revive the agriculture sector with programs to attract youth. The stakeholders who were interviewed acknowledged that climate change has not been given adequate attention. Most stakeholders acknowledged that their focus is on combating drought and maximizing food security, which are intimately related to climate change.

The MoA has observed changes in rainfall trends, including more frequent droughts. The stakeholders admitted that they need to pay close attention to the impacts of climate change in agriculture, particularly for poor communities.

Scenario Modeling Results

The climate models show little change in precipitation in most of Botswana by 2050. The temperature, on the other hand, is shown increasing in all four

GCMs and for all regions of the country, by 1.5° to 3.5°C, thus increasing the evapotranspiration rate, which is already high.

As discussed above, these modeled changes in precipitation and temperature have implications for crop yields: the yields of rainfed maize and sorghum are shown increasing in some districts of Botswana and decreasing in others. Deployment of better farming technologies could increase the yields of both crops, depending on the successful uptake of these technologies by communal farms. However, increases in yields or production are more likely to occur on commercial farms than on subsistence farms, which already lag significantly in crop yields, unless a greater focus is placed on helping subsistence farmers adapt more productive technologies, including irrigation or water harvesting, improved seeds, and increased fertilizer use.

Vulnerability assessments need to be complemented by adaptation strategies designed to offset the impacts of climate change. The government therefore urgently needs to mainstream climate change into its policies and programs if it is to be successful in its efforts to fight poverty, diversify the economy, and improve food security.

Recommendations

We make the following recommendations:

- The government should continue working to improve vulnerability assessments, which should be implemented using models downscaled to the region and the country levels to derive more reliable results with finer resolution of localized impacts.

- There is a need to build capacity in the country to use crop models to assess climate change and other agricultural impacts.

- Research needs to be done, perhaps in conjunction with regional and international institutions, to find crop and livestock strategies that will help farmers adapt to the large projected temperature increases.

- Adaptive measures and strategies should be guided by cost–benefit analyses and by lessons learned.

- Botswana needs to undertake pilot adaptation measures, upscaling the most promising ones.

- The government needs to incorporate climate change into revised policies, plans, strategies, and programs as a way to mainstream climate change into development.

- Policy and program efforts need to be complemented by a forum for information sharing to propagate the most feasible adaptation measures.

- A monitoring and evaluation framework is needed to track the achievement of objectives related to climate change adaptation and to provide feedback into agricultural planning for better sector performance.

Following these recommendations should help Botswana to prepare for many of the challenges that climate change will bring in the coming decades.

References

Bartholome, E., and A. S. Belward. 2005. "GLC2000: A New Approach to Global Land Cover Mapping from Earth Observation Data." *International Journal of Remote Sensing* 26 (9): 1959–1977.

Batisani, N. 2010. "Dynamics of Cultivated Land and Its Association with Rainfall Variability." in *Botswana: Climate Change and Natural Resources Conflicts in Africa*, edited by D. A. Mwiturubani and J.-A. van Wyk. Pretoria, Republic of South Africa: Institute of Security Studies.

Botswana, Central Statistics Office. 2001. *Population and Housing Census*. Gaborone: Government Publishers.

———. 2004. *Household Income and Expenditure Survey (HIES)*. Gaborone: Government Publishers.

Botswana, Ministry of Agriculture. 2010a. *Guidelines for ISPAAD [Integrated Support Programme for Arable Agricultural Development]*. Gaborone: Government Publishers.

———. 2010b. *Guidelines for LIMID [Livestock Management and Infrastructure Development]*. Gaborone: Government Publishers.

———. 2010c. *Guidelines for NAMPAADD [National Master Plan for Arable Agriculture and Dairy Development]*. Gaborone: Government Publishers.

———. 2010d. *Annual Reports 2010*. Gaborone: Government Publishers.

———. Various years. *Early Warning System*. Gaborone: Government Publishers.

Botswana, Ministry of Finance and Development Planning. 2003. *National Development Plan 9*. Gaborone: Government Publishers.

———. 2010. *National Development Plan 10*. Gaborone: Government Publishers.

Botswana, Ministry of Minerals, Energy, and Water Resources. 2009. *Biomass Energy Strategy*. Department of Energy. Gaborone.

Botswana Meat Commission. 2010. *Annual Report*. Gaborone: Botswana Meat Commission.

Botswana Vision 2016 Council. 1997. *Vision 2016.* Gaborone: Government Publishers. www.vision2016.co.bw.

Chipanshi, A. C., R. Chanda, and O. Totolo. 2003. "Vulnerability Assessment of the Maize and Sorghum Crops to Climate Change in Botswana." *Climate Change* 61: 339–360.

Jones, P. G., P. K. Thornton, and J. Heinke. 2009. *Generating Characteristic Daily Weather Data Using Downscaled Climate Model Data from the IPCC's Fourth Assessment.* Project report. Nairobi, Kenya: International Livestock Research Institute.

Lehner, B., and P. Döll. 2004. "Development and Validation of a Global Database of Lakes, Reservoirs, and Wetlands." *Journal of Hydrology* 296 (1–4): 1–22.

Millennium Ecosystem Assessment. 2005. *Ecosystems and Human Well-being: Synthesis.* Washington, DC: Island Press. http://www.maweb.org/en/Global.aspx.

Nelson, G. C., M. W. Rosegrant, A. Palazzo, I. Gray, C. Ingersoll, R. Robertson, S. Tokgoz, et al. 2010. *Food Security, Farming, and Climate Change to 2050: Scenarios, Results, Policy Options.* Washington, DC: International Food Policy Research Institute.

UNEP (United Nations Environment Programme) and IUCN (International Union for the Conservation of Nature). 2009. *World Database on Protected Areas (WDPA): Annual Release.* Accessed 2009. www.wdpa.org/protectedplanet.aspx.

UNPOP (United Nations Department of Economic and Social Affairs–Population Division). 2009. *World Population Prospects: The 2008 Revision.* New York. http://esa.un.org/unpd/wpp/.

World Bank. 2009. *World Development Indicators.* Washington, DC.

———. 2010. *The Costs of Agricultural Adaptation to Climate Change.* Washington, DC.

Chapter 4

LESOTHO

Patrick Gwimbi, Timothy S. Thomas, Sepo Hachigonta, and Lindiwe Majele Sibanda

Lesotho is located on the plateau of southern Africa, with altitudes ranging from about 1,400 meters to more than 3,480 meters above sea level. This position exposes the country to the influences of both the Indian and the Atlantic Oceans, with wide differences in temperature. Annual precipitation is highly variable both temporally and spatially, ranging from 500 millimeters to 760 millimeters. Temperatures are highly variable on diurnal, monthly, and annual time scales, generally ranging between −10° and 30°C. High winds of up to 20 meters per second can sometimes be reached during summer thunderstorms. The variations in topography and the microclimatological influences shape the ecological zones of the country: the lowlands, the foothills, the highlands, and the Senqu River Valley.

Many Basotho (Lesotho's majority ethnic group) pursue rainfed agriculture and are thus highly vulnerable to climate change and variability (Turner et al. 2001; LVAC 2008). The agricultural sector accounts for about 10 percent of the country's gross domestic product (GDP) (Lesotho, Bureau of Statistics 2007b). It is the primary source of income—as well as an important supplementary source of income—for more than half of the population (Lesotho, Bureau of Statistics 2007b; LVAC 2008).

The agricultural sector is subject to multiple shocks and stresses that increase household vulnerability (LVAC 2008). Climate change is one of the pervasive stresses that rural communities have to cope with (LVAC 2002, 2008). The situation is worsened by declining employment opportunities and rising staple food prices that adversely affect household resilience to the shocks brought by climate change (LVAC 2002).

In this chapter we examine the vulnerability of agriculture to climate change in Lesotho and suggest a range of potential options for supporting adaptation efforts. Underlying this general purpose are three key questions:

- What is known about Lesotho's agricultural vulnerability to climate change?

- Where are the knowledge gaps with respect to the challenges that adaptation to climate change poses for agriculture in Lesotho?

- What needs to be done to improve agricultural adaptation to climate change in Lesotho?

The answers to these questions are intended to help policymakers and researchers better understand and anticipate the likely impacts of climate change on agriculture and on vulnerable households in Lesotho and to identify adaptation measures for addressing the challenges of climate change and variability.

The Threat of Climate Change

A number of studies (Lesotho, Ministry of Natural Resources 2000, 2007; LVAC 2008; Matarira 2008) show that that Lesotho is already undergoing changes in climate. The *National Report on Climate Change* (Lesotho, Ministry of Natural Resources 2000) noted that the country is experiencing erratic weather patterns. *Lesotho's National Adaptation Programme of Action (NAPA) on Climate Change* (Lesotho, Ministry of Natural Resources 2007) and the *Lesotho Food Security and Vulnerability Monitoring Report* (LVAC 2008) similarly observed the widespread negative impacts of climate change on agriculture and other livelihoods. Matarira (2008) also noted that temperatures are on a rising trend, while rainfall is on a declining trend, concluding that these trends have implications for the agriculture and water supply of rural communities. Most of these assessments anticipate increased warming accompanied by drier conditions across Lesotho.

The effects of climate change on human health are both direct and indirect. People are exposed directly to changing weather patterns and frequent extreme events; fatalities are associated with severe weather such as snowfalls and droughts, as well as with worsened child undernutrition and both waterborne and vector-borne diseases (Lesotho, Ministry of Natural Resources 2000). Indirectly, climate change affects the quality of water, air, and food, causing changes in the ecosystem, agriculture, industry, human settlements, and the economy (Khanna 2010). Finally, where poor transport conditions make rural areas difficult to access, food prices—already high—are especially sensitive to climate change and variability.

Country Profile

Lesotho is a landlocked, mountainous country surrounded by the Republic of South Africa and covering 30,355 square kilometers Only 13 percent of its land area is deemed suitable for crop production (Mochebelele et al. 1992); the rest consists predominantly of rocky mountains and foothills. The lowest altitude is 1,400 meters above sea level, with the highest peak rising to 3,482 meters (Chakela 1999).

Agroecologically, the country is divided into four zones: the lowlands (17 percent), the foothills (15 percent), the Senqu River Valley (9 percent), and the mountains (59 percent). The lowlands, in the western region of the country, are between 1,400 meters and 1,800 meters above sea level. The lowlands are bordered on the east by the foothills zone, which lies between 1,800 meters and 2,200 meters above sea level. The Senqu River Valley stretches into the southeastern mountainous region of the country. The mountain zone, known as the Maluti, ranges from 2,200 meters to 3,484 meters above sea level. The country's climate is extremely variable; temperatures range from −10°C in winter to 30°C in summer in the lowlands (Chakela 1999). Winters are cold and dry, becoming extremely cold in the highlands, where the mountains are usually snow covered during June, July, and August. Normal annual rainfall averages 750 millimeters but varies considerably among different regions of the country. The highlands (average annual rainfall 760 millimeters) receive most of the rain. The rainy season runs mainly from October to March.

Economic activities are largely confined to the lowlands, the foothills, and the Senqu River Valley; the mountain region is suitable only for grazing and, in recent years, for water and hydropower development (LVAC 2008). The mountains are endowed with the bulk of the natural resources, including abundant water resources, critical biodiversity, and gemstones. However, there has been environmental degradation over the years, attributed to overexploitation, especially by grazing.

Review of Current Trends

Economic and Demographic Indicators

Population

The population of Lesotho, estimated at 1.96 million in 2009, has a declining annual growth rate (WHO 2009). Table 4.1 shows the growth rates from 1960 to 2008. The rate of growth increased from 1.9 percent in 1960–1969 to

TABLE 4.1 Population growth rates in Lesotho, 1960–2008 (percent)

Decade	Total growth rate	Rural growth rate	Urban growth rate
1960–1969	1.9	1.4	11.4
1970–1979	2.2	1.9	5.3
1980–1989	2.2	2.0	4.0
1990–1999	1.6	0.9	5.3
2000–2008	0.8	−0.1	3.8

Source: Authors' calculations, based on *World Development Indicators* (World Bank 2009).

2.2 percent in 1970–1979, remained stable from 1980 to 1989, then declined to 1.6 percent in 1990–1999 and 0.82 percent in 2000–2008.

The decline in population growth rates has been attributed to a combination of factors: the over- and underenumeration of the population in 1986 and 1996, respectively (Lesotho, Ministry of Development Planning 2002); declining fertility rates; increased mortality rates attributed to the HIV/AIDS pandemic; and emigration over the past decade (Lesotho, Bureau of Statistics 2007b).

The fertility rate declined significantly over the past three decades of the 20th century—from a high of 5.4 children per woman in the mid-1970s and 5.3 in the mid-1980s, to 4.1 in the mid-1990s, 4.2 in 2001, and 3.5 in 2004 (Lesotho, Ministry of Health and Social Welfare 2005; Lesotho, Bureau of Statistics 2007b). The decline in fertility has been attributed to a rise in contraception rates (from 23 percent in 1990 to 36.1 percent in 2000), as well as the significant increase in school enrolment due to free primary education (Lesotho, Bureau of Statistics 2010)—delaying marriage for most of the girls attending school (Lesotho, Bureau of Statistics, and UNDP 2006).

Figure 4.1 shows trends in Lesotho's total population and rural population (left axis), as well as the share of the population that is urban (right axis). The percentage of the population that is urban has risen steadily—from 3.4 percent in 1960 to over 25 percent in 2008 (World Bank 2009). The increase in the urban population can be attributed to the lack of opportunities in rural areas compared to the perceived economic opportunities in urban areas. Rural-to-urban migration has been increasing due to declining agriculture (attributed to poor soils, changing climatic conditions, and unimproved farming techniques), along with declining remittances from migrant laborers in South

FIGURE 4.1 Population trends in Lesotho: Total population, rural population, and percent urban, 1960–2008

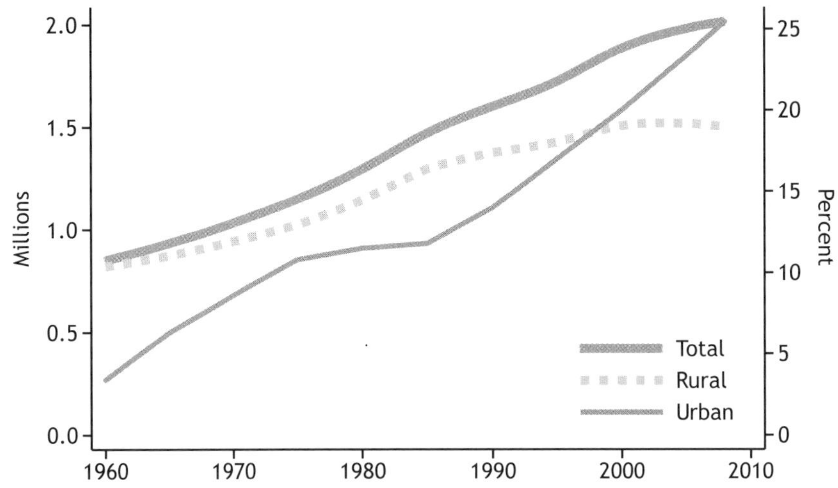

Source: *World Development Indicators* (World Bank 2009).

Africa as the mines cut back on production (Chakela 1999; LVAC 2005; Molapo 2005; Owusu-Ampomah, Naysmith, and Rubincam 2009).

Internal migration from the mountainous eastern region to the northern and western regions of the country also contributes to high population densities (50–2,000 persons per square kilometer) in the lowland and foothill zones (Figure 4.2).

Income

Figure 4.3 shows trends in GDP per capita and the proportion of GDP from agriculture from 1960 to 2008. GDP growth has averaged 3.4 percent in the past decade (Lesotho, Bureau of Statistics, and UNDP 2006) and generally continues to increase.

The launch in 1987–1988 of the Lesotho Highlands Water Project (LHWP)—designed to exploit the Senqu River system by exporting water to South Africa and providing hydroelectricity to the domestic market—marked a major economic transition for the country. Construction of the LHWP stimulated economic growth, generating 9,000 jobs in Phase 1 as well as 1 million US dollars per month at the current exchange rate in royalty payments (Lesotho, Bureau of Statistics, and UNDP 2006).

FIGURE 4.2 Population distribution in Lesotho, 2000 (persons per square kilometer)

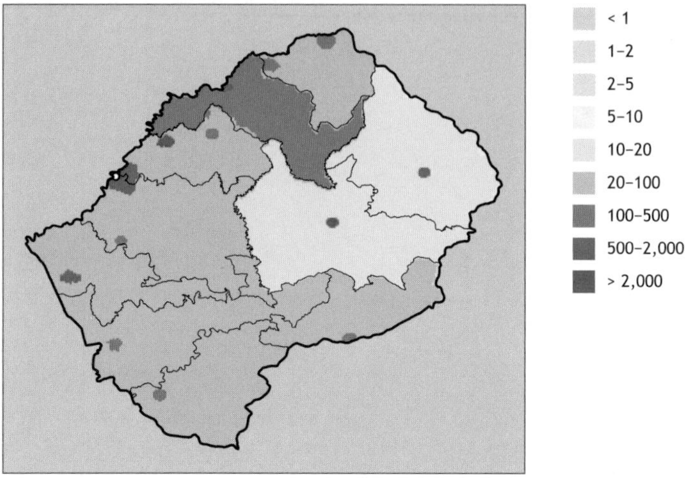

Legend:
- < 1
- 1–2
- 2–5
- 5–10
- 10–20
- 20–100
- 100–500
- 500–2,000
- > 2,000

Source: CIESIN et al. (2004).

The Government of Lesotho draws the majority of its revenue from customs duties, as well as from taxes on the textile and apparel industries, which accounted for 45 percent of the country's GDP in 2007. The GDP in 2006 was $1.4 billion, or $2,879 per capita (Owusu-Ampomah, Naysmith, and Rubincam 2009).

However, despite Lesotho's impressive growth in GDP per capita, the majority of the population has seen little economic gain because most of the companies are foreign owned. The majority of the population remains largely dependent on subsistence agriculture (LVAC 2008). Recent droughts have decreased overall agricultural output, testing households' coping mechanisms and exacerbating food insecurity (FAO 2006; Owusu-Ampomah, Naysmith, and Rubincam 2009).

The contribution of agriculture to GDP has been declining steadily (see Figure 4.3), from more than 50 percent in 1960s to 25 percent in 1980 and then to 17 percent by 2005, and it is currently less than 10 percent (Owusu-Ampomah, Naysmith, and Rubincam 2009). The sector nevertheless has a strong impact on the growth and development of the economy, especially by providing livelihoods for the majority of the population.

The declining contribution of agriculture has been attributed to several factors: low levels of fertilizer application; low and erratic rainfall; hail, frost, and soil erosion; and inadequate strategies to adapt to climate stresses (Chakela 1999; LVAC 2005; Molapo 2005).

FIGURE 4.3 Per capita GDP in Lesotho (constant 2000 US$) and share of GDP from agriculture (percent), 1960–2008

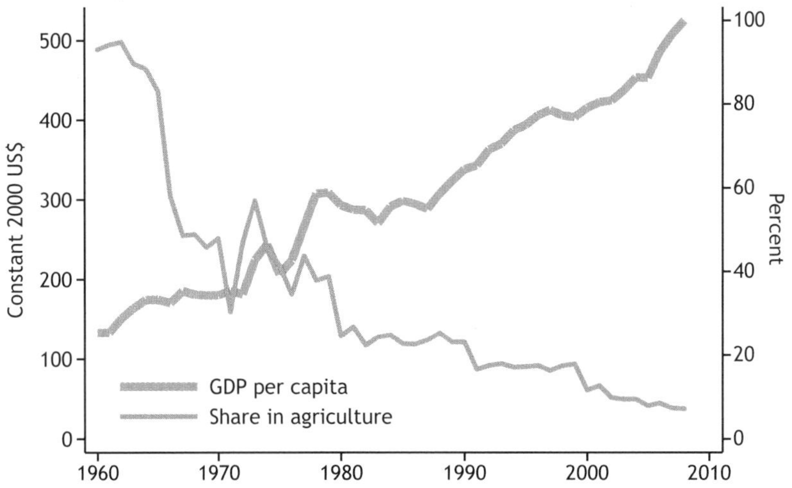

Source: *World Development Indicators* (World Bank 2009).
Note: GDP = gross domestic product; US$ = US dollars.

Vulnerability to Climate Change

In Lesotho, vulnerability to climate change is based on the exposure and sensitivity of eight sectors assessed by the National Climate Change Study Team and the National Adaptation Programme of Action (NAPA). The eight sectors deemed vulnerable to climate change are water, agriculture, rangelands, forestry, soils, health, biodiversity, and Basotho culture. The NAPA report (Lesotho, Ministry of Natural Resources 2007) cites exposure and sensitivity of the geographic space as an indicator of vulnerability. The report delineates three climate change vulnerability zones:

- Zone I—the southern lowlands and the Senqu River Valley.
- Zone II—the mountains.
- Zone III—the (western) lowlands and foothills.

Zone I is regarded as the most vulnerable area in the country, followed by Zone II. Zone I is inhabited mainly by subsistence farmers and small livestock farmers. Zone II is mountainous terrain with minimal land for cultivation. Zone III is especially prone to drought.

FIGURE 4.4 Well-being indicators in Lesotho, 1960–2008

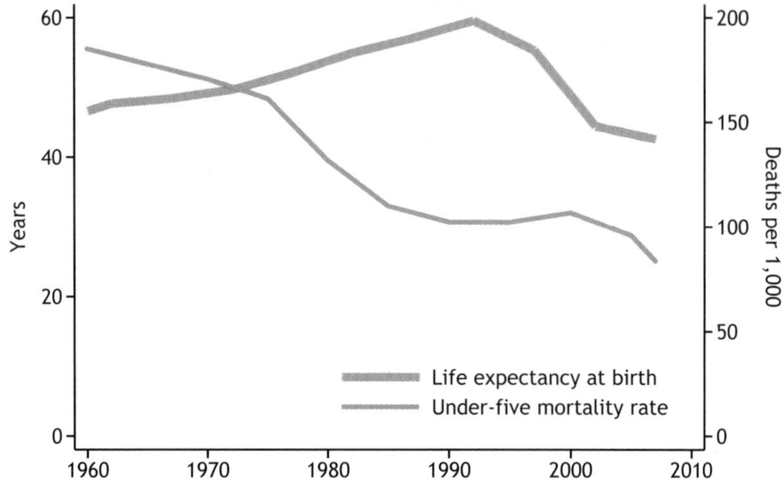

[Chart showing Life expectancy at birth (Years, left axis) and Under-five mortality rate (Deaths per 1,000, right axis) from 1960 to 2010]

Source: *World Development Indicators* (World Bank 2009).

Health

According to the 2004 *Lesotho Demographic and Health Survey,* 20 percent of children under age five were underweight, 38.2 percent chronically malnourished, and 4.3 percent acutely malnourished (Lesotho, Ministry of Health and Social Welfare 2005). Chronic malnutrition overall was a shocking 42 percent in the 2007 *National Nutrition Survey* (Lesotho, FNCO 2009), a major health challenge.

At 42.6 years, life expectancy in Lesotho is among the lowest in the world (Owusu-Ampomah, Naysmith, and Rubincam 2009). Figure 4.4 shows an increase in life expectancy from 1960 through 1990 and a sharp decline thereafter. Similarly, the mortality rate for children under age five shows a gradual decline from 1960 to 1975 and a sharp decline to 1990, followed by a slight increase through 2000 and a steady decline thereafter.

The major factors accounting for improvements in well-being in the 1970s and 1980s were improving food security, health services, and nutrition and hygiene practices. These advances were particularly important in lowering infant mortality; reductions in under-five mortality, in turn, contributed largely to the increased life expectancy. Declining well-being after 1990 was driven by several factors: increasing poverty, food insecurity, and malnutrition, along with the HIV/AIDS pandemic (Lesotho, Bureau of Statistics 2001; WHO 2009).

TABLE 4.2 Education and labor statistics for Lesotho, 1980s, 1990s, and 2000s

Indicator	Year	Percent
Primary school enrollment (percent gross, three-year average)	2006	114.4
Secondary school enrollment (percent gross, three-year average)	2006	37.0
Adult literacy rate	2001	82.2
Percent employed in agriculture	1999	72.3
Percent with vulnerable employment (own farm or day labor)	1987	38.2
Under-five malnutrition (weight for age)	2005	16.6

Source: *World Development Indicators* (World Bank 2009).

Education and Labor

Lesotho's achievements in raising its education and literacy standards have been remarkable (Table 4.2). The adult literacy rate is comparatively high compared to the rest of Africa south of the Sahara, at 82.2 percent; primary and secondary enrollment and school attendance have been increasing due to free primary education (WHO 2009).

The education completion rates, however, are much lower than the enrollment rates. Although a high percentage of children in Lesotho start primary education, as of 2005, only 62 percent of those starting complete all 7 years (World Bank 2009). Children from low-income households are more likely to drop out of school (Sechaba Consultants 2000; LVAC 2003). Poverty forces many rural households to withdraw their children from school to eke out a daily subsistence as part of the households' coping mechanism (LVAC 2003).

The unemployment rate is high, probably even exceeding the official rate of 40–45 percent. Sechaba Consultants (2000) estimated unemployment at 50.8 percent for 1999; NAPA (Lesotho, Ministry of Natural Resources 2007) estimates it at 60 percent. The decline in migrant labor due to retrenchments in the South African mining and textile industries, along with declines in local agriculture, has resulted in many workers losing their jobs.

Poverty

The poverty level in Lesotho remains high, in spite of strong GDP growth in recent years. Poverty is countrywide, with 62 percent of the population living on less than $2 per day (Wood et al. 2010). Lesotho's NAPA report (Lesotho, Ministry of Natural Resources 2007) states that more than 85 percent of the population is exposed to the risks of climate change and is more vulnerable because of crippling poverty. The *2002/03 and 1994/95 Household Budget Survey* (Lesotho, Bureau of Statistics 2007a) classified 66.6 percent of the population as poor and 36.4 percent as very poor, living on less than $1 per day. Poverty is worst in rural areas where farming is dependent on rainfall.

Land Use Overview

Lesotho is generally considered a grassland biome with limited forest cover (Low and Rebello 1996). The country has extensive areas of shrub cover (Figure 4.5); less than 1 percent of the country's total area is forest and woodland. The grassland appears to be deteriorating, however, at an alarming rate due to unsustainable range management practices and frequent droughts (Marake et al. 1998; LVAC 2008). The cultivable land is largely confined to the lowlands and the foothills on the western border and to the Senqu River Valley in the south.

The Maluti alpine region, a unique habitat, hosts approximately 1,375 species of plants, 250 species of birds, 50 species of mammals, and 30 species of amphibians and reptiles (Marake et al. 1998). The entire afro-alpine area, however, is currently under heavy grazing pressure from domestic livestock. This is particularly evident in bogs, which are rapidly losing their hydrologic function due overgrazing and trampling during summer months (Marake et al. 1998). The Maluti mountain range covers 35,000 square kilometers, of which 60 percent is in Lesotho.

Figure 4.6 shows the land classified by the International Union for Conservation of Nature as protected areas in Lesotho. Two areas—the Sehlabathebe Wildlife Sanctuary and National Park and the Masitise Nature Reserve—are protected primarily for ecosystem preservation and recreation. These two areas cover approximately 6,495 hectares.

The availability of and access to transport infrastructure influence access to markets and vulnerability to lack of food availability (Paavola 2003). Lesotho's total road network of 5,000 kilometers heavily favors the lowlands, leaving the rural mountain districts highly inaccessible; the local food supply and prices are influenced by the accessibility of the markets (Lesotho, Ministry of Development Planning 2002). Although arterial roads connect all districts, there are relatively few rural roads connecting villages and towns in the mountain districts. These districts comprise 75 percent of the country's total area and are home to about a quarter of the population. The only railway line in Maseru extends for only 2.6 kilometers and links Lesotho with South Africa.

Travel time to the largest market, Maseru—the only city of more than 500,000 in Lesotho—is more than 5 hours for remote areas, especially those in the mountain zone, which are poorly connected to the transport network

FIGURE 4.5 Land cover and land use in Lesotho, 2000

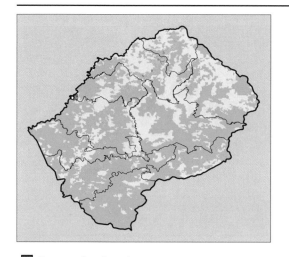

- Tree cover, broadleaved, evergreen
- Tree cover, broadleaved, deciduous, closed
- Tree cover, broadleaved, open
- Tree cover, broadleaved, needle-leaved, evergreen
- Tree cover, broadleaved, needle-leaved, deciduous
- Tree cover, broadleaved, mixed leaf type
- Tree cover, broadleaved, regularly flooded, fresh water
- Tree cover, broadleaved, regularly flooded, saline water
- Mosaic of tree cover/other natural vegetation
- Tree cover, burnt
- Shrub cover, closed-open, evergreen
- Shrub cover, closed-open, deciduous
- Herbaceous cover, closed-open
- Sparse herbaceous or sparse shrub cover
- Regularly flooded shrub or herbaceous cover
- Cultivated and managed areas
- Mosaic of cropland/tree cover/other natural vegetation
- Mosaic of cropland/shrub/grass cover
- Bare areas
- Water bodies
- Snow and ice
- Artificial surfaces and associated areas
- No data

Source: GLC2000 (Bartholome and Belward 2005).

FIGURE 4.6 Protected areas in Lesotho, 2009

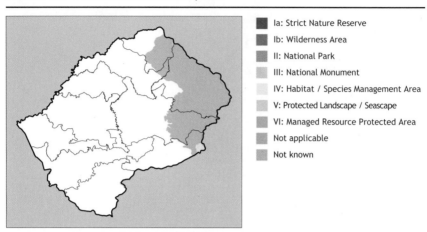

Sources: Protected areas are from the World Database on Protected Areas (UNEP and IUCN 2009). Water bodies are from the World Wildlife Fund's Global Lakes and Wetlands Database (Lehner and Döll 2004).

(Figure 4.7). Poor households, unable to access large markets, are forced to pay high prices for food in the markets closest to them.

Based on the NAPA, supplemented by socioeconomic and biophysical indicators, the following land use sectors are viewed as directly and indirectly vulnerable to climate change.

Water Resources

Lesotho is one of the richest countries in water resources in southern Africa. The vast water resource potential of Lesotho has considerable importance to the economic development of the country. Owing to its abundant and surplus water, Lesotho exports water to its neighbor, South Africa, through the LHWP (Mwangi 2010). This binational project of Lesotho and South Africa is aimed at harnessing the water resources of the highlands of Lesotho for the reciprocal benefit of both countries. It is one of the largest water transfer schemes in the world, and the relevant treaty was signed on October 24, 1986, by both governments. The project consists of four phases. On completion in 2020, five dams, water transfer works, and 200 kilometers of tunnels will have been constructed between the two countries. More than 2 billion cubic meters of water per annum will be transferred from Lesotho to South Africa. The project already earns Lesotho substantial revenues in terms of royalties from water deliveries; for example, between 1996 and 2004 the country earned $225 million.

FIGURE 4.7 Travel time to urban areas of various sizes in Lesotho, circa 2000

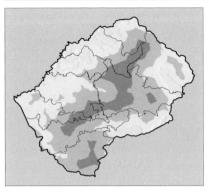
To cities of 500,000 or more people

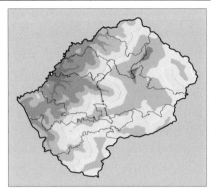
To cities of 100,000 or more people

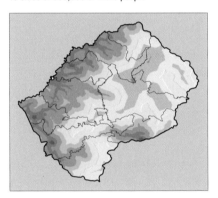
To towns and cities of 25,000 or more people

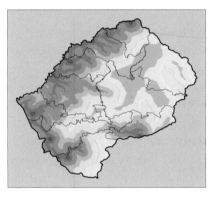
To towns and cities of 10,000 or more people

- Urban location
- < 1 hour
- 1–3 hours
- 3–5 hours
- 5–8 hours
- 8–11 hours
- 11–16 hours
- 16–26 hours
- > 26 hours

Source: Authors' calculations.

Water stress and water scarcity may, however, worsen in the future. Unreliable seasonal flows pose serious risks to water supplies in the country (Lesotho, Ministry of Natural Resources 2000). The *National Report on Climate Change* warns of severe impacts of climate change:

> Lower runoff could translate itself into an ecological disaster, and lead to the closure of many water-based economic and social activities. Dry conditions for most of the year and the resultant lower sub-surface flow would lead to dry springs and wells, lower water tables and higher borehole costs, reduced yields of many water sources, and severe water stress, particularly for the rural population who mainly depend on ground water. [Lesotho, Ministry of Natural Resources 2000, ix]

Agriculture Overview

Agriculture is a key economic sector and a major source of employment in Lesotho: 60–70 percent of the country's labor force obtains supplemental income from agriculture. The sector is characterized by low and declining production due to a number of factors, including climate change and variability.

Crop Production and Consumption

Table 4.3 shows that the total land under cultivation between 2006 and 2008 was 271,000 hectares. The principal crops are maize, sorghum, and wheat. From 2006 to 2008, these three crops accounted for an average of 84.7 percent of the cultivated area: maize at 61.7 percent, sorghum at 13.6 percent, and wheat at 9.4 percent.

The share of area sown with cereals has declined steadily, from 450,000 hectares in 1960 to 300,500 hectares in 1996 (Lesotho, Ministry of Natural Resources 2000) and from over 200,000 hectares in 2004/2005 to just over 150,000 hectares in 2006/2007 (Owusu-Ampomah, Naysmith, and Rubincam 2009). The decline in area under cultivation can be attributed to several factors, among them erratic rainfall and the nonavailability of inputs such as fertilizers (Owusu-Ampomah, Naysmith, and Rubincam 2009).

Maize is the staple foodcrop, contributing 44.4 percent to the daily diet, followed by wheat and potatoes (Table 4.4). Legumes, particularly beans and peas, are widely grown in Lesotho, both for subsistence consumption and as cash crops. These provide major sources of protein in the local diet. Other significant crops include potatoes and vegetables.

TABLE 4.3 Harvest area of leading agricultural commodities in Lesotho, 2006–2008 (thousands of hectares)

Rank	Crop	Percent of total	Harvest area
	Total	100.0	271
1	Maize	61.7	167
2	Sorghum	13.6	37
3	Wheat	9.4	26
4	Beans	8.6	23
5	Potatoes	2.1	6
6	Peas	1.4	4
7	Other fresh vegetables	1.4	4
8	Other fresh fruits	1.4	4
9	Barley	0.3	1
10	Oats	0.0	0

Source: FAOSTAT (FAO 2010).
Note: All values are based on the three-year average for 2006–2008.

TABLE 4.4 Consumption of leading food commodities in Lesotho, 2003–2005 (thousands of metric tons)

Rank	Crop	Percent of total	Food consumption
	Total	100.0	705
1	Maize	44.4	313
2	Wheat	13.2	93
3	Potatoes	12.5	88
4	Sugar	4.1	29
5	Other vegetables	3.5	25
6	Beer	3.3	23
7	Sorghum	2.8	20
8	Fermented beverages	2.5	17
9	Other fruits	2.0	14
10	Oats	1.4	10

Source: FAOSTAT (FAO 2010).
Note: All values are based on the three-year average for 2003–2005.

Adverse weather conditions in recent years have caused significant seasonal food insecurity in Lesotho. Because of declining productivity, loss of productive land through soil erosion, and unfavorable weather conditions, domestic food production covers just about 30 percent of the national food requirements (Mukeere and Dradri 2006). The Food and Agriculture Organization of the United Nations and the World Food Programme estimates cited by Mukeere and Dradri (2006) put the number of people vulnerable to food insecurity in the country at 549,000, constituting about 30 percent of the total population.

Crop Yields

Yields per hectare are highly variable between ecological zones and also vary by crop type and the prevailing climatic conditions. Climatic hazards such as hail, frost, and drought, together with low fertilizer applications, soil erosion, and the use of low-yielding crop varieties, are the main causes of the poor crop yields (LVAC 2008).

The estimated national average yield for sorghum is significantly higher than that of maize. This reversal of the normal pattern has occurred because sorghum is more drought resistant than maize. Winter wheat, which normally starts growing mid-April, makes use of the late rains.

It has been stated that as much as 80 percent of the variability of agricultural production in Lesotho is due to weather conditions, especially for rainfed production systems (Hyden 1996; LMS 1999). Late rainfall may reduce the area planted. For example, between 1990 and 1996 the total area under cultivation fluctuated between 136,500 and 300,500 hectares, down from 450,000 hectares in 1960 (Lesotho, Ministry of Natural Resources 2000). In 2009, only 122,808 hectares had been planted with crops—a decrease of 18,778 hectares (13 percent) from the previous season (World Food Programme 2010). Late rains, early frosts, and erratic hailstorms affect not only the land area planted but also harvests (FAO 2006).

Risks associated with climate change are higher in the southern lowlands, the Senqu River Valley, and the mountains compared to the other zones. The most vulnerable households in these zones are those with assets and livelihoods sensitive to climatic risks and having weak adaptation capacity.

Sharply rising staple food prices and declining employment opportunities adversely affect households' resilience in coping with declining food availability and access. Households' response to these challenges have included coming up with coping strategies such as beer brewing as alternative livelihood strategies (LVAC 2002).

Rangeland and Livestock

Cattle, sheep, and goats are used as vital sources of cash to purchase food when agricultural production is low—when crops fail and wage income drops (LVAC 2002; Turner 2003). Livestock also play many sociocultural roles, serving as food at feasts and burial ceremonies and as bride wealth, sacrifices, and offerings (Rohde et al. 2006). The country's rangelands are primarily used for livestock grazing.

Lesotho's rangelands are in general in poor condition and are declining, with apparent erosion of the topsoil and abundant undesirable vegetation species (Turner 2003). Annual soil loss from rangelands is estimated at 23.4 million metric tons per year (Marake et al. 1998).

Climate change is adding to the challenges posed by rangeland degradation and overstocking. The first *National Report on Climate Change* (Lesotho, Ministry of Natural Resources 2000) noted that due to reduced and delayed precipitation, the country would lose a lot of its nutritious climax grass species and gain a lot of hardy and less nutritious varieties, with serious consequences for livestock productivity. These changes negatively affect the subsector, which contributes 55–65 percent of the agricultural output per year (Lesotho, Ministry of Natural Resources 2000). The combination of generally increasing temperatures and shifting rainfall amounts and patterns has impacts on the carrying capacity of rangelands and the sustainability of the ecosystem. These impacts include changes in the productivity of rainfed forage, reduced water availability, more widespread water shortages, and livestock diseases. Some recent research regarding African grasslands in general, however, suggests that temperature-driven factors might help the grasslands (Scheiter and Higgins 2009).

The Lesotho livestock sector consists of cattle, sheep, goats, horses, donkeys, pigs, and poultry. Livestock are kept for both economic and social reasons. Cattle are raised mostly for subsistence use, including draft power, milk, fuel (dung), and meat.

The major problems facing the sector are rangeland deterioration as a result of overstocking and the consequent overgrazing. The poor grass nutrition as a result of overstocking reduces reproductive rates, milk production, and draft power.

Scenarios for the Future

Economic and Demographic Indicators

Population

Figure 4.8 shows population projections for 2010 to 2050. The optimistic variant shows a declining population after 2030, falling below 2 million by 2050. The population rose from 1,862,275 in 1996 to just 1,880,661 in 2006 (Lesotho, Bureau of Statistics 2010)—a 1 percent increase over 10 years. The optimistic variant is plausible, given the declining population growth trend shown by figures from two successive censuses. Similarly, the United Nations Population Fund, as cited by Owusu-Ampomah, Naysmith, and Rubincam (2009), projects that the population will fall to 1.8 million by 2050, with a negative overall growth rate (−0.3 percent). The HIV/AIDS pandemic, declining fertility, and food insecurity are factors in the low population projections.

Income

Lesotho's GDP per capita is projected to increase by around 50 percent between 2010 and 2025 in all the scenarios, and then to increase at higher rates in all three scenarios, with the optimistic scenario showing very rapid growth (Figure 4.9). The slowdown between 2010 and 2025 in the pessimistic

FIGURE 4.8 Population projections for Lesotho, 2010–2050

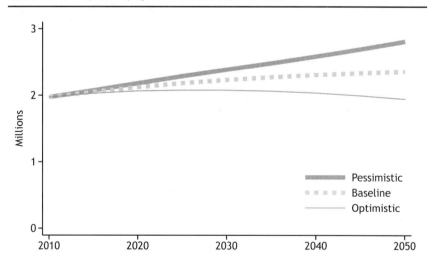

Source: UNPOP (2009).

FIGURE 4.9 Gross domestic product (GDP) per capita in Lesotho, future scenarios, 2010–2050

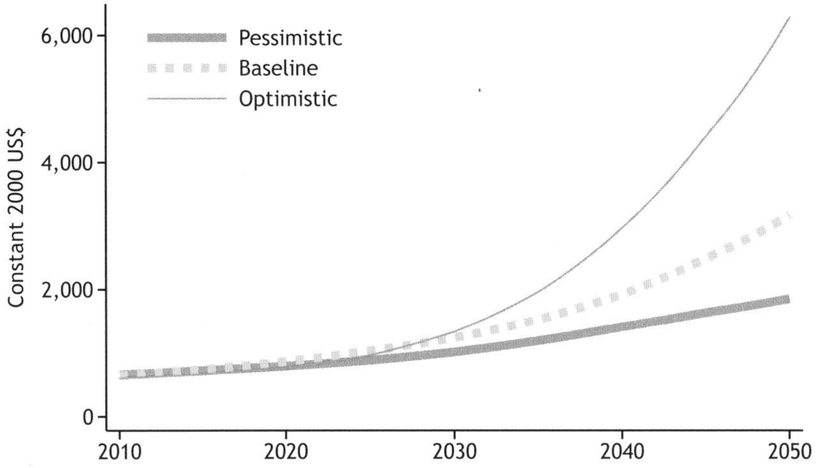

Sources: Computed from GDP data from the World Bank Economic Adaptation to Climate Change project (World Bank 2010), from the Millennium Ecosystem Assessment (2005) reports, and from population data from the United Nations (UNPOP 2009).
Note: US$ = US dollars.

scenario relative to the growth from 1960 to 2010 is due, in part, to the global financial and economic crisis, but high inflation driven by soaring food and oil prices also plays a role (African Development Bank and OECD 2009). The government, however, has continued to strengthen capacity in the textiles sector while exploring other market opportunities in Asia and Africa, and this is going to increase the country's GDP (African Development Bank and OECD 2009). The current trends—shrinking incomes, rising unemployment, and a weak overall resource balance—are all projected to continue, driving the projected static growth to 2025. The optimistic projection assumes that GDP will continue to increase at its past rate, with no change (or possibly an increase) in the contribution of agriculture to GDP.

Biophysical Analysis

Climate Models

The various general circulation models (GCMs) of climate change we used in this study show Lesotho becoming warmer, with diminished precipitation, by 2050. Figure 4.10 shows the projected precipitation deviations

FIGURE 4.10 Changes in mean annual precipitation in Lesotho, 2000–2050, A1B scenario (millimeters)

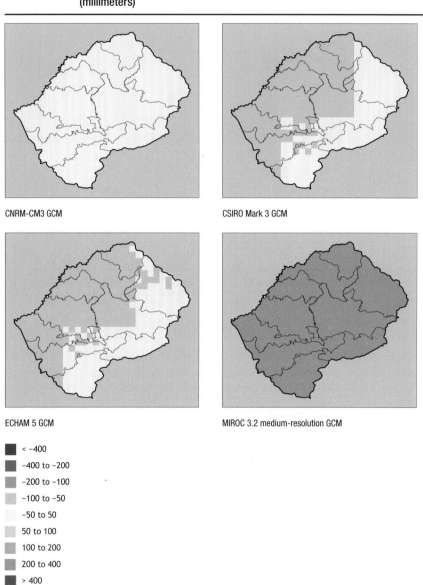

Source: Authors' calculations based on Jones, Thornton, and Heinke (2009).
Notes: A1B = greenhouse gas emissions scenario that assumes fast economic growth, a population that peaks midcentury, and the development of new and efficient technologies, along with a balanced use of energy sources; CNRM-CM3 = National Meteorological Research Center–Climate Model 3; CSIRO = climate model developed at the Australia Commonwealth Scientific and Industrial Research Organisation; ECHAM 5 = fifth-generation climate model developed at the Max Planck Institute for Meteorology (Hamburg); GCM = general circulation model; MIROC = Model for Interdisciplinary Research on Climate, developed by the University of Tokyo Center for Climate System Research.

from normal rainfall levels under the four downscaled climate models used.[1] Only CNRM-CM3 shows minimal change in precipitation throughout the country (varying from −50 to 50 millimeters). CSIRO Mark 3 and ECHAM 5 show a significant decrease in rainfall of between 50 and 100 millimeters per annum in the lowland, foothill, and southern Senqu Valley zones, with little change for the mountain and northern Senqu Valley zones (±50 millimeters). MIROC 3.2 shows severe reductions in rainfall (between −100 and −200 millimeters) for the whole country.

Figure 4.11 shows the change in average daily maximum temperatures in the A1B scenario according to the four GCMs.[2] CNRM-CM3 and ECHAM 5 show the greatest temperature increase—between 2.0° and 2.5°C for the entire country, with pockets of increase as high as 3.0°C. CSIRO Mark 3 shows temperature increases of between 1.0° and 2.0°C throughout the country, with the lower reductions in the mountain zone. MIROC 3.2 similarly shows temperature increases for the whole country of between 1.5° and 2.0°C.

Significant changes in precipitation and temperature could have severe impacts on people's livelihoods and especially on agriculture—particularly in the lowlands, the foothills, and the Senqu Valley, the most densely populated and crop-growing regions of the country. In these zones, increasing temperatures and decreasing precipitation might lead to a substantial decrease in crop harvests.

Crop Models

The Decision Support System for Agrotechnology Transfer (DSSAT) was used to assess the influence of projected climate changes on yields of maize, sorghum, and wheat. Yield changes will depend on the type of crop. Maize yields are projected to decrease as precipitation declines and temperatures increase, whereas sorghum yields are shown making significant gains in some zones due to that crop's drought tolerance.

---

1   CNRM-CM3 is National Meteorological Research Center–Climate Model 3. CSIRO Mark 3 is a climate model developed at the Australia Commonwealth Scientific and Industrial Research Organisation. ECHAM 5 is a fifth-generation climate model developed at the Max Planck Institute for Meteorology in Hamburg. MIROC 3.2 is the Model for Interdisciplinary Research on Climate, developed at the University of Tokyo Center for Climate System Research.
2   The A1B scenario describes a world of very rapid economic growth, low population growth, and rapid introduction of new and more efficient technologies, with moderate resource user with and a balanced use of technologies.

FIGURE 4.11 Change in monthly mean maximum daily temperature in Lesotho for the warmest month, 2000–2050, A1B scenario (°C)

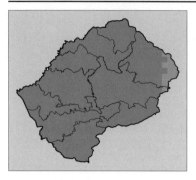

CNRM-CM3 GCM

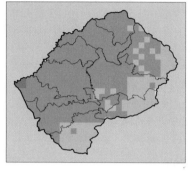

CSIRO Mark 3 GCM

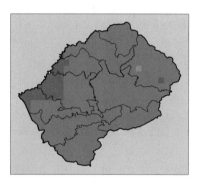

ECHAM 5 GCM

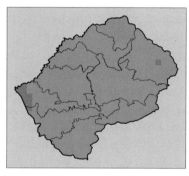

MIROC 3.2 medium-resolution GCM

- < -1
- -1 to -0.5
- -0.5 to 0
- 0 to 0.5
- 0.5 to 1
- 1 to 1.5
- 1.5 to 2
- 2 to 2.5
- 2.5 to 3
- 3 to 3.5
- > 3.5

Source: Authors' calculations based on Jones, Thornton, and Heinke (2009).
Notes: A1B = greenhouse gas emissions scenario that assumes fast economic growth, a population that peaks midcentury, and the development of new and efficient technologies, along with a balanced use of energy sources; CNRM-CM3 = National Meteorological Research Center–Climate Model 3; CSIRO = climate model developed at the Australia Commonwealth Scientific and Industrial Research Organisation; ECHAM 5 = fifth-generation climate model developed at the Max Planck Institute for Meteorology (Hamburg); GCM = general circulation model; MIROC = Model for Interdisciplinary Research on Climate, developed by the University of Tokyo Center for Climate System Research.

All four GCMs show declining maize yields in the lowlands and foothills (Figure 4.12). The southern lowlands zone shows a yield loss of more than 25 percent of baseline. CSIRO Mark 3 and ECHAM 5 show declines of 5–25 percent in much of the northern lowland zone, with decreasing precipitation and increasing temperatures.

In general, a potential maize yield loss—between 5 percent and more than 25 percent—is shown in areas with potential precipitation declines of between 100 and 400 millimeters and temperature increases of between 1.5° and 2.5°C. However, some areas in the mountain zone and at the edges of the foothill zones are shown gaining some ground for maize in all four models. This could be due to reduced frost as temperatures increase and, more generally, to temperatures' becoming warm enough to support maize cultivation. (See Figure 4.12 for the new area projected to be gained.)

Other studies (Matarira 2008; Matarira, Pullanikkatil, and Maletjane 2008) project negative impacts of climate change for maize production. Muller (2009) projects decreasing crop yields in those regions currently constrained by increasing temperatures and declining rainfall. Figure 4.13 shows sorghum yield losses of 5–25 percent in the lowlands in all four GCMs. However, all models show new area gains as well as yield gains of 5–25 percent for the foothills and the Senqu Valley, as well as for scattered parts of the rest of the country. These results suggest that one adaptation option might be growing more drought-tolerant and heat-tolerant sorghum. However, climate change might present some favorable opportunities for Lesotho, because several areas that were previously too cold for cultivation will present themselves as favorable for maize and sorghum. Unplanned migration can lead to environmental damage and might result from farmers' seeing favorable growing conditions in places that are perhaps currently uncultivated while at the same time experiencing losses due to climate change in their present locations. Policymakers should consider whether any of these areas ought to be protected by establishing national parks or other kinds of limited-access areas or whether these areas would be best used by establishing new farms.

All four GCMs show mixed results for wheat—with a yield gain of 5–25 percent projected for most of the lowland regions and a yield loss of 5–25 percent in the other regions (Figure 4.14). ECHAM 5 and MIROC 3.2 show the Senqu Valley and central mountain regions experiencing significant yield reductions of 5–25 percent of baseline.

FIGURE 4.12 Yield change under climate change: Rainfed maize in Lesotho, 2000–2050, A1B scenario

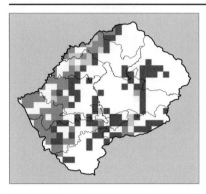

CNRM-CM3 GCM

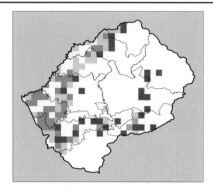

CSIRO Mark 3 GCM

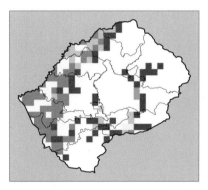

ECHAM 5 GCM

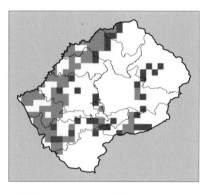

MIROC 3.2 medium-resolution GCM

- Baseline area lost
- Yield loss >25% of baseline
- Yield loss 5–25% of baseline
- Yield change within 5% of baseline
- Yield gain 5–25% of baseline
- Yield gain > 25% of baseline
- New area gained

Source: Authors' calculations.

Notes: A1B = greenhouse gas emissions scenario that assumes fast economic growth, a population that peaks midcentury, and the development of new and efficient technologies, along with a balanced use of energy sources; CNRM-CM3 = National Meteorological Research Center–Climate Model 3; CSIRO = climate model developed at the Australia Commonwealth Scientific and Industrial Research Organisation; ECHAM 5 = fifth-generation climate model developed at the Max Planck Institute for Meteorology (Hamburg); GCM = general circulation model; MIROC = Model for Interdisciplinary Research on Climate, developed by the University of Tokyo Center for Climate System Research.

FIGURE 4.13 Yield change under climate change: Rainfed sorghum in Lesotho, 2000–2050, A1B scenario

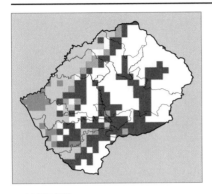

CNRM-CM3 GCM

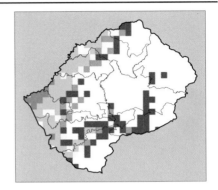

CSIRO Mark 3 GCM

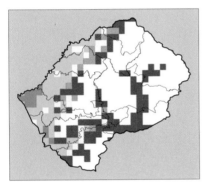

ECHAM 5 GCM

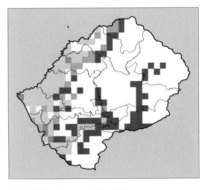

MIROC 3.2 medium-resolution GCM

- Baseline area lost
- Yield loss >25% of baseline
- Yield loss 5–25% of baseline
- Yield change within 5% of baseline
- Yield gain 5–25% of baseline
- Yield gain > 25% of baseline
- New area gained

Source: Authors' calculations.
Notes: A1B = greenhouse gas emissions scenario that assumes fast economic growth, a population that peaks midcentury, and the development of new and efficient technologies, along with a balanced use of energy sources; CNRM-CM3 = National Meteorological Research Center–Climate Model 3; CSIRO = climate model developed at the Australia Commonwealth Scientific and Industrial Research Organisation; ECHAM 5 = fifth-generation climate model developed at the Max Planck Institute for Meteorology (Hamburg); GCM = general circulation model; MIROC = Model for Interdisciplinary Research on Climate, developed by the University of Tokyo Center for Climate System Research.

FIGURE 4.14 Yield change under climate change: Rainfed wheat in Lesotho, 2000–2050, A1B scenario

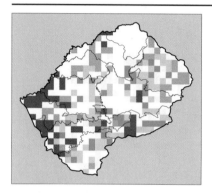

CNRM-CM3 GCM

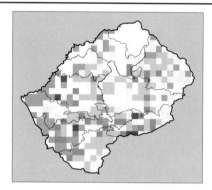

CSIRO Mark 3 GCM

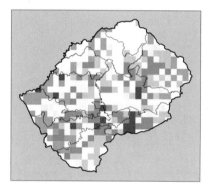

ECHAM 5 GCM

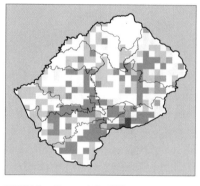

MIROC 3.2 medium-resolution GCM

- ■ Baseline area lost
- ▪ Yield loss >25% of baseline
- □ Yield loss 5–25% of baseline
-   Yield change within 5% of baseline
- ▪ Yield gain 5–25% of baseline
- ■ Yield gain > 25% of baseline
- ■ New area gained

Source: Authors' calculations.

Notes: A1B = greenhouse gas emissions scenario that assumes fast economic growth, a population that peaks midcentury, and the development of new and efficient technologies, along with a balanced use of energy sources; CNRM-CM3 = National Meteorological Research Center–Climate Model 3; CSIRO = climate model developed at the Australia Commonwealth Scientific and Industrial Research Organisation; ECHAM 5 = fifth-generation climate model developed at the Max Planck Institute for Meteorology (Hamburg); GCM = general circulation model; MIROC = Model for Interdisciplinary Research on Climate, developed by the University of Tokyo Center for Climate System Research.

Vulnerability to Climate Change

Lesotho faces a major threat from potential climate change. Drought, land degradation, and loss of biodiversity are the key potential stressors to agriculture. With high unemployment, estimated at 60 percent (Lesotho, Ministry of Natural Resources 2007), most communities are likely to be vulnerable to climate change. The vulnerability of the different sectors is assessed below.

Biophysical Impacts

Agriculture

The climate and crop model scenarios in this study show potential food deficits due to the stresses of decreased rainfall and increased temperature. In the absence of effective adaptation, more land will become fallow and yields will continue to decline. On the other hand, new areas may open up for cultivation, because climate change will likely raise temperatures, which will permit cultivation in areas that were previously too cold.

Water Resources

Climate change scenarios for Lesotho suggest some potential challenges for the water sector. Predictions are that lower level of precipitation due to climate change is likely to result in reduced availability of fresh water. This situation means that a condition of water stress could be reached earlier than predicted. Predicted climate change scenarios for the water sector in Lesotho show a reduction in surface and subsurface runoff owing to the anticipated lower precipitation (Mwangi 2010). Given the current rate of population growth and the necessary levels of service, the projected climate change, and the availability of fresh water, it is estimated that the country will enter a period of water stress, with less than 1,700 cubic meters available per capita per year, and a period of water scarcity, with less than 1,000 cubic meters available per capita per year by the years 2019 and 2062, respectively (Mwangi 2010). This translates to a reduction of slightly more than 60 percent of the water currently available per capita per year. Under climate change, these lower levels of service are likely to be reached earlier than predicted.

Rangeland and Livestock

In the past, chronic drought impeded the recovery of grazed grasses and vegetation in Lesotho, with major destruction of rangelands and severe attenuation of the carrying capacity of pastoral lands. As a result, the number and quality of livestock produced have deteriorated significantly (Lesotho, Ministry of Natural Resources 2007). Future climate change is expected to continue to degrade the already stressed rangelands.

Health

Figure 4.15 shows the impact of future GDP and population scenarios, combined with climate scenarios, on the malnutrition of children under five years of age from 2010 to 2050. Figure 4.16 shows the share of malnourished children. All three scenarios show the *numbers* increasing until 2020. In the pessimistic scenario, malnutrition increases from 68,000 children in 2010 to 100,000 by 2020, thereafter gradually declining to 86,000 by 2050. The baseline scenario shows the number decreasing to 70,000 by 2050, and the optimistic scenario shows it declining to fewer than 50,000.

One of the major causes of illness among children is protein-caloric malnutrition. Figure 4.17 shows the kilocalories available per capita in Lesotho. All three scenarios show the kilocalories per capita decreasing, from more than 2,500—the recommended level for young men—in 2010 to 2,000 by 2030. In the pessimistic scenario, the level then remains unchanged through 2050. This is a case in which positive income effects are negated by the adverse effects of price increases (which are exacerbated by the model's having

FIGURE 4.15 Number of malnourished children under five years of age in Lesotho in multiple income and climate scenarios, 2010–2050

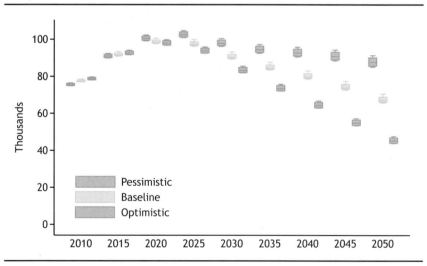

Source: Based on analysis conducted for Nelson et al. (2010).
Note: The box and whiskers plot for each socioeconomic scenario shows the range of effects from the four future climate scenarios.

FIGURE 4.16 Share of malnourished children under five years of age in Lesotho in multiple income and climate scenarios, 2010–2050

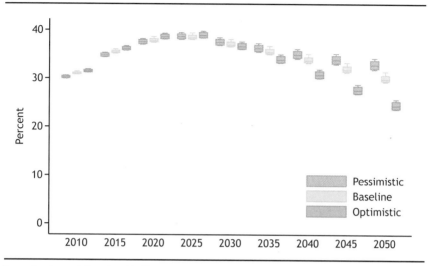

Source: Based on analysis conducted for Nelson et al. (2010).
Note: The box and whiskers plot for each socioeconomic scenario shows the range of effects from the four future climate scenarios.

FIGURE 4.17 Kilocalories per capita in Lesotho in multiple income and climate scenarios, 2010–2050

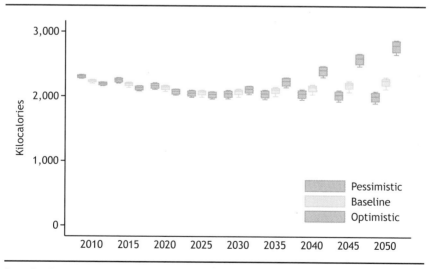

Source: Based on analysis conducted for Nelson et al. (2010).
Note: The box and whiskers plot for each socioeconomic scenario shows the range of effects from the four future climate scenarios.

the own-price elasticities of staple foods set too high, as discussed in Chapter 2). The baseline scenario shows kilocalories per capita increasing to 2,200 by 2050, whereas the optimistic scenario shows an increase to 2,800.

Agriculture

The next three sets of figures show simulation results from the International Model for Policy Analysis of Agricultural Commodities and Trade associated with key agricultural crops (maize, sorghum, and wheat) in Lesotho. Each crop has five graphs showing production, yield, area, net exports, and world price.

In the three scenarios, the area sown with maize remains more or less unchanged, whereas production and yields increase by more than 200 percent between 2010 and 2050 (Figure 4.18). Production, yields, and harvested area under sorghum are also shown to increase greatly (Figure 4.19). All three scenarios show the global price of maize and sorghum increasing, indicating considerable potential for the export of surplus crops. Wheat production, however, will not keep pace with increases in domestic demand, and Lesotho will have to import the deficit (Figure 4.20).

Increases in maize production will be highly dependent on increases in productivity because the area under cultivation will remain unchanged. Increases in cereal productivity will depend, in turn, on high levels of inputs as GDP increases, improving farmers' income. Using the low population growth rate scenario, showing fewer people by 2050, an expected decrease in consumption of cereals will be expected, leading to more exports. If this downward trend in population growth materializes, there is a potential for Lesotho to achieve the 2050 food security projections. In addition, it should be noted that the low population projection could be attributed to high levels of HIV/AIDS and emigration.

Policy Response to the Challenges of Climate Change

Lesotho is working with the global community to address the challenges posed by climate change. It has ratified the United Nations Framework Convention on Climate Change (UNFCCC) and the Kyoto Protocol and has submitted its first *National Report on Climate Change* in 2000. In 2006 Lesotho participated in the preparation of the NAPA report. The key objectives of NAPA include identification of regions and communities vulnerable to climate change, assessing the impact of climate change on community livelihoods, and identifying and prioritizing adaptation activities for the vulnerable zones.

FIGURE 4.18 Impact of changes in GDP and population on maize in Lesotho, 2010–2050

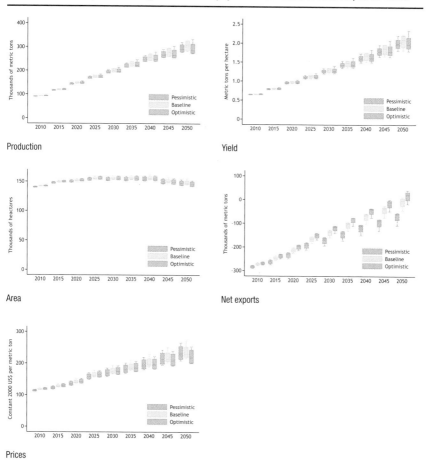

Source: Based on analysis conducted for Nelson et al. (2010).
Notes: The box and whiskers plot for each socioeconomic scenario shows the range of effects from the four future climate scenarios. GDP = gross domestic product; US$ = US dollars.

NAPA includes a prioritized list of adaptation projects, such as enhancements of agricultural productivity (including irrigation systems), early warning systems for droughts and water resource development, and research and development.

Current Programs

In addition, Lesotho has put in place several measures and strategies to deal with the adverse effects of extreme climatic events. A Disaster Management

FIGURE 4.19 Impact of changes in GDP and population on sorghum in Lesotho, 2010–2050

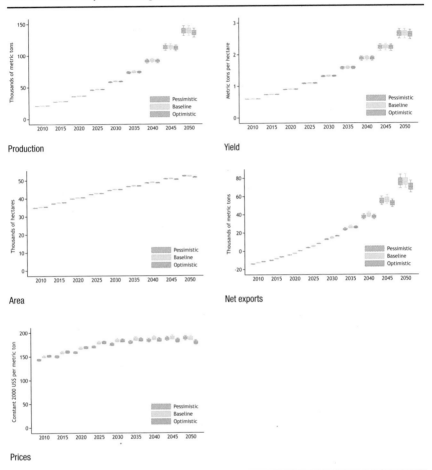

Source: Based on analysis conducted for Nelson et al. (2010).
Notes: The box and whiskers plot for each socioeconomic scenario shows the range of effects from the four future climate scenarios. GDP = gross domestic product; US$ = US dollars.

Authority was established to handle extreme weather and other natural disasters. Nongovernmental organizations and faith groups run similar programs.

Lesotho is currently preparing its second Convention on Initial National Communications (INCs) report. Current climate-related projects include the early warning project for agriculture, designed to develop a climate change policy for Lesotho and to strengthen early warning systems, as well as efforts to capacitate individuals, institutions, and communities with technical

FIGURE 4.20 Impact of changes in GDP and population on wheat in Lesotho, 2010–2050

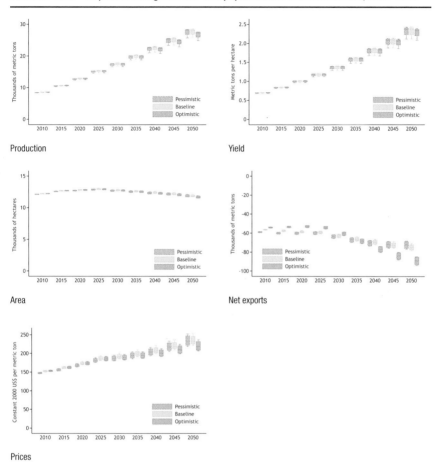

Source: Based on analysis conducted for Nelson et al. (2010).
Notes: The box and whiskers plot for each socioeconomic scenario shows the range of effects from the four future climate scenarios. GDP = gross domestic product; US$ = US dollars.

knowledge, skills, information, and resources through the Africa Adaptation Program. Lesotho has also received a Japanese grant to develop a program to address disasters caused by climate change.

Implementation Obstacles

Most of the strategies proposed by NAPA to address the impacts of climate change are valid today; communities would be much less vulnerable if these strategies were already being implemented. Factors constraining the implementation of the strategies are not likely to change significantly given the budgetary constraints facing the country.

Despite the efforts being made to address the challenges of climate change adaptation, significant gaps and disconnects still exist.

- **Policy:** Climate change has yet to be internalized into Lesotho's major policy planning documents by the government. An analysis of national development plans, poverty reduction strategies, sectoral strategies, and other documents in climate-sensitive sectors shows that even where climate change is mentioned, specific operational guidance is generally lacking.

- **Public awareness:** Climate change is a new concept in Lesotho, especially at grassroots levels. NAPA (Lesotho, Ministry of Natural Resources 2007) highlights that a key barrier to implementing programs for climate change adaptation is the lack of awareness of the potential impact of climate change on people's livelihoods and the adaptation options available. There is also a lack of awareness at the institutional level.

- **Human resources:** The lack of technical human resources on climate change issues across the development sectors of the government is one of the challenges facing Lesotho. Climate change expertise seems to be limited, at the government level, to the Department of Meteorological Services in the Ministry of Natural Resources. That department has only limited leverage over other sectors' activities, guidelines, and programs.

- **Government financial capacity:** On its own, Lesotho has low financial capacities to address the challenges of climate change adaptation. Adaptation will require substantial funding, and this is one of the major challenges facing the country.

Conclusions and Policy Recommendations

Agriculture is the lifeline for the majority of Lesotho's population, although it accounts for just 10 percent of the country's GDP. Agriculture employs 60–70 percent of the country's labor force. The modeled climate scenarios show that the country will be adversely affected in terms of its food production and water supply. Yet, as noted in this chapter, there appear to be opportunities with regard to food production in areas that have not previously been able to support cultivation due to cold temperatures. Poverty, health risks, and limited education pose additional challenges to implementing adaptive policies. In some rural areas, where poor road networks hamper access to markets, food prices are highly sensitive to climate variability.

In response to these challenges, Lesotho has committed itself to tackling the impacts of climate change by ratifying the UNFCCC and the Kyoto Protocol. In 2000, in keeping with its obligations under the INCs, Lesotho submitted its *National Report on Climate Change*. In addition, in 2006 NAPA prepared a prioritized list of adaptation projects, including the enhancement of agricultural productivity, irrigation, the introduction of drought-resistant varieties, early warning systems for droughts, water resource development, and research and development.

Despite these efforts, there are still gaps and disconnects in mainstreaming climate change issues, increasing awareness through outreach programs, creating institutional frameworks for action, and adapting drought-resistant crop varieties. The costs of adapting to climate change are high for Lesotho to self-finance adaptation efforts.

We make the following recommendations:

- The Government of Lesotho needs to incorporate adaptations to address climate change as part of its long-term planning and development programs, along with budget allocations.

- Raising awareness of the impacts of climate change should be given priority and adequate funding.

- Capacity building in knowledge and skills related to climate change is needed across all the sectors of development and in learning institutions across different disciplines.

- Smallholder irrigation projects targeting vulnerable communities are needed.

- Drought-tolerant and heat-tolerant crop varieties and hardy livestock need to be developed and promoted.

- Information capacity on climate change monitoring and early warning systems needs to be strengthened.

- Functioning communications technologies need to be made available to poor and inaccessible communities to allow broad access to seasonal weather forecasts and early warning systems.

Small nations are often challenged institutionally because of expectations that they function in the same capacity as larger nations but with fewer personnel. Climate change presents large institutional challenges because of the magnitude of the changes expected. The government could perhaps improve capacity by seeking to learn from the adaptation plans of other countries, which could provide a template for their own use. This also may suggest supporting roles for international and regional agencies.

References

African Development Bank and OECD (Organisation for Economic Co-operation and Development). 2009. "African Economic Outlook." Accessed July 17, 2010. http://ls.china-embassy.org/eng/xwdt/P020100715196012967784.pdf.

Bartholome, E., and A. S. Belward. 2005. "GLC2000: A New Approach to Global Land Cover Mapping from Earth Observation Data." *International Journal of Remote Sensing* 26 (9): 1959–1977.

Chakela, Q. K. 1999. *State of Environment in Lesotho*. Maseru: National Environment Secretariat (NES), Ministry of Environment, Gender, and Youth Affairs, Government of Lesotho.

CIESIN (Center for International Earth Science Information Network), Columbia University, IFPRI (International Food Policy Research Institute), World Bank, and CIAT (Centro Internacional de Agricultura Tropical). 2004. *Global Rural–Urban Mapping Project (GRUMP), Alpha Version: Population Density Grids*. Palisades, NY, US: Socioeconomic Data and Applications Center (SEDAC), Columbia University. http://sedac.ciesin.columbia.edu/gpw.

FAO (Food and Agriculture Organization of the United Nations). 2006. *Assessment of 2005/06 Agricultural Production in Lesotho*. Accessed July 1, 2010. www.sadc.int/fanr/aims/rvaa/Documents/Lesotho/2006%20Lesotho%202006%20Agricultural%20Season%20Assessment.pdf.

———. 2010. FAOSTAT. Accessed December 22, 2010. http://faostat.fao.org/site/573/default.aspx#ancor.

Hyden, L. 1996. "Meteorological Droughts and Rainfall Variability in the Lesotho Lowlands." Licentiate thesis, Division of Hydraulic Engineering, Department of Civil and Environmental Engineering, Royal Institute of Technology, Stockholm.

Jones, P. G., P. K. Thornton, and J. Heinke. 2009. *Generating Characteristic Daily Weather Data Using Downscaled Climate Model Data from the IPCC's Fourth Assessment*. Project report. Nairobi, Kenya: International Livestock Research Institute.

Khanna, S. 2010. "Climate Change and Oral Health: Current Challenges and Future Scope." *International Journal of Environmental Science and Development* 1 (2): 190–192.

Lehner, B., and P. Döll. 2004. "Development and Validation of a Global Database of Lakes, Reservoirs, and Wetlands." *Journal of Hydrology* 296 (1–4): 1–22.

Lesotho, Bureau of Statistics. 2001. *Demography Survey 2001.* Maseru.

———. 2007a. *2002/03 and 1994/95 Household Budget Survey: Analytical Report.* Maseru.

———. 2007b. *2006 Lesotho Census of Population and Housing Preliminary Results Report.* Maseru.

———. 2010. *National and Sub-national Population Projections.* Maseru.

Lesotho, Bureau of Statistics, and UNDP (United Nations Development Programme). 2007. *National Human Development Report 2006.* Maseru.

Lesotho, FNCO (Food and Nutrition Coordination Office). 2009. *Lesotho National Nutrition Survey November–December 2007.* Maseru.

Lesotho, Ministry of Development Planning. 2002. "Lesotho's Country Report." Paper prepared for the fourth meeting of the follow-up committee on the implementation of the DND and the ICPD–PA in Yaoundé, Cameroon, January 28–31.

Lesotho, Ministry of Health and Social Welfare. 2005. *Lesotho Demographic and Health Survey 2004.* Maseru.

Lesotho, Ministry of Natural Resources. 2000. *National Report on Climate Change.* Maseru.

———. 2007. *Lesotho's National Adaptation Programme of Action (NAPA) on Climate Change.* Maseru.

LMS (Lesotho Meteorological Service). 1999. *Vulnerability Assessment Report–UNEP/GEF Climate Change Study in Lesotho.* Project GF/2200-96-16: 156. Maseru.

Low, A. B., and A. G. Rebelo. 1996. *Vegetation of South Africa, Lesotho, and Swaziland.* Pretoria, Republic of South Africa: Department of Environmental Affairs and Tourism.

LVAC (Lesotho Vulnerability Assessment Commission). 2002. *Food Security Assessment Report.* Maseru: Disaster Management Authority.

———. 2003. *Lesotho Livelihoods-Based Vulnerability Assessment (LBVA).* Maseru.

———. 2005. *Annual Vulnerability Monitoring Report.* Maseru.

———. 2008. *Lesotho Food Security and Vulnerability Monitoring Report.* Maseru.

Marake, M., C. Mokhoku, M. Majoro, and N. Mokitimi. 1998. *Global Change and Subsistence Rangelands in Southern Africa: Resource Variability, Access, and Use in Relation to Rural Livelihoods and Welfare; A Preliminary Report and Literature Review for Lesotho.* Roma, Lesotho: National University of Lesotho.

Matarira, C. H. 2008. "Climate Variability and Change: Implications for Food—Poverty Reduction Strategies in Lesotho." Paper presented at the International Conference on Natural Resource Management, Climate Change, and Economic Development in Africa: Issues, Opportunities and Challenges in Nairobi, Kenya, September 15–17, 2008.

Matarira, C. H., D. Pullanikkatil, and M. Maletjane. 2008. "Climate Variability and Change: Vulnerability of Communities in Senqu Ecological Zone, Lesotho." Paper presented at the conference Adapting to Changes in the Orange River Basin: Can Adaptive Water Resources Management Make a Difference? at the National University of Lesotho, Roma, October 23–24.

Millennium Ecosystem Assessment. 2005. *Ecosystems and Human Well-being: Synthesis*. Washington, DC: Island Press. http://www.maweb.org/en/Global.aspx.

Mochebelele, M. T., N. L. Mokitimi, M. T. Ngqaleni, G. G. Storey, and B. M. Swallow. 1992. *Agricultural Marketing in Lesotho*. Johannesburg: International Development Research Centre.

Molapo, L. 2005. "Urban Water Provision in Maseru (Lesotho): A Geographical Analysis." Masters thesis, University of the Free State, Bloemfontein, Republic of South Africa.

Mukeere, B., and S. Dradri. 2006. *Food Aid, Food Production and Food Markets in Lesotho: An Analytical Review*. Rome: World Food Programme.

Muller, C. 2009. *Climate Change Impact on Sub-Saharan Africa: An Overview and Analysis of Scenarios and Models*. Discussion paper. Bonn, Germany: Deutsches Institut für Entwicklungspolitik.

Mwangi, O. G. 2010. "Climate Change, Hydro Politics, and Security in Lesotho." In *Climate Change and Natural Resources Conflicts in Africa*, edited by D. A. Mwiturubani and J. van Wyk. Accessed January 8, 2011. www.iss.co.za/pubs/Monographs/170/Mono170.pdf.

Nelson, G. C., M. W. Rosegrant, A. Palazzo, I. Gray, C. Ingersoll, R. Robertson, S. Tokgoz, et al. 2010. *Food Security, Farming, and Climate Change to 2050: Scenarios, Results, Policy Options*. Washington, DC: International Food Policy Research Institute.

Owusu-Ampomah, K., S. Naysmith, and C. Rubincam. 2009. *"Emergencies" in HIV and AIDS-Affected Countries in Southern Africa: Shifting the Paradigm in Lesotho*. Maseru: National AIDS Commission.

Paavola, J. 2003. "Vulnerability to Climate Change in Tanzania: Sources, Substance, and Solutions." Paper presented at the inaugural workshop of Southern Africa Vulnerability Initiative (SAVI) in Maputo, Mozambique, June 19–21.

Rohde, R. F., N. M. Moleele, M. Mphale, N. Allsopp, R. Chanda, M. T. Hoffman, L. Magole, and E. Young. 2006. "Dynamics of Grazing Policy and Practice: Environmental and Social Impacts in Three Communal Areas of Southern Africa." *Environmental Science and Policy* 9: 302–316.

Scheiter, Simon, and Steven I. Higgins. 2009. "Impacts of Climate Change on the Vegetation of Africa: An Adaptive Dynamic Vegetation Modelling Approach." *Global Change Biology* 15 (9): 2224–2246.

Sechaba Consultants. 2000. *Poverty and Livelihoods in Lesotho, 2000: More Than a Mapping Exercise*. Maseru.

Turner, S. D. 2003. *The Southern African Food Crisis: Lesotho Literature Review.* Maseru: CARE Lesotho.

Turner, S. D., et al. 2001. *Livelihoods in Lesotho.* Maseru: CARE Lesotho.

UNEP (United Nations Environment Programme) and IUCN (International Union for Conservation of Nature). 2009. *World Database on Protected Areas (WDPA) Annual Release 2009.* Accessed January 8, 2011. http://www.wdpa.org/protectedplanet.aspx.

UNPOP (United Nations Department of Economic and Social Affairs–Population Division). 2009. *World Population Prospects: The 2008 Revision.* New York. http://esa.un.org/unpd/wpp/.

WHO (World Health Organization). 2009. *Country Cooperation Strategy 2008–2013.* Maseru, Lesotho.

Wood, S., G. Hyman, U. Deichmann, E. Barona, R. Tenorio, Z. Guo, S. Castano, O. Rivera, E. Diaz, and J. Marin. 2010. "Sub-national Poverty Maps for the Developing World Using International Poverty Lines: Preliminary Data Release." Accessed January 18, 2011. http://povertymap.info (password protected). Some also available at http://labs.harvestchoice.org/2010/08/poverty-maps/.

World Bank. 2009. *World Development Indicators.* Washington, DC.

———. 2010. *The Costs of Agricultural Adaptation to Climate Change.* Washington, DC.

World Food Programme. 2010. *Lesotho Food Security Monitoring System Quarterly Bulletin.* May. Accessed January 25, 2011. http://documents.wfp.org/stellent/groups/public/documents/ena/wfp221556.pdf.

Chapter 5

MALAWI

John D. K. Saka, Pickford Sibale, Timothy S. Thomas,
Sepo Hachigonta, and Lindiwe Majele Sibanda

Malawi is located in the eastern part of southern Africa between latitude 9°22′ and 17°7′ South and between longitude 32°40′ and 35°55′ East. Its total area is 118,483 square kilometers, of which 94,275 square kilometers is land, while 24,208 square kilometers is water. Malawi is a landlocked country bordering Tanzania, Mozambique, and Zambia and does not have direct access to the Indian Ocean.

The topography of the country is highly varied; the Great Rift Valley runs from north to south through the country, containing Lake Malawi, and the landscape around the valley consists of large plateaus at an elevation of around 800–1,200 meters but with peaks as high as 3,000 meters. The climate of the country is tropical, but its high elevation means that the temperatures are relatively cool.

Lake Malawi is Africa's third largest lake and the world's eleventh largest and covers much of the country. It measures about 550 kilometers by 15–80 kilometers and occupies a deep Rift Valley trough that cuts through the country along a north–south line. The lake's surface elevation is about 474 meters, and its deepest point is 230 meters below sea level.

Lake Malawi and the Shire River are part of the Great Rift Valley, and on either side of the rift are escarpments. In the west, the highlands include the Nyika (highest elevation 2,607 meters), Viphya (2,058 meters), and Dedza (2,198 meters) Plateaus; in the east they include the Shire Highlands (1,774 meters), the Zomba Plateau (2,087 meters), and the Mangochi and Namizimu Hills (1,796 meters). The eastern highlands continue northward into Mozambique and Tanzania. Behind the rift-edge highlands the land descends gently to the Central African Plateau, which, at elevations around 1,000 meters, covers the Lilongwe and Kasungu Plains. The country's lowest elevation of about 37 meters is on the Rift Valley floor at the extreme south, while Mulanje Mountain, an ancient volcanic plug standing on the plateau to the southeast, at 3,050 meters, is the highest point in central Africa.

Climate and Climate Change in Malawi

The great variations in Malawi's landscape result in wide spatial differences in climate. The vast water surface of Lake Malawi has a cooling effect, but because of its low elevation, the margins of the lake generate long hot seasons and high humidity, with a mean annual temperature of 24°C (75°F).

The climate is tropical continental but is significantly moderated by the effects of Lake Malawi, high altitudes, and proximity to the influence of westerly frontal systems that move eastward around the South African coast. There are two distinct seasons: the rainy season (October–April) and the dry season (May–October). The latter is further subdivided into two parts: (1) cool and wet (May–August) and (2) hot and dry (September–October).

Climate Change and the Vulnerability of Agriculture

Agriculture is the backbone of Malawi's economy, providing more than 50 percent of the country's gross domestic product (GDP). Both crop and livestock production depend on rainfall as the main source of water supply because less than 5 percent of arable land is under irrigation. In the past 30 years, Malawi has experienced variability and unpredictability in its seasonal rainfall.

High variability in rainfall could imply recurrent drought conditions in lower-rainfall zones (e.g., in the Shire Valley region). Malawi could therefore subsequently experience failure of the more desired foodcrops and pasturage. These changes are expected to cause many shifts in food production. Most crops are sensitive to changes in climate conditions, including alterations in temperature, moisture, and carbon dioxide levels. Furthermore, major changes in climate influence populations of beneficial organisms and pests and alter their effectiveness in agricultural ecosystems. Although there will be gains in certain crops in some regions of the world, the overall global impacts of climate change on agriculture, especially rainfed agriculture, are expected to be negative, threatening global food security. These impacts are (1) direct, on crops and livestock productivity domestically; (2) indirect, on the availability and prices of food domestically and in international markets; and (3) indirect, on income from agricultural production at both the farm and the country levels.

This chapter concerns the outcomes of various general circulation models (GCMs) relating to Malawi's agricultural vulnerability to climate change and how various policies and programs influence the capacity of the national systems to address the vulnerability of agriculture to climate change.

Review of Current Trends

Climate Change Concepts

In the Fourth Assessment Report of the Intergovernmental Panel on Climate Change (IPCC), Working Group 1 defines *climate* as "average weather, usually described in terms of the mean and variability of temperature, precipitation, and wind over a period of time ranging from months to millions of years (the classical period is 30 years)" (IPCC 2007).

GCMs model the physics and chemistry of the atmosphere, its interactions with oceans and land surface, and greenhouse gas levels. Several GCMs have been developed independently around the world. In this study, GCMs together with crop and socioeconomic models were used to simulate the interactions between humans and their surroundings. Precipitation and temperature levels in the A1B scenario were obtained from four GCMs: CNRM-CM3, CSIRO Mark 3, ECHAM 5, and MIROC 3.2.[1] These data were used with Decision Support Software for Agrotechnology Transfer (DSSAT) crop models to assess the likely impact of climate change on yield. Additionally, the International Model for Policy Analysis of Agricultural Commodities and Trade developed by the International Food and Policy Research Institute was used to estimate the impact of GDP and population, together with climate change, on the various agriculture outcomes: yield in metric tons per hectare; crop area, net exports; and world price for the crop.

Economic and Demographic Indicators

Population

The 2008 Malawi Population and Housing Census revealed a total population of nearly 13.1 million and a national annual population growth rate of 2.8 percent (World Bank 2009). The population increased from 9.9 million in 1998, so the rate of the overall population increase was 32 percent in 10 years. The population 18 years of age and over was 6.2 million, of whom 3.2 million were female and 3 million were male. In Figure 5.1, the total and rural population

---

1 The A1B scenario is a greenhouse gas emissions scenario that assumes fast economic growth, a population that peaks midcentury, and the development of new and efficient technologies, along with a balanced use of energy sources. CNRM-CM3 is National Meteorological Research Center–Climate Model 3. CSIRO is a climate model developed at the Australia Commonwealth Scientific and Industrial Research Organisation. ECHAM 5 is a fifth-generation climate model developed at the Max Planck Institute for Meteorology in Hamburg. MIROC is the Model for Interdisciplinary Research on Climate, developed at the University of Tokyo Center for Climate System Research.

FIGURE 5.1 Population trends in Malawi: Total population, rural population, and percent urban, 1960–2008

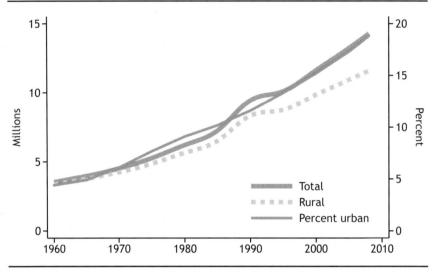

Source: *World Development Indicators* (World Bank 2009).

counts (left axis) and the share of the population that was urban (right axis) are provided.

In 2008 the urban population of Malawi constituted 18.8 percent of the estimated population (World Bank 2009), up from 8.5 percent in 1994. In such urban areas, employment, water services, electricity, postal services, bus services, and businesses are most prominent. The relative population growth rates and urbanization are shown in Table 5.1. The decade with the highest growth rate was 1980–1989 (4.3 percent), and this was largely due to a major influx of refugees from the war in Mozambique. The decade with the lowest

TABLE 5.1 Population growth rates in Malawi, 1960–2008 (percent)

Decade	Total growth rate	Rural growth rate	Urban growth rate
1960–1969	2.4	2.3	5.6
1970–1979	3.2	2.9	7.3
1980–1989	4.3	4.0	6.7
1990–1999	1.8	1.4	4.5
2000–2008	2.6	2.0	5.2

Source: Authors' calculations, based on *World Development Indicators* (World Bank 2009).

growth rate was 1990–1999 (1.80 percent), which coincided with the repatriation of the refugees back to Mozambique after the war.

Malawi has witnessed rapid urbanization in the past 50 years, with urban growth rates higher than rural growth rates. Much of the urbanization has been due to youth looking for employment and business opportunities. The highest rate of urbanization was seen during 1970–1979 (at 7.3 percent), and the lowest rate was seen during 1990–1999 (at 4.50 percent).

One of the United Nations (UN) Millennium Development Goals (MDGs) on education, especially MDG 2, calls for universal access to basic education by 2015. Malawi has not made significant progress toward achieving this goal. The program is off track; only about 38 percent of the 91 percent of children who enroll in primary school complete Standard 8 (Malawi, Ministry of Economic Planning and Development 2006). Further, at the completion of Standard 8, most girls drop out of schools for various reasons. Policymakers should ensure that programs are in place to keep girls in school beyond Standard 8.

Despite the high rate of growth in the urban population, the rural population is still growing at a rate of 2 percent per annum. This indicates that agriculture will continue to be important in the economy of Malawi for the foreseeable future.

The population density in south and central Malawi is higher than in the north (Figure 5.2). The actual population density figures from the last census are 185 persons per square kilometer for the south, 154 for the center, and 63 for the north (Malawi, National Statistics Office 2008). The geographic distribution of the population in Malawi is shown in Figure 5.2, based on data available in 2000.

Income

Malawi is one of the least developed countries in the world, and therefore poverty remains a key challenge in the country. Figure 5.3 shows the GDP per capita for 1960–2009. There was a steady increase in GDP per capita of about $30 per decade from 1960 to about 1980; thereafter, no significant growth has been seen. However, significant progress has been made in tackling food insecurity. During the past decade, the country has witnessed increased household food security (Malawi, Ministry of Agriculture and Food Security 2006). Although overall well-being remains low, it is improving, to the extent that the country's UN Human Development Index score was 0.493 in 2009, which gave Malawi a ranking of 160 out of 182 countries; it had moved up from 164 out of 177 countries in 2007/2008. The share of agriculture in GDP has

FIGURE 5.2 Population distribution in Malawi, 2000 (persons per square kilometer)

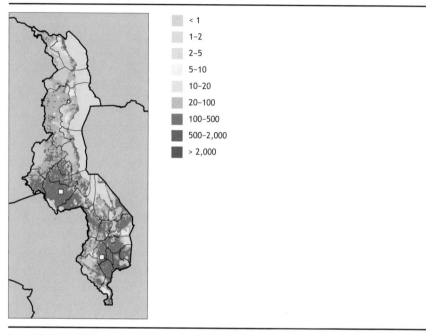

Source: CIESIN et al. (2004).

fallen through the years, from around 50 percent in 1960 to around 35 percent in 2008 (see Figure 5.3).

The country's education enrolment and labor statistics are shown in Table 5.2. The data show that the primary school gross enrollment rate in Malawi is very high (116.5 percent). However, the secondary school enrollment rate is low (28.3 percent). This is due to the high dropout rates among girls, largely due to early marriages and family pressures to remain at home. Malawi needs an educated population to manage climate shocks. This is particularly important because the agriculture sector employs 90 percent of the rural population, of which 80 percent are employed in off-farm activities or as day laborers.

Additional well-being indicators for Malawi are shown in Figure 5.4. Over the past 50 years, life expectancy at birth marginally improved, from 37 years in 1960 to about 50 years in 1992, before dropping to 45 years in 2002. This drop has been attributed to the HIV/AIDS pandemic. Life expectancy has since improved, to about 48 years in 2005, and it is currently rising.

The under-five mortality rate has significantly dropped, to nearly 100 deaths per 1,000 births as of 2010, from more than 360 in 1960. This is at least in part

FIGURE 5.3 Per capita GDP in Malawi (constant 2000 US$) and share of GDP from agriculture (percent), 1960–2008

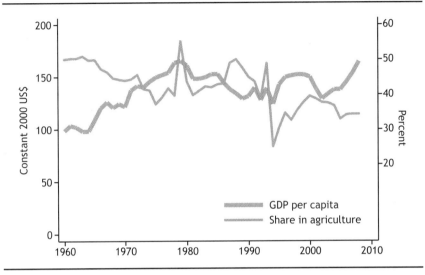

Source: *World Development Indicators* (World Bank 2009).
Note GDP = gross domestic product; US$ = US dollars.

because several government programs—for immunization, provision of bed nets, and improvement of nutrition—are under way, checking malaria and other diseases. This trend needs to be sustained.

In Figure 5.5, the proportion of the population living on less than $2 per day is given. The highest concentration of poor people live in the southern (about 49.7 percent) and central (about 33.9 percent) regions, which are also the most densely populated rural areas (Malawi, Ministry of Agriculture and Food Security 2006).

TABLE 5.2 Education and labor statistics for Malawi, 2000s

Indicator	Year	Value (percent)
Primary school enrollment (percent gross, three-year average)	2007	116.5
Secondary school enrollment (percent gross, three-year average)	2007	28.3
Adult literacy rate	2007	71.8
Percent employed in agriculture	2007	90.0
Percent with vulnerable employment (own farm or day labor)	2007	80.0
Under-five malnutrition (weight for age)	2005	18.4

Source: World *Development Indicators* (World Bank 2009).

FIGURE 5.4 Well-being indicators in Malawi, 1960–2008

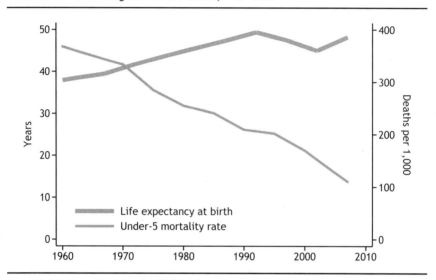

Source: *World Development Indicators* (World Bank 2009).

FIGURE 5.5 Poverty in Malawi, circa 2005 (percentage of population below US$2 per day)

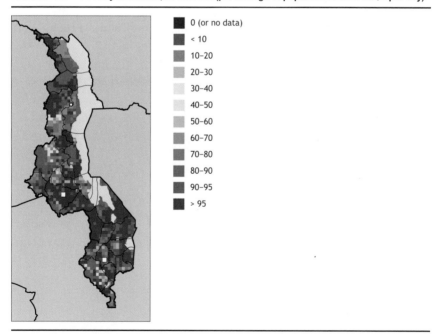

Source: Wood et al. (2010).
Note: Based on 2005 US$ (US dollars) and on purchasing power parity value.

FIGURE 5.6 Land cover and land use in Malawi, 2000

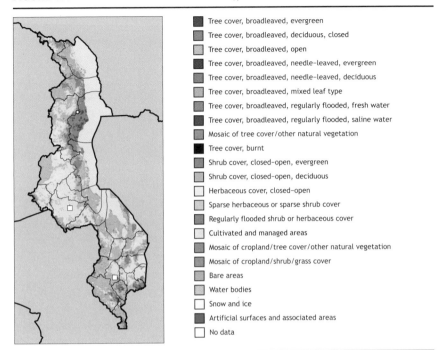

Source: GLC2000 (Bartholome and Belward 2005).

Land Use Overview

The land cover and land use in Malawi are shown in Figure 5.6. In the 1900s, prior to colonization, almost 100 percent of the land area in Malawi was covered with forests. The forest cover has decreased drastically due to increased human activity. For example, between 1972 and the 1990s, forest resources decreased by 41 percent, representing an average annual loss of 2.4 percent (Malawi SDNP 1998). This was due to several factors, including the increasing growth of urban settlements, overdependence on forest products as sources of energy, and the clearing of forests for agriculture-related activities. Further, in the tobacco sector, the processing of flue-cured tobacco by most commercial estates continued to contribute significantly to environmental degradation. The loss of biomass due to the harvesting of wood to provide fuel for tobacco curing is estimated at almost 85,000 cubic meters per year (Malawi, Ministry of Lands and Housing 2002). Over a 20-year period (1972–1992), Malawi's forest resources were reduced by more than half (57 percent) of their size, with an estimated annual deforestation rate of 2.8 percent (Haarstad 2009).

Approximately 35 percent of the land area of Malawi is now covered by forests. The country has between 5,000 and 6,000 plant species in both montane and lowland ecosystems. According to the Red Data Lists Assessment for Malawi, 114 plant species are endemic, 5 species are critically endangered, and 14 species are endangered (IUCN 2010).

In Malawi the Land Act classifies three major forms of tenure, namely public, private, and customary (Malawi, Ministry of Lands and Housing 2002). The adoption of the land tenure system will facilitate farmers' investment in land improvement and the improvement of agricultural technologies such as agroforestry and reforestation.

Agricultural activities in Malawi are located in all the rural areas where there is cultivable land. Malawi has 3 million hectares under cultivation, with more than 95 percent of this land under rainfed agriculture. The average size of land holdings per household in Malawi is 1.2 hectares, while the average land per capita is 0.33 hectare (Malawi Government and World Bank 2006). The rainfed nature of smallholder farming makes agricultural production prone to adverse weather conditions such as droughts and floods. Drought years have most often resulted in poor crop yields and sometimes in total crop failure, leading to serious food shortages, hunger, and malnutrition. Flooding also disrupts food production, destroys household and community assets, and causes loss of life for both livestock and people.

In Malawi, places where no cultivation is allowed are protected areas, including parks and reserves (Figure 5.7). These locations provide important protection for fragile environmental areas, which may also be important for the tourism industry. The national parks are Nyika, Kasungu, Lengwe, and Liwonde; the game reserves are Nkhota Kota, Majete, Dzalanyama, and Vwaza Marches. Other protected areas are Lake Chilwa and Mulanje Mountain, which are designated as conservation and world heritage sites by the United Nations Educational, Scientific, and Cultural Organization. The parks and reserves are useful as environmental protection areas, acting as water catchment areas for several important rivers in Malawi. The parks and reserves are also useful areas for the protection and conservation of wildlife and for tourism.

The major climatic hazard that affects wildlife in the protected areas is drought, which affects animal reproduction systems and migratory habits. Extended drought also leads to the deaths of animals. For example, the 1979/1980 drought resulted in the deaths of Nyala antelope in Lengwe National Park and in the migration of most of the animals from the reserve (Malawi, EAD 2006).

FIGURE 5.7 Protected areas in Malawi, 2009

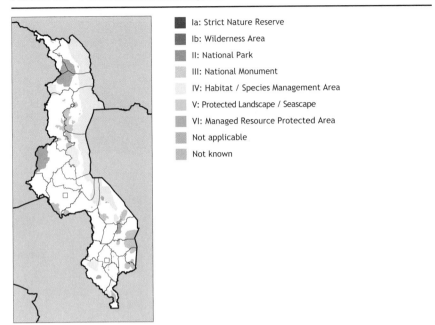

Sources: Protected areas are from the World Database on Protected Areas (UNEP and IUCN 2009). Water bodies are from the World Wildlife Fund's Global Lakes and Wetlands Database (Lehner and Döll 2004).

Figure 5.8 shows the travel times to urban areas of various sizes, which provide potential markets for agricultural products. The importance of transport costs should be considered in policy planning and execution when considering potential for agricultural expansion. That is, if fertile but unused land is far from markets, it represents potential land for expansion only if transportation infrastructure is put in place and if the land does not conflict with the preservation priorities seen in Figure 5.7.

Malawi has four major cities: Blantyre and Zomba in the south, Lilongwe in the center of the country, and Mzuzu in the north. Malawi also has the municipal towns of Luchenza and Mangochi in the south, Kasungu and Salima in the center, and Karonga in the north. There are more trading centers in the south, followed by the center; the north has the fewest. It follows, therefore, that in the northern part of the country, more people take longer than 11–16 hours to travel to big cities, which are largely found in the center and the south. The southern part of the country is more urbanized than the rest of the country.

FIGURE 5.8 Travel time to urban areas of various sizes in Malawi, circa 2000

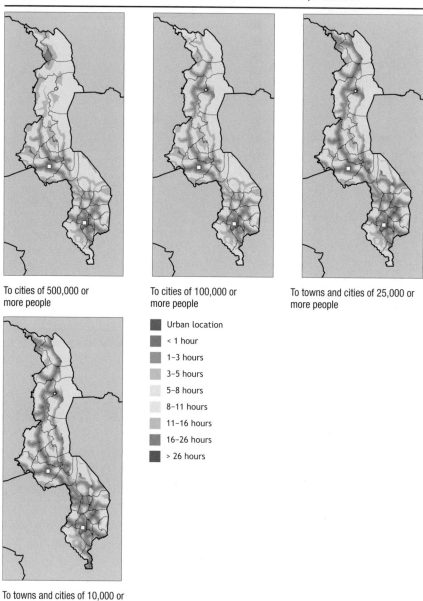

To cities of 500,000 or more people

To cities of 100,000 or more people

To towns and cities of 25,000 or more people

- Urban location
- < 1 hour
- 1–3 hours
- 3–5 hours
- 5–8 hours
- 8–11 hours
- 11–16 hours
- 16–26 hours
- > 26 hours

To towns and cities of 10,000 or more people

Source: Authors' calculations.

Agriculture Overview

In Tables 5.3–5.5, the key agricultural commodities of Malawi are analyzed in terms of area harvested, value of production, and consumption of food commodities.

Maize is the staple foodcrop, but because of its low productivity, it is not surprising that the total acreage allocated to this food commodity is very large (Figure 5.9). The root and tuber crops (cassava and potatoes) have a high allocation, 362,000 hectares, while the legumes and pulses have a slightly higher allocation. Table 5.4 shows that the production value of root and tuber crops is significantly higher than that of cereals (maize and rice) and the commercial crops (tobacco and sugarcane). This could be in part because the value of cassava for home consumption might be overestimated. Table 5.5 reveals that the consumption of root and tuber crops supplements that of maize as a staple for the majority of households. The country has recently witnessed the increased growing of rice, which is an important staple and is exported to neighboring countries. It should be included in the basket of food security foods for the country.

TABLE 5.3 Harvest area of leading agricultural commodities in Malawi, 2006–2008 (thousands of hectares)

Rank	Crop	Percent of total	Harvest area
	Total	100.0	3,389
1	Maize	45.0	1,525
2	Groundnuts	7.6	256
3	Beans	7.5	254
4	Potatoes	5.6	189
5	Cassava	5.1	173
6	Pigeon peas	4.7	160
7	Tobacco	4.1	139
8	Chickpeas	2.8	95
9	Cowpeas	2.4	80
10	Sorghum	2.2	73

Source: FAOSTAT (FAO 2010).
Note: All values are based on the three-year average for 2006–2008.

TABLE 5.4 Value of production of leading agricultural commodities in Malawi, 2005–2007 (millions of US$)

Rank	Crop	Percent of total	Value of production
	Total	100.0	2,541.8
1	Cassava	21.8	553.6
2	Maize	19.6	499.0
3	Potatoes	16.9	429.7
4	Sugarcane	6.9	174.4
5	Tobacco	4.5	114.5
6	Bananas	3.8	97.2
7	Groundnuts	3.7	94.8
8	Beans	3.5	88.1
9	Rice	2.5	64.7
10	Plantains	2.5	62.5

Source: FAOSTAT (FAO 2010).
Note: All values are based on the three-year average for 2005–2007. US$ = US dollars.

TABLE 5.5 Consumption of leading food commodities in Malawi, 2003–2005 (thousands of metric tons)

Rank	Crop	Percent of total	Food consumption
	Total	100.0	6,237
1	Maize	26.1	1,631
2	Cassava	22.6	1,408
3	Potatoes	20.3	1,268
4	Bananas	5.0	312
5	Plantains	4.2	261
6	Other fruits	3.8	236
7	Other vegetables	3.2	198
8	Sugar	2.5	156
9	Fermented beverages	2.3	145
10	Other pulses	1.8	112

Source: FAOSTAT (FAO 2010).
Note: All values are based on the three-year average for 2003–2005.

FIGURE 5.9 Yield (metric tons per hectare) and harvest area density (hectares) for rainfed maize in Malawi, 2000

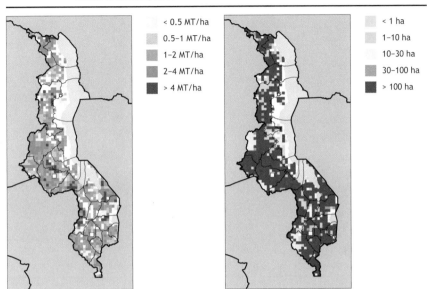

Source: SPAM (Spatial Production Allocation Model) (You and Wood 2006; You, Wood, and Wood-Sichra 2006, 2009).
Note: ha = hectare; MT/ha = metric tons per hectare.

The estimated yield and growing areas for key crops are shown in Figures 5.9–5.13. The yields of rainfed maize in Malawi range from 1.5 metric tons per hectare to 3 metric tons per hectare (see Figure 5.9). These are comparatively low yields when compared to what is possible at research stations. Figure 5.10 shows that the yields of rainfed cassava in Malawi are over 10 metric tons per hectare in most areas. Most of the cassava is grown along the shores of lakes and in many other parts of Malawi, where it has recently been popularized as an important food security crop. The production of rainfed cotton is low (Figure 5.11), with yields ranging from 0.5 to 2 metric tons per hectare, whereas for groundnuts the yields are between 0.5 and 1 metric ton per hectare (Figure 5.12). The yields of rainfed beans in Malawi are generally low due to insect and pest infestations during the growing season (Figure 5.13). Thus yields as low as 0.25–1.0 metric ton per hectare are normal on most smallholder farms.

FIGURE 5.10   Yield (metric tons per hectare) and harvest area density (hectares) for rainfed cassava in Malawi, 2000

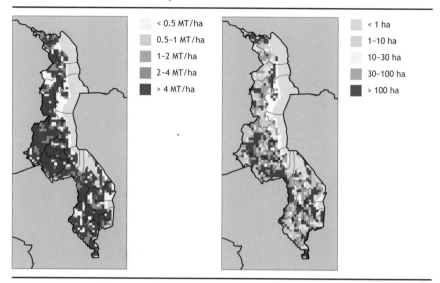

Source: SPAM (Spatial Production Allocation Model) (You and Wood 2006; You, Wood, and Wood-Sichra 2006, 2009).
Note: ha = hectare; MT/ha = metric tons per hectare.

FIGURE 5.11   Yield (metric tons per hectare) and harvest area density (hectares) for rainfed cotton in Malawi, 2000

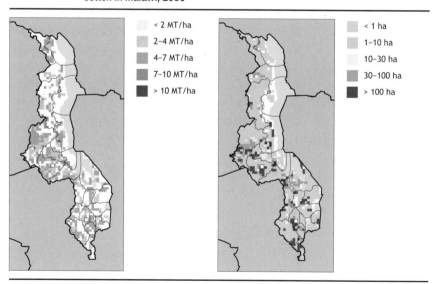

Source: SPAM (Spatial Production Allocation Model) (You and Wood 2006; You, Wood, and Wood-Sichra 2006, 2009).
Note: ha = hectare; MT/ha = metric tons per hectare.

MALAWI 127

FIGURE 5.12 Yield (metric tons per hectare) and harvest area density (hectares) for rainfed groundnuts in Malawi, 2000

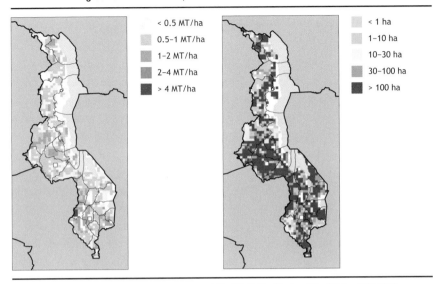

Source: SPAM (Spatial Production Allocation Model) (You and Wood 2006; You, Wood, and Wood-Sichra 2006, 2009).
Note: ha = hectare; MT/ha = metric tons per hectare.

FIGURE 5.13 Yield (metric tons per hectare) and harvest area density (hectares) for rainfed beans in Malawi, 2000

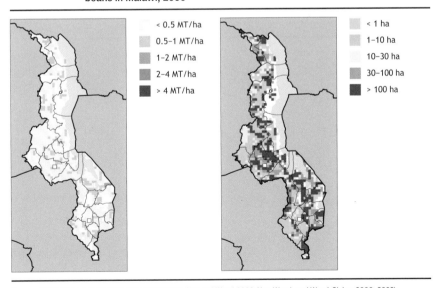

Source: SPAM (Spatial Production Allocation Model) (You and Wood 2006; You, Wood, and Wood-Sichra 2006, 2009).
Note: ha = hectare; MT/ha = metric tons per hectare.

Scenarios for the Future

Economic and Demographic Indicators

Population

The population projections by the UN Population Division (UNPOP 2009) through 2050 are shown in Figure 5.14. All three population projections, based on the high, medium, and low variants, show that Malawi's population is projected to increase significantly, doubling to 31 million people between 2010 and 2050 using the low variant and increasing to 40 million people using the high variant. These projections are reasons for concern and will pose major challenges for the country with its limited land area. For example, to avert hunger, food production will have to increase rapidly and efficiently, with many people moving out of agriculture to follow other livelihood endeavors. There will also be great pressure on the forests for fuelwood unless alternative energy sources are found and used. Managing population growth within acceptable limits consistent with the food resource base is a priority for the country. This entails investing in the education of girls and providing various incentives supportive of small family sizes.

Income

The analysis of income based on three scenarios of GDP per capita for Malawi is provided in Figure 5.15. The optimistic scenario projects an

FIGURE 5.14 Population projections for Malawi, 2010–2050

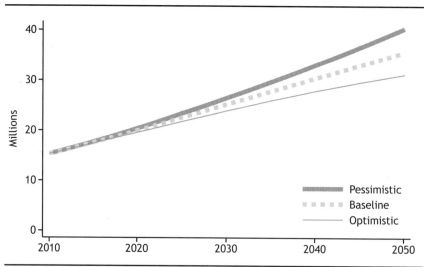

Source: UNPOP (2009).

increase in GDP per capita to almost $2,500 by 2050. Policymakers need to redouble their efforts to ensure that Malawi's population does not grow to the projected levels. In contrast, the pessimistic and baseline scenarios project that incomes will reach a much lower level by the year 2050, $650 and $740, respectively.

Biophysical Analysis

Climate Models

The precipitation changes projected for Malawi in the four climate models are indicated in Figure 5.16. The CNRM-CM3 and CSIRO Mark 3 models project that the country's mean annual precipitation will remain the same except in the CNRM model for the northern region, which will experience an increase of 50–100 millimeters by 2050. The ECHAM 5 model shows the mean annual precipitation decreasing in all of Malawi except the northern region. In contrast, the MIROC 3.2 model shows an increased mean annual precipitation level ranging from 200 to 400 millimeters for the northern and central

FIGURE 5.15 Gross domestic product (GDP) per capita in Malawi, future scenarios, 2010–2050

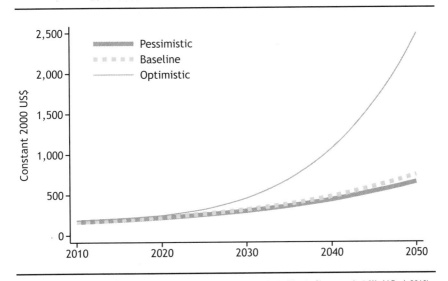

Sources: Computed from GDP data from the World Bank Economic Adaptation to Climate Change project (World Bank 2010), from the Millennium Ecosystem Assessment (2005) reports, and from population data from the United Nations (UNPOP 2009).
Note: US$ = US dollars.

FIGURE 5.16 Changes in mean annual precipitation in Malawi, 2000–2050, A1B scenario (millimeters)

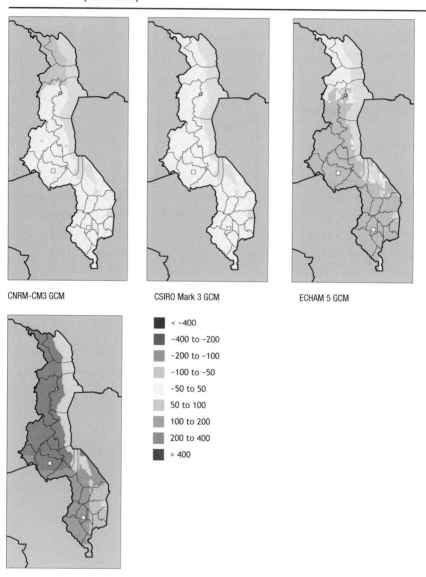

Source: Authors' calculations based on Jones, Thornton, and Heinke (2009).
Notes: A1B = greenhouse gas emissions scenario that assumes fast economic growth, a population that peaks midcentury, and the development of new and efficient technologies, along with a balanced use of energy sources; CNRM-CM3 = National Meteorological Research Center–Climate Model 3; CSIRO = climate model developed at the Australia Commonwealth Scientific and Industrial Research Organisation; ECHAM 5 = fifth-generation climate model developed at the Max Planck Institute for Meteorology (Hamburg); GCM = general circulation model; MIROC = Model for Interdisciplinary Research on Climate, developed by the University of Tokyo Center for Climate System Research.

regions of Malawi and from 50 to 200 millimeters for most of the southern region by the year 2050. The precipitation maps highlight the uncertainties associated with modeling precipitation using GCMs.

Figure 5.17 shows projected annual changes in the mean maximum daily temperature for the warmest month in Malawi by 2050. The CSIRO Mark 3 model projects a relatively modest increase in temperature of between 1° and 1.5°C. The MIROC 3.2 model projects that for the northernmost part, but for the rest of the country it projects an increase of 1.5°–2°C. The CNRM-CM3 model predicts an even warmer future, with temperatures increasing in almost the entire country by 2°–2.5°C. But the ECHAM 5 model predicts the warmest future of all, with the temperature in the central part of Malawi to rise by 2.5°–3°C while that in the rest of the country is to rise by 2°–2.5°C. The projected increase in temperature would result in higher levels of evapotranspiration and reduced moisture, a problem particularly for crops and varieties that are not heat tolerant.

Crop Models

The DSSAT crop modeling system was used to compare future yields by 2050 using four GCMs with the baseline yield (with an unchanged climate). Figure 5.18 shows rainfed maize yields declining by about 5–25 percent of baseline for most parts of the northern and central regions of Malawi in the CNRM-CM3 GCM; for most parts of the Shire Highlands in the south, yields are shown increasing by more than 25 percent of baseline by 2050. The CSIRO Mark 3 GCM shows a brighter prediction for maize in the northern and central regions, with most areas showing a gain in yield of 5–25 percent. The prediction is not as optimistic for the southern region. Although there are areas with projected yield gains of more than 25 percent, there are also large areas showing a 5–25 percent decline in rainfed maize yields. The ECHAM 5 GCM shows rainfed maize yields for most parts of Malawi decreasing by more than 25 percent of baseline, while the MIROC 3.2 GCM is similar in expectations to the CSIRO Mark 3 GCM. The ECHAM 5 GCM is the most pessimistic model, showing the greatest temperature increases and precipitation decreases of the four GCMs considered. It also reveals a reduction in rainfed maize yields, varying from 5 percent to more than 25 percent of baseline yields, except for Mwanza and Neno areas, where yield gains of greater than 25 percent of baseline yields are expected by 2050. The CSIRO model in Figure 5.18 shows significant gains in yields throughout most of the country.

FIGURE 5.17 Change in monthly mean maximum daily temperature in Malawi for the warmest month, 2000–2050, A1B scenario (°C)

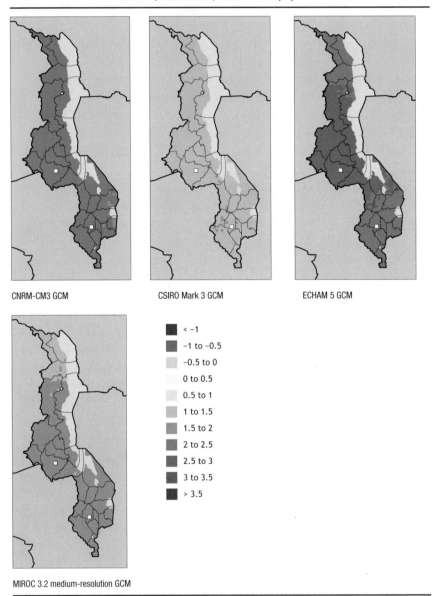

Source: Authors' calculations based on Jones, Thornton, and Heinke (2009).

Notes: A1B = greenhouse gas emissions scenario that assumes fast economic growth, a population that peaks midcentury, and the development of new and efficient technologies, along with a balanced use of energy sources; CNRM-CM3 = National Meteorological Research Center–Climate Model 3; CSIRO = climate model developed at the Australia Commonwealth Scientific and Industrial Research Organisation; ECHAM 5 = fifth-generation climate model developed at the Max Planck Institute for Meteorology (Hamburg); GCM = general circulation model; MIROC = Model for Interdisciplinary Research on Climate, developed by the University of Tokyo Center for Climate System Research.

MALAWI 133

FIGURE 5.18 Yield change under climate change: Rainfed maize in Malawi, 2000–2050, A1B scenario

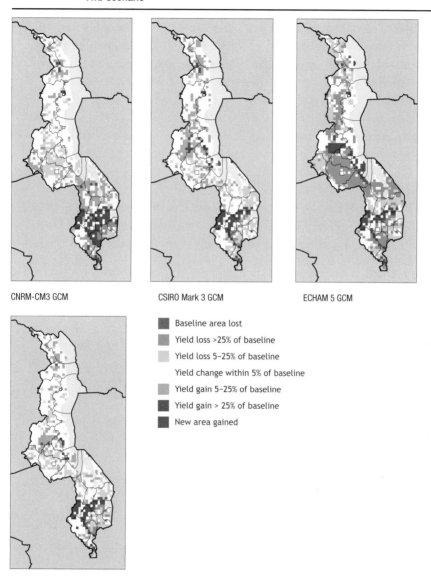

CNRM-CM3 GCM   CSIRO Mark 3 GCM   ECHAM 5 GCM

- Baseline area lost
- Yield loss >25% of baseline
- Yield loss 5–25% of baseline
- Yield change within 5% of baseline
- Yield gain 5–25% of baseline
- Yield gain > 25% of baseline
- New area gained

MIROC 3.2 medium-resolution GCM

Source: Authors' calculations based on Jones, Thornton, and Heinke (2009).
Notes: A1B = greenhouse gas emissions scenario that assumes fast economic growth, a population that peaks midcentury, and the development of new and efficient technologies, along with a balanced use of energy sources; CNRM-CM3 = National Meteorological Research Center–Climate Model 3; CSIRO = climate model developed at the Australia Commonwealth Scientific and Industrial Research Organisation; ECHAM 5 = fifth-generation climate model developed at the Max Planck Institute for Meteorology (Hamburg); GCM = general circulation model; MIROC = Model for Interdisciplinary Research on Climate, developed by the University of Tokyo Center for Climate System Research.

Vulnerability to Climate Change

The projected impact of future GDP and population scenarios on the number of malnourished children under age five is shown in Figure 5.19; Figure 5.20 shows the share of children who are malnourished. The pessimistic and baseline scenarios give similar estimates, topping out at around 1,150,000 in 2030. But the rise from around 850,000 in 2010 does not reflect an increase in the percentage of children malnourished, because the population will also be rising at a similar pace during those years. The optimistic scenario shows a steep decline in the numbers of malnourished children after 2025. The baseline scenario shows a slight decline after 2025, with the number of malnourished children under age five declining very little after 2030.

The decreased numbers of malnourished children are highly correlated with better caloric intake for the general population (Figure 5.21). The availability of kilocalories per capita remains the same under the three scenarios until 2030, when differences emerge. The higher-GDP, low-population scenario shows the availability of kilocalories per capita increasing to nearly 2,800 by 2050. The baseline and pessimistic scenarios show about

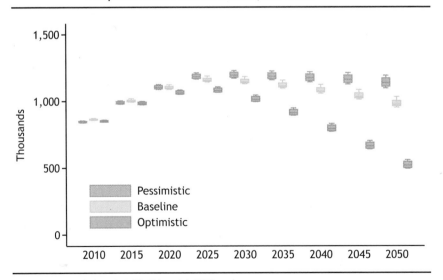

FIGURE 5.19 Number of malnourished children under five years of age in Malawi in multiple income and climate scenarios, 2010–2050

Source: Based on analysis conducted for Nelson et al. (2010).
Note: The box and whiskers plot for each socioeconomic scenario shows the range of effects from the four future climate scenarios.

MALAWI 135

FIGURE 5.20 Share of malnourished children under five years of age in Malawi in multiple income and climate scenarios, 2010–2050

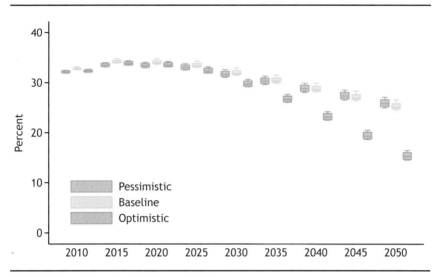

Source: Based on analysis conducted for Nelson et al. (2010).
Note: The box and whiskers plot for each socioeconomic scenario shows the range of effects from the four future climate scenarios.

FIGURE 5.21 Kilocalories per capita in Malawi in multiple income and climate scenarios, 2010–2050

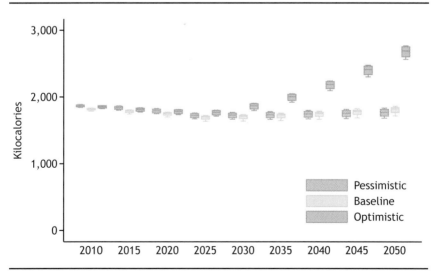

Source: Based on analysis conducted for Nelson et al. (2010).
Note: The box and whiskers plot for each socioeconomic scenario shows the range of effects from the four future climate scenarios.

1,800 kilocalories per capita available in Malawi by the year 2050. In these two scenarios, the large price increases seem to negate the positive effects of the increases in income. Chapter 2, however, discusses the fact that the model's own-price elasticities for staple crops were set too high, amplifying the price effects. Nonetheless, the difference in outcomes between the optimistic scenario and the other two scenarios shows the importance of putting in place the right policies to control population growth and improve broad-based economic growth for Malawi.

Agricultural Outcomes

The results in Figure 5.22 indicate that the total acreage allocated for the production of maize will remain almost the same in the three scenarios from 2010 to 2050. The maize yield will rise by over 15 percent between 2010 and 2030, then remain flat or decline slightly thereafter. Production will follow yield very closely. However, net exports will decrease considerably due to increased population growth by 2050, despite increased world prices for the commodity. It is very important for Malawi to ensure stable maize yields over the 40-year period to achieve stable production levels and meet food security needs. The production levels must be associated with an appropriate annual area for growing maize.

In Figure 5.23, the production of cassava and other roots and tubers and the associated yield per hectare are shown to increase from 2010 to 2050 in the three scenarios, with an increase in yield of almost 50 percent projected. However, there will be some drop in acreage and huge drops in net exports in the country by 2050 in all scenarios. Despite the increase in the world price of cassava and other roots and tuber crops, the model projects that Malawi will be importing these commodities. Malawi also needs to invest in growing and promoting cassava and other root crops in the food security chain. Increasing the yield potential will result in improved productivity during the 40-year period, irrespective of the scenarios.

In contrast with the production of maize and the tropical root crops, including cassava, cotton production over the 40-year period is projected to significantly increase (Figure 5.24), irrespective of the scenarios. The total production and yield per hectare for cotton will more than double by 2050 in the three scenarios, despite only a slight increase in the area dedicated to cotton due to land scarcity in Malawi. However, the net exports will more than triple

FIGURE 5.22 Impact of changes in GDP and population on maize in Malawi, 2010–2050

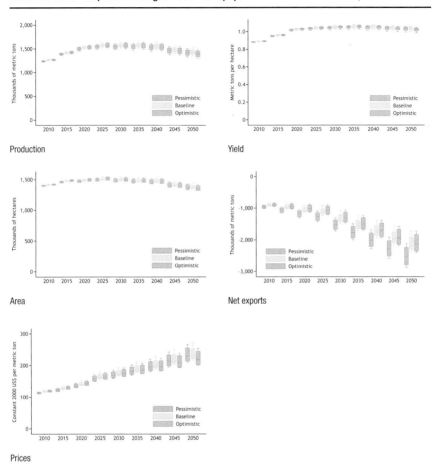

Source: Based on analysis conducted for Nelson et al. (2010).
Notes: The box and whiskers plot for each socioeconomic scenario shows the range of effects from the four future climate scenarios. GDP = gross domestic product; US$ = US dollars.

following the rising world prices of cotton, which will generate higher rates of foreign exchange through exports while keeping the area dedicated to cotton production fairly constant. This cash crop constitutes an important strategy for addressing the impact of climate change in Malawi. The country thus needs to intensify its cotton growing to generate increased foreign exchange earnings under the climate change conditions expected from now until 2050.

FIGURE 5.23 Impact of changes in GDP and population on cassava in Malawi, 2010–2050

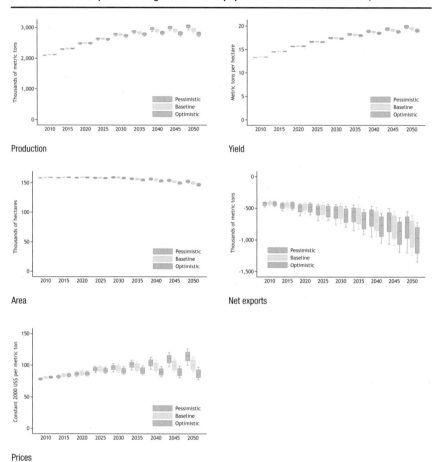

Source: Based on analysis conducted for Nelson et al. (2010).
Notes: The box and whiskers plot for each socioeconomic scenario shows the range of effects from the four future climate scenarios. GDP = gross domestic product; US$ = US dollars.

It is evident from considering the three crop systems that the crops show different responses to climate change in Malawi. Given the country's small landholdings and fast-growing population, the inclusion in the analysis of legumes and other valuable crops to show the impacts of climate change on them would offer various policy options. Further, the impact of climate change on other components of farming systems needs elucidation—for instance,

FIGURE 5.24 Impact of changes in GDP and population on cotton in Malawi, 2010–2050

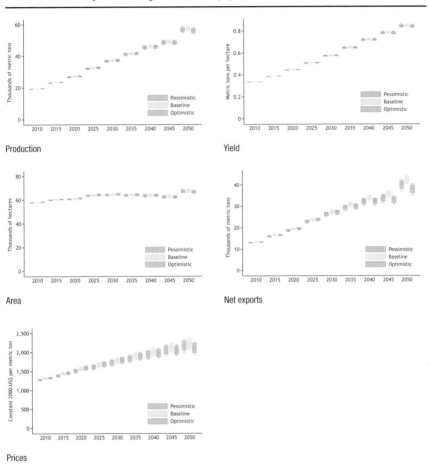

Source: Based on analysis conducted for Nelson et al. (2010).
Notes: The box and whiskers plot for each socioeconomic scenario shows the range of effects from the four future climate scenarios. GDP = gross domestic product; US$ = US dollars.

how climate change influences the occurrence of pests and diseases that affect crop and livestock productivity.

National Consultation in Malawi

A consultative workshop was held with stakeholders during the course of this study in Malawi. During this workshop, several issues were raised, as outlined in Table 5.6.

TABLE 5.6 Climate change issues and recommendations from stakeholder consultations in Malawi

Issue	Observations	Recommendations
Policy on climate change	No policy on climate change exists; the National Adaptation Programme of Action and others policy instruments include some components on climate change.	Malawi needs a policy on climate change, and the current efforts by the United Nations Development Programme to this end should be fully encouraged.
	A policy on risk assessment for mitigating and adapting to climate change needs to be completed.	It is important that a plan be completed soon given the frequent episodes of climate change events.
	Risk assessment and management were included in the national Relief and Preparedness Plan but not developed. The plan is presently being developed.	
	The relationship between vulnerability and climate change definitions and their clear association should be elucidated.	Stakeholders and farming communities should be well informed about the relationship between the two terms.
	It is important to include the role of indigenous knowledge in combating climate change in adaptation policies.	The use and value of indigenous knowledge in combating climate change should be shared among the communities and strongly promoted and included in the policy framework.
Models	The data inputs into the models were not clear; such information is important to inform the outputs of the modeling.	Details on the input variables should be clearly provided and information on the reliability of the models included in the methodology section.
	The agriculture output predictions were restricted to a few crops; inclusion of more crops is required.	Future modeling should include more crops and be used to inform the site-specific recommendations.
	The model results do not seem to be site- and area-specific.	Modeling should ensure that each vulnerable site and area has interventions that are clear and implementable.

Strengthening Local Agricultural Innovation Systems in Malawi

Institutions in Malawi are implementing various programs and projects aimed at reducing the vulnerabilities of communities to climate risks. These programs are contributing to (1) improving community resilience to climate change through the development of sustainable livelihoods, (2) restoring forests in the Upper Middle and Lower Shire Valley Catchments to reduce siltation and associated water flow problems, (3) improving agricultural production under erratic rains and changing climatic conditions, (4) enhancing

TABLE 5.6 *(Continued)*

Issue	Observations	Recommendations
Agriculture outcomes		
Innovative coping strategies	Subsistence agriculture does not offer attractive opportunities for addressing impacts of climate change (i.e., adaptation measures). In the absence of mechanization, this sector would remain ineffective.	The government should proactively implement the Agriculture Sector–Wide Approach component on the commercialization of agriculture, including the Greenbelt Initiative; the medium- and large-scale farmers are likely to ensure food productivity and thus better food security. Mechanization of the agriculture sector should also be aggressively pursued.
	Site-specific associated risks are not known; this information would inform planning and the implementation of coping strategies dependent on the site and area in Malawi.	Malawi should quickly develop its policy on the risk assessment of climate change events.
Impact of climate change on farming systems	Climate change exacerbates outbreaks of diseases and pests that reduce crop yields; this was not included in assessing agriculture's vulnerability to climate change.	Inclusion of the effects of diseases and pests in the various scenarios on crop productivity and yields is important to show the true picture.
	Few crops were included as case studies.	The next exercise should be expanded to include other crops, such as legumes.
National capacity building	The modeling undertaken by the International Food Policy Research Institute and national consultants used only the outputs.	FANRPAN should support human capital development in modeling the various climate scenarios and thus provide better-informed policy advice.
		Basic data inputs into the modeling should be shared with national partners, and these should include evidence-based experience in the use and handling of the models.

Source: Authors.
Note: FANRPAN = Food, Agriculture, and Natural Resources Policy Analysis Network.

Malawi's preparedness to cope with droughts and floods, and (5) improving climatic monitoring to enhance Malawi's early warning capabilities and decisionmaking and its sustainable use of Lake Malawi and Lake Shore area resources (Malawi, EAD 2006).

We briefly describe two projects that concentrate on knowledge and understanding of how various communities are adapting to climatic disasters in the realization that policy- and decisionmakers need to understand the context and strategies of farmers and other stakeholders in agriculture for coping and adapting to extreme climatic conditions. The diverse farming environments and complexities affecting peoples' livelihoods require localized innovation to enhance and sustain agriculture productivity.

Several studies have been done on the risk that climate change poses to people's livelihoods in Malawi (Maro and Majule 2009). A study by ActionAID International (Phiri and Saka 2008) investigated the knowledge and strategies used by stakeholders in agricultural innovation systems in Chikhwawa and Salima to adapt to climate change and variability. It is interesting that communities in both studies are aware of the impacts of climate change. They have experienced both droughts and floods that have negatively affected crop growth and household livelihoods and have thus developed local adaptation strategies. The studies confirm that increased droughts and floods exacerbate poverty levels, leaving households trapped in a cycle of poverty and vulnerability. The two projects have catalyzed processes for two-way communication and engagement among stakeholders and have thus contributed to strengthening the capacities of farmers and other stakeholders to adapt to climate change.

Conclusions and Policy Recommendations

Climate change continues to threaten the livelihoods and food security of many communities in Malawi. Poor communities, particularly women, girls, and the elderly, are most affected. Malawi, like other countries, experiences climate change and weather extremes that have an impact on the agriculture sector, affecting agricultural productivity and thus reducing food security and exacerbating chronic hunger.

The climate models used in this study have indicated various trends for which Malawi has to prepare. The population of the country is estimated to increase to 40 million by 2050 if the right policies are not put in place now. This population size will have serious repercussions for food security and livelihoods in such a small country as Malawi, especially for the most vulnerable people. The biophysical models are clear on one thing: Malawi will experience increased temperatures, ranging from 1° to 3.0°C (depending on the model). Higher temperatures will have negative effects on agriculture in the country. The models do not agree with each other regarding changes in annual precipitation. Two of the four models indicate that the mean annual precipitation will remain essentially the same; one model predicts a decline in precipitation and another predicts an increase. Increased precipitation, however, means that there will be an increased likelihood of floods in the most vulnerable areas of the country, which can destroy crops, livestock, human life, and property. The crop models provide mixed trends in crop yields for maize; by 2050, two of the models show that most parts of the country are likely to experience yield reduction, while one of the models shows the exact opposite, and one had very mixed results. Therefore, expanding the diversification of crops to ones such as roots and tubers, rice, and nontraditional cash crops including cotton might be a strategy that offers greater security for the country. However, alternative strategies include investing in research and extension to find varieties of maize that are more heat tolerant and also possibly more drought tolerant than the varieties that are currently used.

Further, an analysis of existing policies reveals that Malawi has several policies that have a bearing on climate change; these have the potential of facilitating local adaptation efforts by communities. Their joint proactive implementation through the National Adaptation Plan for Action offers great opportunities for Malawi to adequately manage the threats and impacts of climate change. However, as of now the country does not have a policy on climate change, which should be a linchpin for implementing coherent programs; the present initiatives of the UN Development Programme and the government toward its development are timely.

Policymakers should consider doing the following:

1. Create an enabling policy environment for better management of the effects of climate change on the socioeconomic development of the country, including agriculture and sustainable livelihoods. Specifically, the country should
   a. develop a clear policy and implementation plan on climate change;
   b. develop a disaster risk policy and implementation plan that will inform better management of site-specific floods and droughts in Malawi;
   c. implement the Agriculture Sector–Wide Approach component of agriculture commercialization;
   d. include other crops in the modeling for a basket of wider choice; and
   e. establish the relationship between vulnerability and climate change.

2. Enhance the capacity of rural communities to adapt to climate change. Such programs should include
   a. expanding access to agricultural inputs,
   b. facilitating smallholders' access to credit,
   c. establishing a weather-based insurance program,

d. establishing the capability of providing seasonal weather predictions,
   e. expanding the quality and reach of extension services so that smallholder farmers will have technical assistance available to them more frequently,
   f. increasing investment in agricultural research for the development and testing of plant varieties suitable to dealing with climate change threats, and
   g. improving linkages with regional and international organizations regarding climate change adaptation and agricultural research and extension.

3. Strengthen the capacity and coordination of grassroots institutions such as the district, area, and village committees so that they can better provide support services to rural communities, which are the most affected by climate change events, especially during floods and droughts.

With the additional policies suggested here, Malawian farmers will be in a much better position to adapt to changes in the climate in coming decades.

References

Bartholome, E., and A. S. Belward. 2005. "GLC2000: A New Approach to Global Land Cover Mapping from Earth Observation Data." *International Journal of Remote Sensing* 26 (9): 1959–1977.

CIESIN (Center for International Earth Science Information Network), Columbia University, IFPRI (International Food Policy Research Institute), World Bank, and CIAT (Centro Internacional de Agricultura Tropical). 2004. *Global Rural–Urban Mapping Project (GRUMP), Alpha Version: Population Density Grids.* Palisades, NY, US: Socioeconomic Data and Applications Center (SEDAC), Columbia University. http://sedac.ciesin.columbia.edu/gpw.

FAO (Food and Agriculture Organization of the United Nations). 2010. FAOSTAT. Accessed December 22, 2010. http://faostat.fao.org/site/573/default.aspx#ancor.

Haarstad, J. 2009. *Norwegian Environmental Action Plan—Baseline Study, 2009.* Oslo: Norwegian Agency for Development Cooperation. Accessed February 21, 2012. www.norad.no/en/tools-and-publications/publications/.

IPCC (Intergovernmental Panel on Climate Change). 2007. "Regional Climate." In *Climate Change 2007: The Physical Science Basis,* chapter 11. Contribution of Working Group I to the Fourth Assessment Report of the Intergovernmental Panel on Climate Change. Geneva.

IUCN (International Union for Conservation of Nature). 2010. "IUCN Red List of Threatened Species: Version 2010." Accessed March 12, 2012. www.iucnredlist.org.

Jones, P. G., P. K. Thornton, and J. Heinke. 2009. *Generating Characteristic Daily Weather Data Using Downscaled Climate Model Data from the IPCC's Fourth Assessment.* Project report. Nairobi, Kenya: International Livestock Research Institute.

Lehner, B., and P. Döll. 2004. "Development and Validation of a Global Database of Lakes, Reservoirs, and Wetlands." *Journal of Hydrology* 296 (1–4): 1–22.

Malawi, EAD (Environmental Affairs Department). 2006. *National Adaptation Programme of Action (NAPA).* Zomba: Government Press.

Malawi, Ministry of Agriculture and Food Security. 2006. *Ministry of Agriculture and Food Security Strategic Plan.* Lilongwe.

Malawi, Ministry of Economic Planning and Development. 2006. *National Policy on Early Childhood Development.* Lilongwe.

Malawi, Ministry of Lands and Housing. 2002. *Malawi National Land Policy.* Lilongwe. Accessed March 22, 2011. www.cepa.org.mw/documents/legislation/D_Malawi_National_Land_Policy_2002.pdf.

Malawi Government and World Bank. 2006. Malawi Poverty and Vulnerability Assessment: Investing in Our Future. Vol. 2. Lilongwe. Accessed May 12, 2011. www.aec.msu.edu/fs2/mgt/caadp/malawi_pva_draft_052606_final_draft.pdf.

Malawi SDNP (Sustainable Development Network Programme). 1998. "State of Environment Report (SOER)." Accessed October 21, 2010. www.sdnp.org.mw/enviro/soe_report/chapter_4.html.

Maro, P. S., and A. E. Majule, eds. 2009. *Strengthening Local Agricultural Innovations to Adapt to Climate Change in Botswana, Malawi, South Africa, and Tanzania.* Gaborone, Botswana: Southern African Development Community. Accessed March 3, 2012. www.sadc.int/fanr/agricresearch/icart/inforesources/AdaptingtoClimateChange.pdf.

Millennium Ecosystem Assessment. 2005. *Ecosystems and Human Well-being: Synthesis.* Washington, DC: Island Press. http://www.maweb.org/en/Global.aspx.

Nelson, G. C., M. W. Rosegrant, A. Palazzo, I. Gray, C. Ingersoll, R. Robertson, S. Tokgoz, et al. 2010. *Food Security, Farming, and Climate Change to 2050: Scenarios, Results, Policy Options.* Washington, DC: International Food Policy Research Institute.

Phiri, I. M. G., and A. R. Saka. 2008. "The Impact of Changing Environmental Conditions on Vulnerable Communities of the Shire Valley, Southern Malawi." In *The Future of Drylands: International Scientific Conference on Desertification and Drylands Research, Tunis, Tunisia, 19–21 June 2006,* edited by C. Lee and T. Schaaf, chapter 7. New York: Springer-Verlag.

UNEP (United Nations Environment Programme) and IUCN (International Union for Conservation of Nature). 2009. *World Database on Protected Areas (WDPA) Annual Release 2009.* Accessed January 8, 2011. http://www.wdpa.org/protectedplanet.aspx.

UNPOP (United Nations Department of Economic and Social Affairs–Population Division). 2009. *World Population Prospects: The 2008 Revision.* New York. http://esa.un.org/unpd/wpp/.

Wood, S., G. Hyman, U. Deichmann, E. Barona, R. Tenorio, Z. Guo, S. Castano, O. Rivera, E. Diaz, and J. Marin. 2010. "Sub-national Poverty Maps for the Developing World Using International Poverty Lines: Preliminary Data Release." Accessed January 18, 2011. http://povertymap.info (password protected). Some also available at http://labs.harvestchoice.org/2010/08/poverty-maps/.

World Bank. 2009. *World Development Indicators.* Washington, DC.

———. 2010. *The Costs of Agricultural Adaptation to Climate Change.* Washington, DC.

You, L., and S. Wood. 2006. "An Entropy Approach to Spatial Disaggregation of Agricultural Production." *Agricultural Systems* 90 (1–3): 329–347.

You, L., S. Wood, and U. Wood-Sichra. 2006. "Generating Global Crop Distribution Maps: From Census to Grid." Paper presented at the International Association of Agricultural Economists Conference in Brisbane, Australia, August 11–18.

———. 2009. "Generating Plausible Crop Distribution and Performance Maps for Sub-Saharan Africa Using a Spatially Disaggregated Data Fusion and Optimization Approach." *Agricultural Systems* 99 (2–3): 126–140.

Chapter 6

MOZAMBIQUE

Genito A. Maure, Timothy S. Thomas, Sepo Hachigonta, and Lindiwe Majele Sibanda

Africa, in general, is vulnerable to climate change, mostly due to its dependence on agriculture. Mozambique is a prime example. Agriculture is an important sector of the country's economy, and, as indicated by the 2007 FAO country factsheet, around 80 percent of the population (about 19.4 million people in 2007) is employed by this sector, contributing almost 23 percent of the country's gross domestic product (GDP). The same report indicates that 20 of the 128 districts in the country are highly prone to drought, 30 to flooding, and 7 to both risks, which affect about 43 percent of the population overall.

The population of Mozambique has doubled since its independence in 1975, though its growth has not been constant. Despite improvement in the under-five mortality rate, the country's well-being indicators are far below the world average, and Mozambique is ranked as one of the poorest countries in the world. From 1992 to 2010, after the end of the civil war and with improvement of economic conditions, the GDP grew.

The Mozambican National Adaptation Programme of Action (NAPA) for agriculture (Mozambique, MICOA 2007a) contains the following critically important adaptation goals:

- Contribute to self-sufficiency and food security through the promotion of simple agroprocessing technologies for the conservation of food and seeds.

- Increase agricultural productivity through the installation of irrigation systems and exploration of renewable energy sources in agriculture.

- Guarantee a supply of raw materials for industry through the encouragement of local production of seeds and amelioration of the existing road network.

- Promote and support the development of families, cooperatives, the private sector, and job creation though small-scale financing of associations among farmers to engage in small-scale businesses.

- Stimulate increases in commercial agricultural production through encouragement of the cultivation of cash crops.

- Promote agroindustrial development in rural areas to add value to Mozambique's agricultural products both for the national market and for export.

- Identify and promote the good practices of agriculture intensification programs, especially in the most affected areas, through the use of modified crops that are drought tolerant and have a short growing cycle.

- Establish the use of renewable energy for agricultural purposes to reduce costs.

- Intensify mechanized agriculture.

This chapter serves as a first step toward implementing the NAPA for agriculture by identifying specific potential impacts of climate change for this sector.

Review of Current Trends

Economic and Demographic Indicators

Population

Figure 6.1 shows the total and rural population of Mozambique (left axis) as well as the share of the population that is urban (right axis). The population has increased rapidly over the past 30 years, almost doubling since the country's independence in 1975. Figure 6.1 and Table 6.1 show that population growth has not been constant. From 1970 to 1979, the growth rate was around 2.5 percent compared to 2.1 percent in the previous decade. The 1980s then experienced a noticeable drop, from 2.5 percent to 1.0 percent—the lowest growth rate since independence. This was probably due to people's fleeing the nation during the civil war. In the 1990s the population growth rate increased again, reaching 3.1 percent. Since 2000 the annual growth rate has been around 2.0 percent.

There has been rapid urbanization in Mozambique since 1960, with the urban population increasing faster than the rural population (see Table 6.1). The highest urban growth rate was observed in the 1970s (10.7 percent), coinciding with Mozambique's independence, and the 1990s (6.9 percent), coinciding with the end of civil war, which severely disrupted the agricultural sector and the transport sector. The sharp decline in the late 1980s reflects the overall toll of the civil war. Currently both the urban and the rural populations are steadily

FIGURE 6.1 Population trends in Mozambique: Total population, rural population, and percent urban, 1960–2008

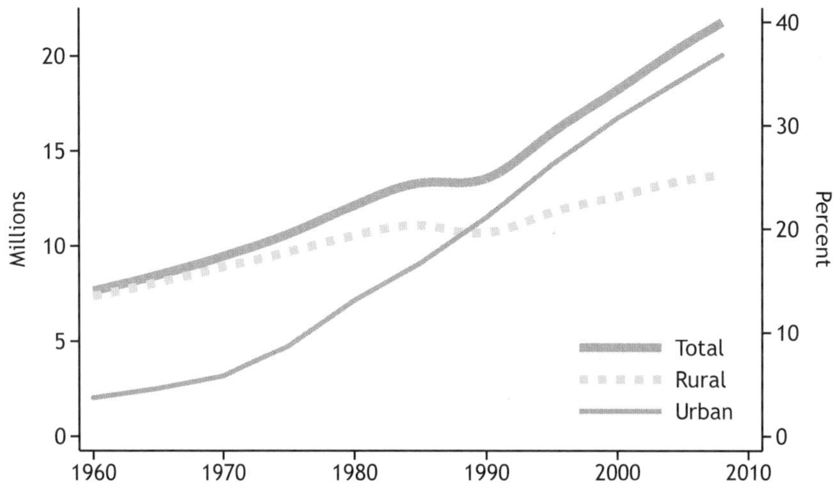

Source: *World Development Indicators* (World Bank 2009).

increasing. Despite these variations, around 65 percent of the population is still rural, making agriculture the country's main economic activity.

Figure 6.2 shows the geographic distribution of the population in Mozambique. The population is mostly concentrated in Zambézia and Nampula Provinces in central–northern Mozambique, as well as along the coast elsewhere. Niassa, the largest province, has the least dense population. In Gaza Province, in the south—a province severely hit by droughts (in the west) and floods (in the east)—the population is concentrated mostly in the floodplains of the Limpopo River.

TABLE 6.1 Population growth rates in Mozambique, 1960–2008 (percent)

Decade	Total growth rate	Rural growth rate	Urban growth rate
1960–1969	2.1	1.9	6.6
1970–1979	2.5	1.7	10.7
1980–1989	1.0	0.1	5.8
1990–1999	3.1	1.8	6.9
2000–2008	2.3	1.1	4.5

Source: Author's calculations based on *World Development Indicators* (World Bank 2009).

FIGURE 6.2 Population distribution in Mozambique, 2000 (persons per square kilometer)

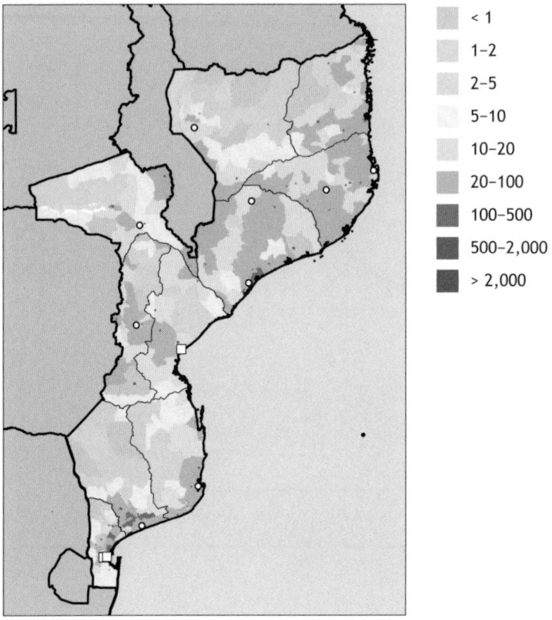

	< 1
	1–2
	2–5
	5–10
	10–20
	20–100
	100–500
	500–2,000
	> 2,000

Source: CIESIN et al. (2004).

Income

Figure 6.3 shows trends in GDP per capita as well as the proportion of GDP from agriculture, obtained from the World Bank's *World Development Indicators* (2009). Agriculture is included because of its importance as a sector vulnerable to the impacts of climate change. From 1981 to 1986, the country's GDP declined due to civil war; it declined again in 1991 due to economic adjustments. From 1992 to 2010, the GDP has been growing. From 1989 to 2000 the share of agriculture in GDP declined from about 45 percent to about 28 percent. From 2000 to 2010, agriculture contributed about a quarter to overall GDP, even though about 70 percent of the population lives on subsistence farming.

Vulnerability to Climate Change

Table 6.2 provides some data on several indicators of a population's vulnerability and resiliency to economic and natural shocks: level of education, literacy, and nutrition status. The three-year gross primary school enrollment for 2007 shows a value of 111 percent, meaning that at least 11 percent of students at the primary level were not the appropriate age for primary school

FIGURE 6.3 Per capita GDP in Mozambique (constant 2000 US$) and share of GDP from agriculture (percent), 1960–2008

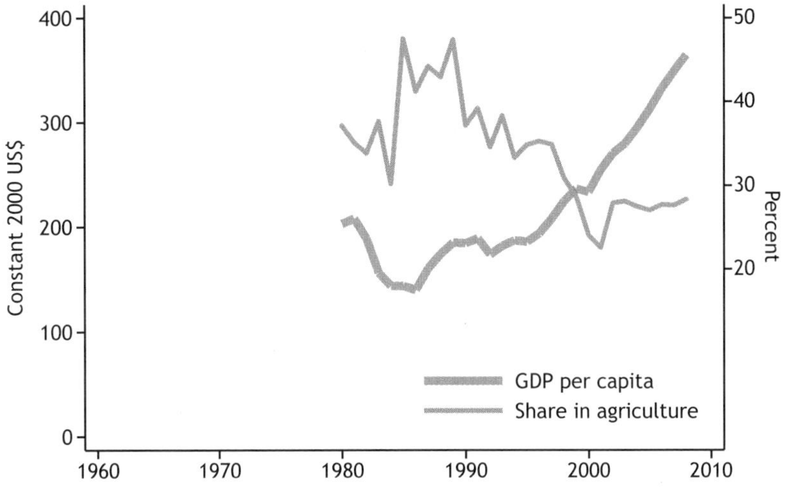

Source: *World Development Indicators* (World Bank 2009).
Note: GDP = gross domestic product; US$ = US dollars.

enrollment. Secondary school enrollment is worse, with only 18.3 percent of youth enrolled. This low percentage was partly due to the involvement of young members of the population in subsistence activities, to the detriment of school. The adult literacy rate in 2007 was still below 50 percent of the total adult population. In 2003 the level of under-five malnutrition was very high—around a fifth of the under-five population.

Figure 6.4 shows two noneconomic indicators associated with poverty: life expectancy and under-five mortality. The under-five mortality rate in Mozambique has been declining over the past 40 years, driven by improvements

TABLE 6.2 Education and nutrition statistics for Mozambique, 2000s

Indicator	Year	Value (percent)
Primary school enrollment (percent gross, three-year average)	2007	111.0
Secondary school enrollment (percent gross, three-year average)	2007	18.3
Adult literacy rate	2007	44.4
Under-five malnutrition (weight for age)	2003	21.2

Source: *World Development Indicators* (World Bank 2009).

FIGURE 6.4 Well-being indicators in Mozambique, 1960–2008

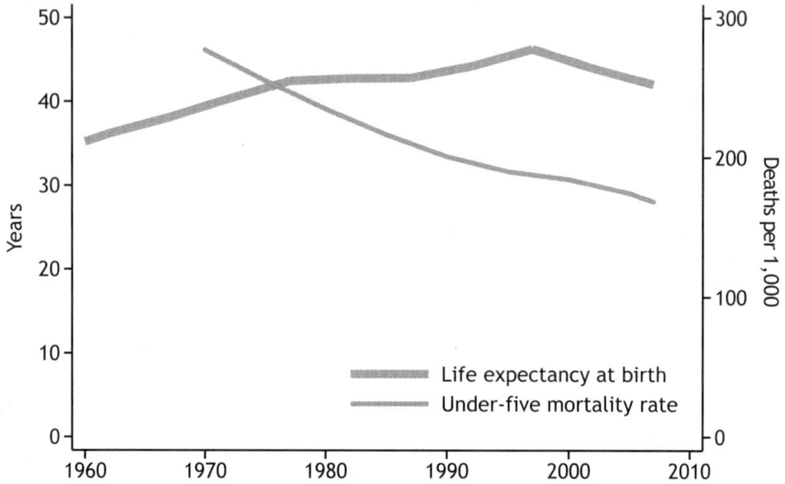

[Chart showing Life expectancy at birth (Years, left axis) and Under-five mortality rate (Deaths per 1,000, right axis) from 1960 to 2010]

Source: *World Development Indicators* (World Bank 2009).

in health services (quality and access) throughout the country, especially in rural areas with previously unmet medical needs. Life expectancy at birth increased in the period from the 1960s to 1996–1998 but has been decreasing since that time, driven by the high incidence of HIV/AIDS, vector-borne diseases, and malnutrition. Mozambique ranks far worse than the global average on both indicators.

Figure 6.5 shows the proportion of the population living on less than $2 per day, a poverty indicator that affects more than 50 percent of the population across the country. The provinces of Inhambane and Sofala are the poorest, followed by Niassa, Tete, and Gaza. Cabo-Delgado, in the north, has the lowest levels of poverty, in part because most of its population lives near the coast and therefore relies on food sources not as widely available in the rest of the country, such as fisheries, and in part because Pemba is becoming more important as a tourist destination. Rural areas far from the major urban centers tend to be poorer, implying that the areas that rely solely on agriculture are the poorest.

Land Use Overview

Figure 6.6 shows land cover and land use in Mozambique as of 2000, including the distribution of forest and agricultural activities. Niassa, Zambézia, and Nampula Provinces, in the north, have the best land for agriculture; Maputo,

FIGURE 6.5 Poverty in Mozambique, circa 2005 (percentage of population below US$2 per day)

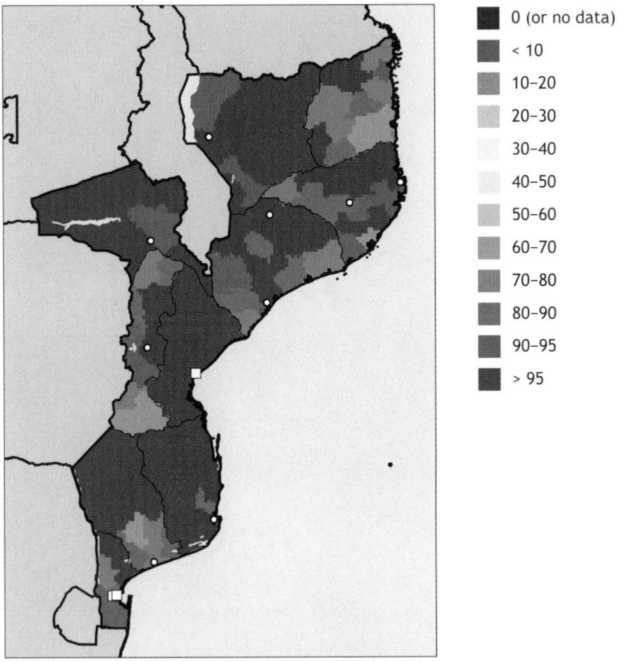

Legend:
- 0 (or no data)
- < 10
- 10–20
- 20–30
- 30–40
- 40–50
- 50–60
- 60–70
- 70–80
- 80–90
- 90–95
- > 95

Source: Wood et al. (2010).
Note: Based on 2005 US$ (US dollars) and on purchasing power parity value.

Gaza, and Inhambane in the south have poor, dry land. The land in the northern and western regions of central Mozambique is mostly covered by shrubs, while the central and southern regions are covered mainly by trees. The coastal part of central Mozambique is covered by trees that are regularly flooded by fresh water and in other cases affected by salt intrusion.

Figure 6.7 shows the locations of protected areas, including parks and reserves. These locations provide important protection for fragile environmental areas, which may also be important for tourism. About 16 percent of the country is protected, either under national legislation or under the International Union for Conservation of Nature classification (Mozambique, MICOA 2007b). Those areas may be sources of income for local communities if they benefit from tourism activities and job opportunities. On the other hand, protection may have a negative local impact if it limits the land usable for agriculture.

FIGURE 6.6 Land cover and land use in Mozambique, 2000

■ Tree cover, broadleaved, evergreen
□ Tree cover, broadleaved, deciduous, closed
□ Tree cover, broadleaved, open
■ Tree cover, broadleaved, needle-leaved, evergreen
□ Tree cover, broadleaved, needle-leaved, deciduous
□ Tree cover, broadleaved, mixed leaf type
□ Tree cover, broadleaved, regularly flooded, fresh water
■ Tree cover, broadleaved, regularly flooded, saline water
□ Mosaic of tree cover/other natural vegetation
■ Tree cover, burnt
▨ Shrub cover, closed-open, evergreen
□ Shrub cover, closed-open, deciduous
□ Herbaceous cover, closed-open
□ Sparse herbaceous or sparse shrub cover
■ Regularly flooded shrub or herbaceous cover
□ Cultivated and managed areas
■ Mosaic of cropland/tree cover/other natural vegetation
■ Mosaic of cropland/shrub/grass cover
■ Bare areas
□ Water bodies
□ Snow and ice
■ Artificial surfaces and associated areas
□ No data

Source: GLC2000 (Bartholome and Belward 2005).

FIGURE 6.7 Protected areas in Mozambique, 2009

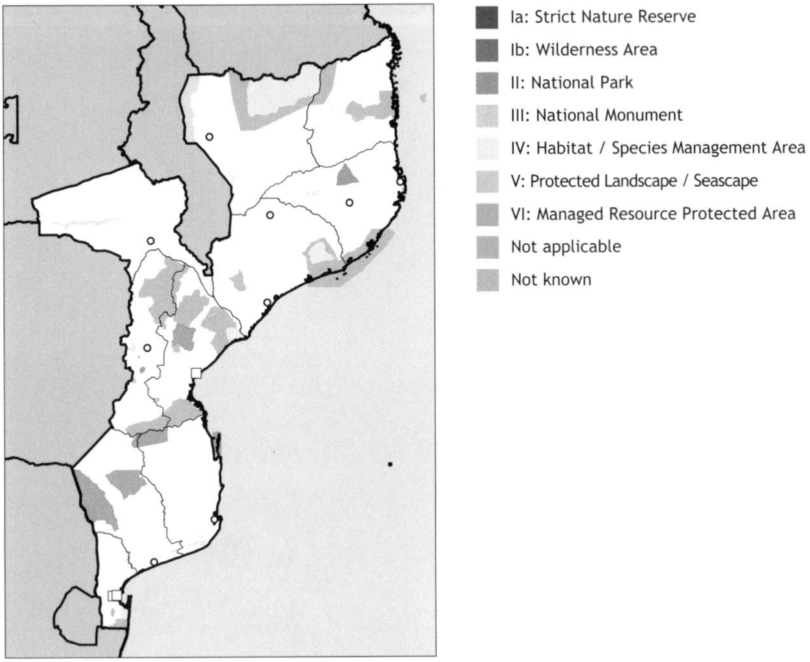

- Ia: Strict Nature Reserve
- Ib: Wilderness Area
- II: National Park
- III: National Monument
- IV: Habitat / Species Management Area
- V: Protected Landscape / Seascape
- VI: Managed Resource Protected Area
- Not applicable
- Not known

Sources: Protected areas are from the World Database on Protected Areas (UNEP and IUCN 2009). Water bodies are from the World Wildlife Fund's Global Lakes and Wetlands Database (Lehner and Döll 2004).

Figure 6.8 shows travel times to urban areas of various sizes, which are potential markets for agricultural products. The road network in the country is very weak; apart from roads to major cities, most of the country does not have paved roads for transporting agricultural products to distribution areas. The provinces of Niassa, Cabo-Delgado, Nampula, and Tete—agricultural areas in the north—have the longest travel times to the main cities, followed by Inhambane and Gaza in the south.

Less travel time is needed along the coastline, which has the best road network and large population concentrations. The roads in the south allow for reduced travel times between towns and cities, from less than an hour to three hours. In contrast, in the western region of Tete Province (close to the border with Zambia and Zimbabwe) a bit more than a day is required to travel between neighboring cities or towns of fewer than 10,000 people, as in northern Mozambique.

FIGURE 6.8 Travel time to urban areas of various sizes in Mozambique, circa 2000

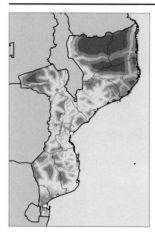

To cities of 500,000 or more people

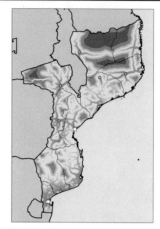

To cities of 100,000 or more people

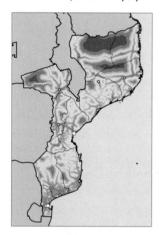

To towns and cities of 25,000 or more people

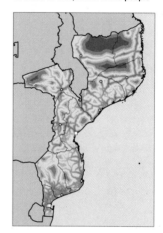

To towns and cities of 10,000 or more people

- Urban location
- < 1 hour
- 1–3 hours
- 3–5 hours
- 5–8 hours
- 8–11 hours
- 11–16 hours
- 16–26 hours
- > 26 hours

Source: Authors' calculations.

Agriculture Overview

Tables 6.3–6.5 show key agricultural commodities ranked by area harvested, value of harvest, and amount of food for human consumption (by weight) for recent years. Maize and cassava are the dominant crops, representing almost 50 percent of the area harvested. As Table 6.4 shows, cassava and maize make up approximately 65 percent of the total value of production in Mozambique. Although maize occupies more land than cassava, it ranks second in value. Tobacco comes third, at 5.7 percent of the total value of crops; it is grown in very specific areas of the country (Government of Mozambique, Ministry of Agriculture 2005). Table 6.5 shows that cassava, maize, and wheat represent more than 75 percent of agricultural food consumption; cassava alone is more than 50 percent of the total. Note that the third most widely consumed commodity, wheat, is not produced domestically.

The next two figures show the estimated yield and growing areas of key crops. Maize is grown throughout the country, with the highest yields in the provinces of Niassa, Tete, and Maputo (Figure 6.9). Cassava, on the other hand, appears to be grown mostly in the provinces of Cabo-Delgado, Nampula, and Zambézia, as well as along the coast in Inhambane and in

TABLE 6.3 Harvest area of leading agricultural commodities in Mozambique, 2006–2008 (thousands of hectares per year)

Rank	Crop	Percent of total	Harvest area
	Total	100.0	4,666
1	Maize	31.5	1,471
2	Cassava	18.3	852
3	Other pulses	9.2	430
4	Seed cotton	7.5	350
5	Sorghum	7.3	342
6	Groundnuts	6.3	295
7	Sugarcane	3.6	167
8	Rice	3.5	163
9	Castor oil seed	3.1	145
10	Other oilseeds	1.9	90

Source: FAOSTAT (FAO 2010).
Note: All values are based on the three-year average for 2006–2008.

TABLE 6.4 Value of production of leading agricultural commodities in Mozambique, 2005–2007 (millions of US$)

Rank	Crop	Percent of total	Value of production
	Total	100.0	1,852.5
1	Cassava	53.9	997.8
2	Maize	10.4	193.4
3	Tobacco	5.7	105.7
4	Potatoes	3.3	62.1
5	Coconuts	3.0	54.8
6	Groundnuts	2.2	40.8
7	Bananas	2.1	39.0
8	Pineapples	2.1	38.2
9	Sorghum	2.0	36.4
10	Seed cotton	1.4	26.4

Source: FAOSTAT (FAO 2010).
Note: All values are based on the three-year average for 2005–2007. US$ = US dollars.

TABLE 6.5 Consumption of leading food commodities in Mozambique, 2003–2005 (thousands of metric tons)

Rank	Crop	Percent of total	Food consumption
	Total	100.0	8,268
1	Cassava	56.1	4,641
2	Maize	14.0	1,155
3	Wheat	5.2	430
4	Rice	4.0	334
5	Sorghum	3.5	291
6	Other fruits	2.2	182
7	Other pulses	2.1	176
8	Sugar	1.7	137
9	Other vegetables	1.2	97
10	Bananas	1.0	84

Source: FAOSTAT (FAO 2010).
Note: All values are based on the three-year average for 2003–2005.

FIGURE 6.9 Yield (metric tons per hectare) and harvest area density (hectares) for rainfed maize in Mozambique, 2000

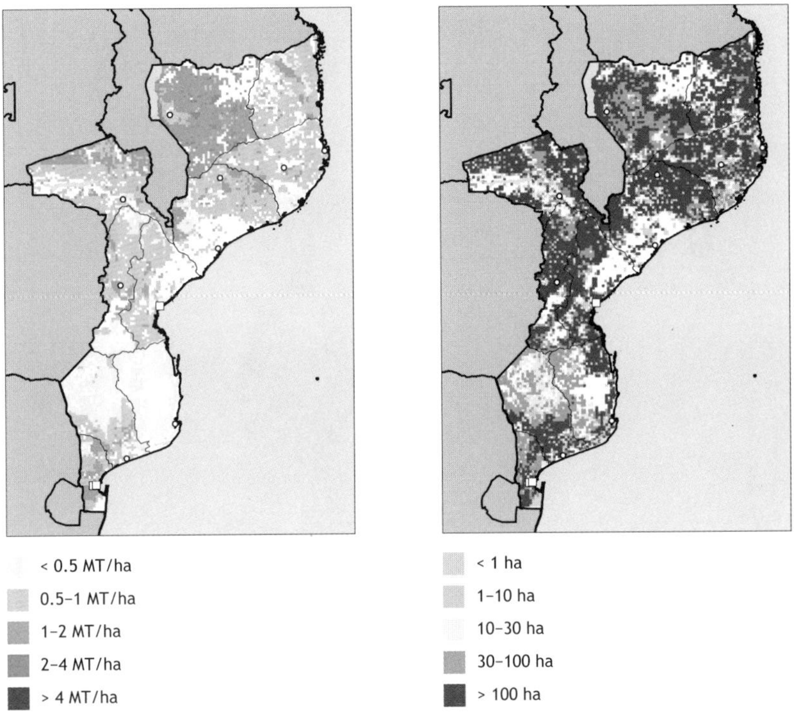

< 0.5 MT/ha	< 1 ha
0.5–1 MT/ha	1–10 ha
1–2 MT/ha	10–30 ha
2–4 MT/ha	30–100 ha
> 4 MT/ha	> 100 ha

Source: SPAM (Spatial Production Allocation Model) (You and Wood 2006; You, Wood, and Wood-Sichra 2006, 2009).
Note: ha = hectare; MT/ha = metric tons per hectare.

central Manica (Figure 6.10). The highest yields are noted in places where the acreage of cassava appears to be limited: in Tete and northern Sofala.

Scenarios for the Future

Economic and Demographic Scenarios

Population

Figure 6.11 shows the population projections of the UN Population Division through 2050 (UNPOP 2009). The population projections for 2050 are about 38 million for the low variant, 43 million for the medium variant, and 50 million for the high variant. All projections show steady population growth, implying an increasing demand for food and other resources.

FIGURE 6.10 Yield (metric tons per hectare) and harvest area density (hectares) for rainfed cassava in Mozambique, 2000

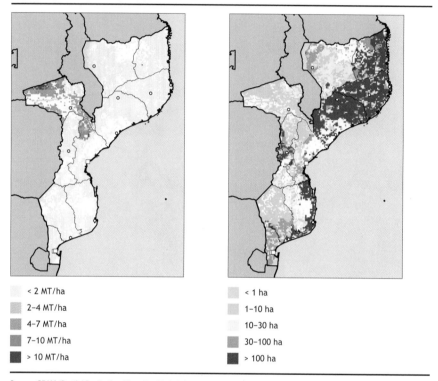

< 2 MT/ha
2–4 MT/ha
4–7 MT/ha
7–10 MT/ha
> 10 MT/ha

< 1 ha
1–10 ha
10–30 ha
30–100 ha
> 100 ha

Source: SPAM (Spatial Production Allocation Model) (You and Wood 2006; You, Wood, and Wood-Sichra 2006, 2009).
Note: ha = hectare; MT/ha = metric tons per hectare.

Income

Figure 6.12 presents three overall scenarios for GDP per capita derived by combining three GDP scenarios with the three population scenarios of Figure 6.11 (based on United Nations population data). The optimistic scenario combines high GDP growth with low population growth, the baseline scenario combines medium GDP growth with medium population growth, and the pessimistic scenario combines low GDP growth with high population growth. (The agricultural modeling in the next section uses these scenarios as well.)

The three scenarios all show a very moderate increase until around 2027, when the optimistic scenario shows a much more rapid increase than the others, which should overtake the baseline scenario by around 2037. Even the pessimistic scenario, however, shows GDP almost doubling by 2030.

FIGURE 6.11 Population projections for Mozambique, 2010–2050

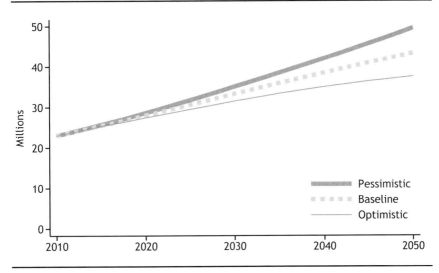

Source: UNPOP (2009).

FIGURE 6.12 Gross domestic product (GDP) per capita in Mozambique, future scenarios, 2010–2050

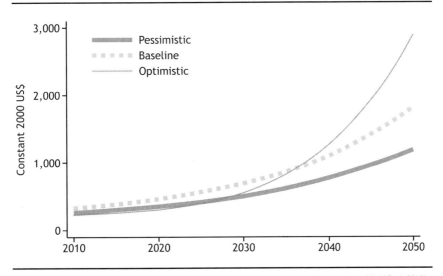

Sources: Computed from GDP data from the World Bank Economic Adaptation to Climate Change project (World Bank 2010), from the Millennium Ecosystem Assessment (2005) reports, and from population data from the United Nations (UNPOP 2009).
Note: US$ = US dollars.

Biophysical Analysis

Climate Models

Figure 6.13 shows projected annual precipitation changes under the four downscaled global circulation models (GCMs) in the A1B scenario.[1] CNRM-CM3 and CSIRO Mark 3 show very little change in rainfall across the country, with slight variations more noticeable in the CSIRO Mark 3 GCM, with a reduction in rainfall in the eastern part of Inhambane Province and an increase in part of Tete Province. ECHAM 5 shows less rainfall for the entire country, with the reduction greater in southern Mozambique, with rainfall reaching almost 200 millimeters in places. MIROC 3.2 shows little change over most of the coastal area and the southern part of the country, but in the northern and northwestern parts, away from the coast, there is a predicted increase in rainfall exceeding 200 millimeters in a few places.

Figure 6.14 shows changes in maximum temperature for the month with the highest mean daily maximum temperature. All four models show an increase in temperature, though the degree of change differs between models. CSIRO predicts the least change, with most of the country in the range of 1°–1.5°C, though changes in parts of the south go almost as high as 2°C. The ECHAM model shows an area with the highest change of all four models, with an increase of more than 2.5°C for much of Tete Province. The other two models fall in between these extremes, with the CNRM-CM3 GCM predictions slightly hotter than those of the MIROC 3.2 GCM.

Crop Models

The Decision Support Software for Agrotechnology Transfer (DSSAT) crop modeling system was used to compare the yields for 2050 to the baseline yields assuming an unchanged 2000 climate. The results for rainfed maize are mapped in Figure 6.15. They are geographically varied, with some areas showing gains in yield and others showing losses. Between the GCMs, we note that some results differ (compare the results for the northern parts of the country in CNRM-CM3 to those in ECHAM 5), and others are in agreement

---

1 The A1B scenario is a greenhouse gas emissions scenario that assumes fast economic growth, a population that peaks midcentury, and the development of new and efficient technologies, along with a balanced use of energy sources. CNRM-CM3 is National Meteorological Research Center–Climate Model 3. CSIRO is a climate model developed at the Australia Commonwealth Scientific and Industrial Research Organisation. ECHAM 5 is a fifth-generation climate model developed at the Max Planck Institute for Meteorology in Hamburg. MIROC is the Model for Interdisciplinary Research on Climate, developed at the University of Tokyo Center for Climate System Research.

FIGURE 6.13 Change in mean annual precipitation in Mozambique, 2000–2050, A1B scenario (millimeters)

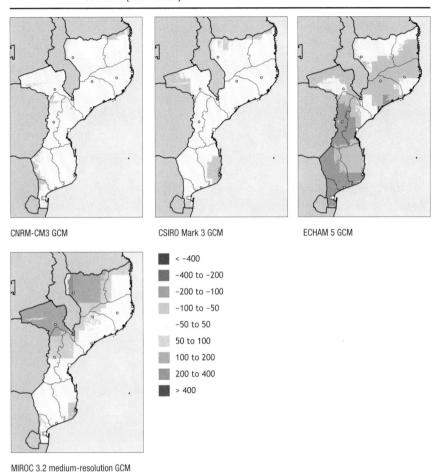

Source: Authors' calculations based on Jones, Thornton, and Heinke (2009).
Notes: A1B = greenhouse gas emissions scenario that assumes fast economic growth, a population that peaks midcentury, and the development of new and efficient technologies, along with a balanced use of energy sources; CNRM-CM3 = National Meteorological Research Center–Climate Model 3; CSIRO = climate model developed at the Australia Commonwealth Scientific and Industrial Research Organisation; ECHAM 5 = fifth-generation climate model developed at the Max Planck Institute for Meteorology (Hamburg); GCM = general circulation model; MIROC = Model for Interdisciplinary Research on Climate, developed by the University of Tokyo Center for Climate System Research.

FIGURE 6.14 Change in monthly mean maximum daily temperature in Mozambique for the warmest month, 2000–2050, A1B scenario (°C)

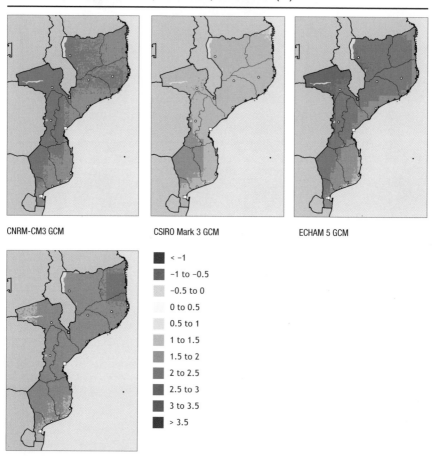

CNRM-CM3 GCM   CSIRO Mark 3 GCM   ECHAM 5 GCM

MIROC 3.2 medium-resolution GCM

- < −1
- −1 to −0.5
- −0.5 to 0
- 0 to 0.5
- 0.5 to 1
- 1 to 1.5
- 1.5 to 2
- 2 to 2.5
- 2.5 to 3
- 3 to 3.5
- > 3.5

Source: Authors' calculations based on Jones, Thornton, and Heinke (2009).
Notes: A1B = greenhouse gas emissions scenario that assumes fast economic growth, a population that peaks midcentury, and the development of new and efficient technologies, along with a balanced use of energy sources; CNRM-CM3 = National Meteorological Research Center–Climate Model 3; CSIRO = climate model developed at the Australia Commonwealth Scientific and Industrial Research Organisation; ECHAM 5 = fifth-generation climate model developed at the Max Planck Institute for Meteorology (Hamburg); GCM = general circulation model; MIROC = Model for Interdisciplinary Research on Climate, developed by the University of Tokyo Center for Climate System Research.

FIGURE 6.15 Yield change under climate change: Rainfed maize in Mozambique, 2000–2050, A1B scenario

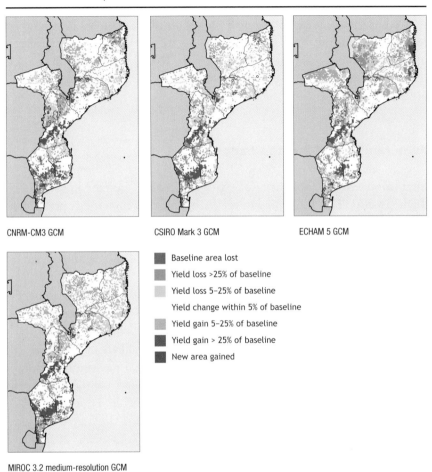

Source: Authors' calculations based on Jones, Thornton, and Heinke (2009).
Notes: A1B = greenhouse gas emissions scenario that assumes fast economic growth, a population that peaks midcentury, and the development of new and efficient technologies, along with a balanced use of energy sources; CNRM-CM3 = National Meteorological Research Center–Climate Model 3; CSIRO = climate model developed at the Australia Commonwealth Scientific and Industrial Research Organisation; ECHAM 5 = fifth-generation climate model developed at the Max Planck Institute for Meteorology (Hamburg); GCM = general circulation model; MIROC = Model for Interdisciplinary Research on Climate, developed by the University of Tokyo Center for Climate System Research.

(similarities are noted between CSIRO Mark 3 and CNRM-CM3, especially in the south).

Comparing the results of Figure 6.15 to those of the National Disasters Management Institute (INGC 2009a, 2009b), we see that our results appear to be slightly more pessimistic than those of the INGC, particularly in the ECHAM model, which seems to predict more yield reduction than the other models.

Vulnerability to Climate Change

The results from running the DSSAT crop model were used as input for the International Model for Policy Analysis of Agricultural Commodities and Trade (IMPACT), which computes global agricultural commodity prices and output by country and region. IMPACT was run with four climate model and scenario combinations. In particular, we used the CSIRO and the MIROC models, with the A1B and the B1 scenarios.[2] Those four combinations were run for each of the three GDP per capita scenarios.

In addition to agricultural predictions, IMPACT also produces scenarios of the number of malnourished children under the age of five, as well as the available kilocalories per capita. Figure 6.16 shows the impact of future GDP and population scenarios on the number of malnourished children under age five; Figure 6.17 shows the share. All scenarios show a moderate increase in the number of malnourished children from 2010 to 2025 and reductions thereafter, with sharper reductions in the optimistic and baseline scenarios. In the pessimistic scenario malnutrition peaks in 2023, at around 1 million, then declines to around 800,000 in 2050. In the baseline scenario malnutrition peaks around 2020 and then declines to 520,000 by 2050. In the optimistic scenario malnutrition peaks slightly later than in the baseline scenario and slightly earlier than in the pessimistic scenario, then declines sharply to fewer than 250,000 by 2050.

Figure 6.18 shows the average number of kilocalories per day available to each person in Mozambique. The level remains more or less constant under the pessimistic scenario, at about 1,800 kilocalories per capita through 2030, then rises to around 2,050 kilocalories per capita. The baseline scenario shows a constant level to 2025 and an increase thereafter. The level in the optimistic scenario is similar to that in the baseline scenario but rises more sharply,

---

2 B1 is a greenhouse gas emissions scenario that assumes a population that peaks midcentury (like A1B) but with rapid changes toward a service and information economy and the introduction of clean and resource-efficient technologies.

FIGURE 6.16 Number of malnourished children under five years of age in Mozambique in multiple income and climate scenarios, 2010–2050

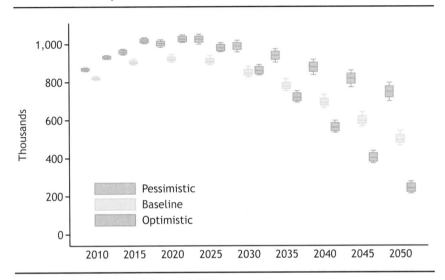

Source: Based on analysis conducted for Nelson et al. (2010).
Note: The box and whiskers plot for each socioeconomic scenario shows the range of effects from the four future climate scenarios.

FIGURE 6.17 Share of malnourished children under five years of age in Mozambique in multiple income and climate scenarios, 2010–2050

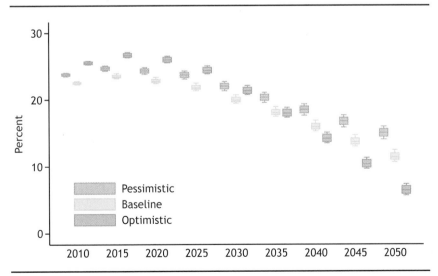

Source: Based on analysis conducted for Nelson et al. (2010).
Note: The box and whiskers plot for each socioeconomic scenario shows the range of effects from the four future climate scenarios.

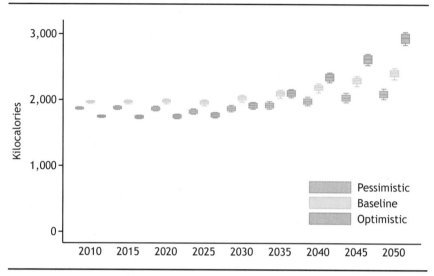

FIGURE 6.18 Kilocalories per capita in Mozambique in multiple income and climate scenarios, 2010–2050

Source: Based on analysis conducted for Nelson et al. (2010).
Note: The box and whiskers plot for each socioeconomic scenario shows the range of effects from the four future climate scenarios.

showing an increase by 2050 to 2,900 kilocalories, well above the 2,000 kilocalories per day considered healthy for an active adult.

Agricultural Outcomes

The next two sets of figures show simulation results from the IMPACT model associated with the two key agricultural crops in Mozambique. Each featured crop has five graphs: production, yield, area, net exports, and world price.

The production and yield of maize are shown increasing from 2010 to 2050 (Figure 6.19). Net exports are expected to increase to a plateau in 2025, followed by a slow decrease after 2040 but with a high variance between GCMs. The world price of maize is shown increasing throughout.

The production and yield of cassava are shown increasing slowly to 2025 and then declining slightly (Figure 6.20). The flatlining and eventual decline of the production and yield of cassava drive a decline in net exports: net exports of cassava are shown increasing slightly to 2015, followed by a reduction to 2050, and finally becoming negative. The world price of cassava

FIGURE 6.19 Impact of changes in GDP and population on rainfed maize in Mozambique, 2010–2050

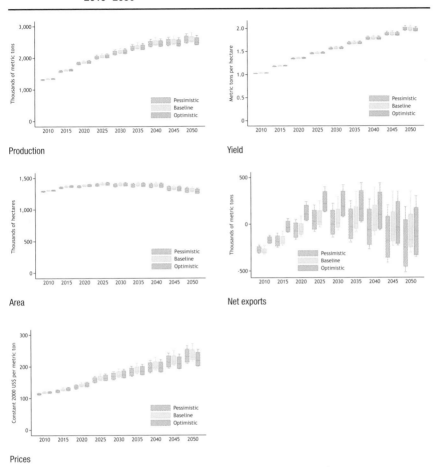

Source: Based on analysis conducted for Nelson et al. (2010).
Notes: The box and whiskers plot for each socioeconomic scenario shows the range of effects from the four future climate scenarios. GDP = gross domestic product; US$ = US dollars.

is shown increasing through 2040 in all scenarios. In the pessimistic scenario, cassava prices continue to increase. In the baseline scenario the prices appear constant from 2040 to 2050, whereas the optimistic scenario shows a minor decline from 2040 to 2050.

Because cassava is one of the main crops consumed in the country (see Figure 6.20), efforts should be made to develop alternative food sources or improved cassava varieties that are adapted to the changing climate.

FIGURE 6.20 Impact of changes in GDP and population on rainfed cassava in Mozambique, 2010–2050

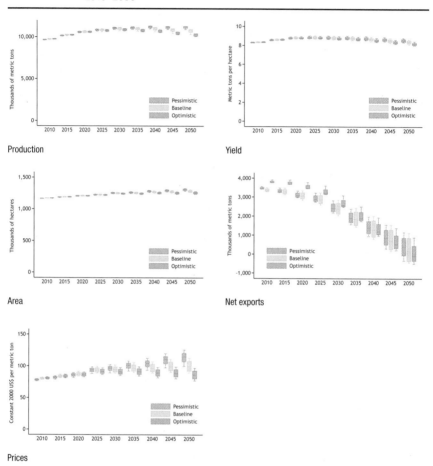

Source: Based on analysis conducted for Nelson et al. (2010).
Notes: The box and whiskers plot for each socioeconomic scenario shows the range of effects from the four future climate scenarios. GDP = gross domestic product; US$ = US dollars.

Future Research Directions

On November 23, 2010, a national meeting was held in Maputo to share the results of the initial version of this chapter and to obtain feedback from the relevant stakeholders, including governmental, nongovernmental, and educational institutions. The meeting produced the following recommendations for future research:

- Include rice yields in the simulations because, according to recent government policies, rice production will be the driver of the Mozambican economy.
- Include agroclimatic tests to predict the yields or suitability of different varieties of the most relevant crops.
- Overlap, when possible, the maps of future scenarios for the different variables and different crops to illustrate the weights of results.
- Clarify whether the supply and demand of neighboring countries are taken into account in a country study.

Conclusions and Policy Recommendations

This chapter takes a first step toward implementing the Mozambican NAPA for agriculture by identifying specific potential impacts of climate change for the agricultural sector:

- The climate scenarios generated for southern Africa show that temperatures are likely to rise in Mozambique and that precipitation patterns will change in time, location, and intensity. There was disagreement between the models, with two showing very little change, one showing a substantial reduction in rainfall, and one showing an increase in rainfall.
- For northern Mozambique the models show better future conditions for agriculture. However, the weak road network may hamper any attempt to use this region as a source of food for the rest of the country.
- Rainfed agriculture faces risks from a changing climate. The IMPACT model results show very little change in cassava productivity over the entire period 2010–2050, though maize yield may double over the same period of time.

We recommend the following actions be taken:

- Enhance the road network in the north of Mozambique so that agricultural production can be distributed throughout the country.
- Contribute to self-sufficiency and food security by finding alternative crops such as sorghum.
- Find alternative cassava varieties suitable to the changing climate.

In the context of low incomes, increased population pressure on natural systems, and a weak road network, the stress on agriculture in the country poses a potentially severe problem. Under these circumstances, the need to implement the NAPA for agriculture is urgent. The highest priority must be given to the objective "Strengthen capacities of agricultural producers to cope with climate change."

References

Bartholome, E., and A. S. Belward. 2005. "GLC2000: A New Approach to Global Land Cover Mapping from Earth Observation Data." *International Journal of Remote Sensing* 26 (9): 1959–1977.

CIESIN (Center for International Earth Science Information Network), Columbia University, IFPRI (International Food Policy Research Institute), World Bank, and CIAT (Centro Internacional de Agricultura Tropical). 2004. *Global Rural–Urban Mapping Project (GRUMP), Alpha Version: Population Density Grids.* Palisades, NY, US: Socioeconomic Data and Applications Center (SEDAC), Columbia University. http://sedac.ciesin.columbia.edu/gpw.

FAO (Food and Agriculture Organization of the United Nations). 2007. *Mozambique Fact Sheet.* FAO: Rome. Accessed November 21, 2010. www.fao.org/fileadmin/templates/tc/tce/pdf/Mozambique_factsheet.pdf.

———. 2010. FAOSTAT. Accessed December 22, 2010. http://faostat.fao.org/site/573/default.aspx#ancor.

Government of Mozambique, Ministry of Agriculture (MINAG). 2005. *Proposta do quadro legal e institucional sobre biosegurança em Moçambique.* Accessed January 17, 2012. www.unep.org/biosafety/files/MZNBFrepPT.pdf.

INGC (National Disasters Management Institute). 2009a. *Main Report: INGC Climate Change Report: Study on the Impact of Climate Change on Disaster Risk in Mozambique,* edited by K. Asante, R. Brito, G. Brundrit, P., Epstein, A. Fernandes, M. R. Marques, A. Mavume, M. Metzger, A. Patt, A. Queface, R. Sanchez del Valle, M. Tadross, and R. Brito. Maputo: Mozambique Ministry of State Administration.

———. 2009b. *Synthesis Report: INGC Climate Change Report: Study on the Impact of Climate Change on Disaster Risk in Mozambique,* edited by B. van Logchemand and R. Brito. Maputo: Mozambique Ministry of State Administration.

Jones, P. G., P. K. Thornton, and J. Heinke. 2009. *Generating Characteristic Daily Weather Data Using Downscaled Climate Model Data from the IPCC's Fourth Assessment.* Project report. Nairobi, Kenya: International Livestock Research Institute.

Lehner, B., and P. Döll. 2004. "Development and Validation of a Global Database of Lakes, Reservoirs, and Wetlands." *Journal of Hydrology* 296 (1–4): 1–22.

Millennium Ecosystem Assessment. 2005. *Ecosystems and Human Well-being: Synthesis*. Washington, DC: Island Press. http://www.maweb.org/en/Global.aspx.

Mozambique, MICOA (Ministry of Coordination of Environmental Affairs). 2007a. *National Adaptation Programme of Action*. Maputo.

———. 2007b. *4th National Report on Implementation of the Convention on Biological Diversity in Mozambique*. Maputo. www.convambientais.gov.mz/index.php?option=com_docman&task=doc_download&gid=5.

Nelson, G. C., M. W. Rosegrant, A. Palazzo, I. Gray, C. Ingersoll, R. Robertson, S. Tokgoz, et al. 2010. *Food Security, Farming, and Climate Change to 2050: Scenarios, Results, Policy Options*. Washington, DC: International Food Policy Research Institute.

UNEP (United Nations Environment Programme) and IUCN (International Union for the Conservation of Nature). 2009. *World Database on Protected Areas (WDPA): Annual Release 2009*. Accessed 2009. www.wdpa.org/protectedplanet.aspx.

UNPOP (United Nations Department of Economic and Social Affairs–Population Division). 2009. *World Population Prospects: The 2008 Revision*. New York. http://esa.un.org/unpd/wpp/.

Wood, S., G. Hyman, U. Deichmann, E. Barona, R. Tenorio, Z. Guo, S. Castano, O. Rivera, E. Diaz, and J. Marin. 2010. "Sub-national Poverty Maps for the Developing World Using International Poverty Lines: Preliminary Data Release." Accessed May 6, 2010. http://povertymap.info. Some also available at http://labs.harvestchoice.org/2010/08/poverty-maps/.

World Bank. 2009. *World Development Indicators*. Washington, DC.

———. 2010. *The Costs of Agricultural Adaptation to Climate Change*. Washington, DC.

You, L., and S. Wood. 2006. "An Entropy Approach to Spatial Disaggregation of Agricultural Production." *Agricultural Systems* 90 (1–3): 329–347.

You, L., S. Wood, and U. Wood-Sichra. 2006. "Generating Global Crop Distribution Maps: From Census to Grid." Paper presented at the International Association of Agricultural Economists Conference in Brisbane, Australia, August 11–18.

———. 2009. "Generating Plausible Crop Distribution and Performance Maps for Sub-Saharan Africa Using a Spatially Disaggregated Data Fusion and Optimization Approach." *Agricultural Systems* 99 (2–3): 126–140.

Chapter 7

SOUTH AFRICA

Peter Johnston, Timothy S. Thomas, Sepo Hachigonta,
and Lindiwe Majele Sibanda

In examining agricultural vulnerability to climate change in South Africa, we see that an important factor is the enormous existing socioeconomic disparity in access to resources, poverty levels, and capacities to adapt. Recent research results suggest that the South African farming sector is characterized by medium-level exposure risk, coupled with medium to high levels of social vulnerability (Gbetibouo and Ringler 2009).

South Africa is unique in southern Africa: from a climatological perspective, it has a steep rainfall gradient from west to east, as well as three different rainfall regimes; from a developmental perspective, it has a highly developed industrial and commercial infrastructure. Agricultural production is mostly commercial, with only 11 percent of the land arable—and much of that is located in marginally viable areas. Less than 20 percent of the country's production is from small-scale agriculture.

Review of Current Trends

Economic and Demographic Indicators

Population

Figure 7.1 shows trends for the total and rural populations of South Africa (left axis), as well as the share of the population that is urban (right axis). The figure provides additional information concerning rates of population growth. The graph shows that, although the total population is increasing steadily, the rate of growth decreased around 2000. This finding is backed up by the numbers in Table 7.1. The period 2000–2008 shows much slower growth than in previous periods, at 1.20 percent; this may be related to the fact that a population census was not taken during that period, but it may also reflect the impact of HIV/AIDS on mortality rates. Urban growth rates reflect increased trends toward urbanization. Although rural informal agriculture may be in decline,

FIGURE 7.1 Population trends in South Africa: Total population, rural population, and percent urban, 1960–2008

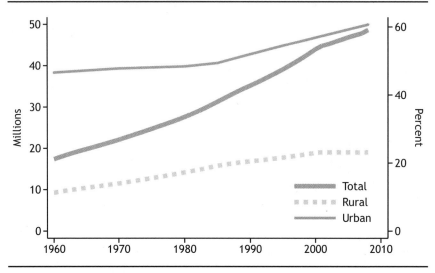

Source: *World Development Indicators* (World Bank 2009).

urban agriculture may be increasing. The latter holds little potential for commercial production. Rural agriculture, when shifting from subsistence to commercial, would generally use less labor and more mechanization along with efficiencies of scale.

Figure 7.2 shows the geographic distribution of population in South Africa.

Income

Figure 7.3 shows trends in gross domestic product (GDP) per capita and the proportion of GDP from agriculture. The decline in the contribution to GDP from agriculture reflects the increasing contributions from other sectors in the

TABLE 7.1 Population growth rates in South Africa, 1960–2008 (percent)

Decade	Total growth rate	Rural growth rate	Urban growth rate
1960–1969	2.4	2.1	2.6
1970–1979	2.2	2.1	2.3
1980–1989	2.5	1.8	3.2
1990–1999	2.2	1.1	3.1
2000–2008	1.2	0.0	2.0

Source: *World Development Indicators* (World Bank 2009).

FIGURE 7.2 Population distribution in South Africa, 2000 (persons per square kilometer)

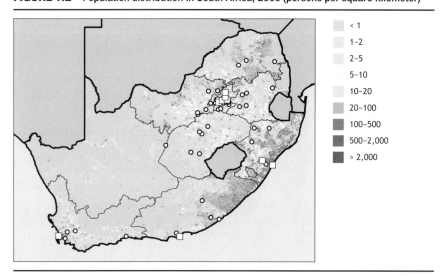

Source: CIESIN et al. (2004).

FIGURE 7.3 Per capita GDP in South Africa (constant 2000 US$) and share of GDP from agriculture (percent), 1960–2008

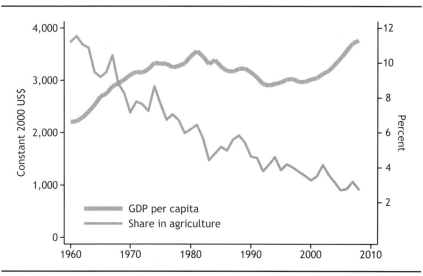

Source: *World Development Indicators* (World Bank 2009).
Note: GDP = gross domestic product; US$ = US dollars.

economy. The overall increase in GDP per capita since 1990 should indicate a growing improvement in personal well-being.

Vulnerability to Climate Change

Table 7.2 provides some data on indicators of a population's vulnerability or resiliency to economic shocks: level of education, literacy, and concentration of labor in poorer or less dynamic sectors. The figures for South Africa appear to indicate a healthy rate of literacy and education, but the values include a significant number of non-school-age learners currently in the education system. Similarly, although the adult literacy level appears high, it does not accurately reflect existing adult education levels. Although 65 percent of whites over 20 years old and 40 percent of ethnic Indians have completed at least high school, this figure is only 14 percent for black South Africans and 17 percent for the colored population.

Figure 7.4 shows two noneconomic correlates of poverty: life expectancy and under-five mortality. Life expectancy has decreased since 1990, and child mortality has also decreased. Figure 7.5 shows the HIV infection rate. This has influenced life expectancy. A positive sign is the leveling off of the rate of infection in the past decade.

Figure 7.6 shows the distribution of the population living on less than $2 per day (the criterion for poverty). This indicator is based on actual income and ignores social grants and remittances. It is measured according to magisterial and provincial boundaries and does not reflect overall population density. It gives a good indication of the locations of the poorest members of the population, who reside mostly in rural areas. The World Bank (2012) calculates that 31.3 percent of the total population of South Africa was living on less than $2 per day in 2009 (2009 data measured at 2005 prices).

TABLE 7.2 Education and labor statistics for South Africa, 2007

Indicator	Year	Value (percent)
Primary school enrollment (percent gross, three-year average)	2007	102.5
Secondary school enrollment (percent gross, three-year average)	2007	97.1
Adult literacy rate (percent)	2007	88.0
Percent employed in agriculture	2007	8.8
Percent with vulnerable employment (own farm or temporary labor)	2007	2.7

Source: *World Development Indicators* (World Bank 2009).

FIGURE 7.4 Well-being indicators in South Africa, 1960–2008: Life expectancy and under-five mortality rate

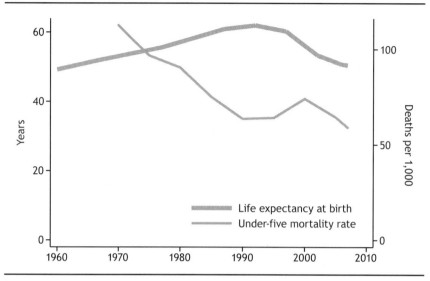

Source: *World Development Indicators* (World Bank 2009).

FIGURE 7.5 Well-being indicators in South Africa: Prevalence of HIV infection, 1990–2008

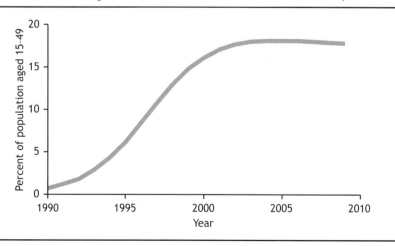

Source: *World Development Indicators* (World Bank 2009).

FIGURE 7.6 Poverty in South Africa, circa 2005 (percentage of population below US$2 per day)

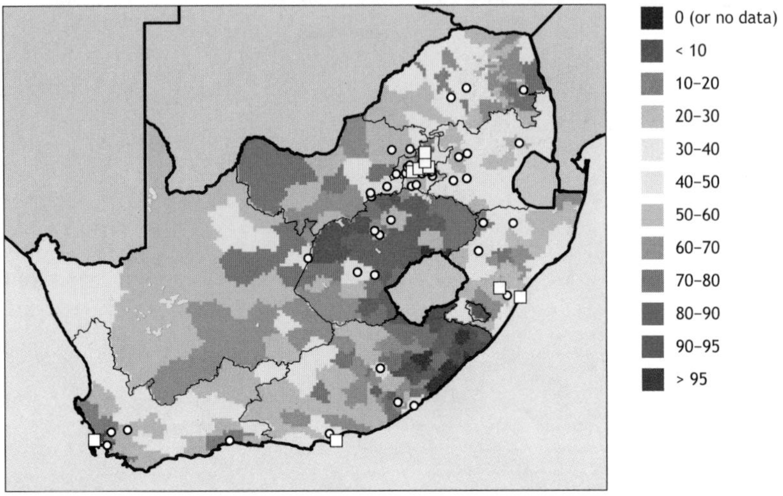

[Map of South Africa showing poverty percentages with legend: 0 (or no data), < 10, 10–20, 20–30, 30–40, 40–50, 50–60, 60–70, 70–80, 80–90, 90–95, > 95]

Source: Wood et al. (2010).
Note: Based on 2005 US$ (US dollars) and on purchasing power parity value.

Land Use Overview

Figure 7.7 shows the natural land cover and agricultural land use in South Africa as of 2000. The natural land cover has been mostly replaced by field crops, forests, and orchards where arable land is present; elsewhere, natural vegetation is used for grazing.

Figure 7.8 shows the locations of protected areas, including parks and reserves. These locations provide important protection for fragile environments, which may also be important for the tourism industry. South Africa enjoys the third-highest level of biodiversity in the world. Although the country covers only 2 percent of the world's land area, nearly 10 percent of the world's plant species and 7 percent of its reptiles, birds, and mammals are found there. South Africa's marine life is similarly diverse, partly as a result of the extreme contrast between the water masses on the east and west coasts. Over 10,000 plant and animal species—almost 15 percent of the coastal species known worldwide—are found in South African waters, with about 12 percent of these occurring nowhere else. By May 2008, about 5.9 percent of South Africa's land surface area was under formal conservation through the system of national and provincial protected areas, whereas approximately

FIGURE 7.7 Land cover and land use in South Africa, 2000

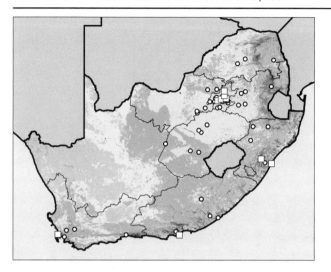

- Tree cover, broadleaved, evergreen
- Tree cover, broadleaved, deciduous, closed
- Tree cover, broadleaved, open
- Tree cover, broadleaved, needle-leaved, evergreen
- Tree cover, broadleaved, needle-leaved, deciduous
- Tree cover, broadleaved, mixed leaf type
- Tree cover, broadleaved, regularly flooded, fresh water
- Tree cover, broadleaved, regularly flooded, saline water
- Mosaic of tree cover/other natural vegetation
- Tree cover, burnt
- Shrub cover, closed-open, evergreen
- Shrub cover, closed-open, deciduous
- Herbaceous cover, closed-open
- Sparse herbaceous or sparse shrub cover
- Regularly flooded shrub or herbaceous cover
- Cultivated and managed areas
- Mosaic of cropland/tree cover/other natural vegetation
- Mosaic of cropland/shrub/grass cover
- Bare areas
- Water bodies
- Snow and ice
- Artificial surfaces and associated areas
- No data

Source: GLC2000 (Bartholome and Belward 2005).

FIGURE 7.7 Protected areas in South Africa, 2009

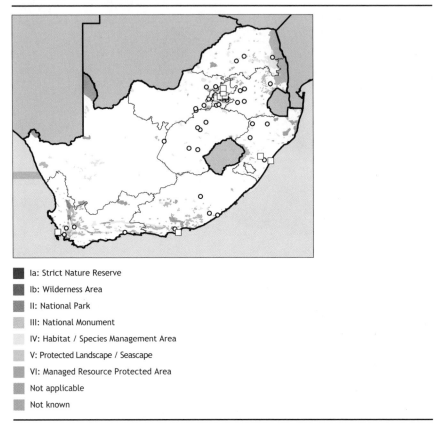

- Ia: Strict Nature Reserve
- Ib: Wilderness Area
- II: National Park
- III: National Monument
- IV: Habitat / Species Management Area
- V: Protected Landscape / Seascape
- VI: Managed Resource Protected Area
- Not applicable
- Not known

Sources: Protected areas are from the World Database on Protected Areas (UNEP and IUCN 2009). Water bodies are from the World Wildlife Fund's Global Lakes and Wetlands Database (Lehner and Döll 2004).

12 percent was under game farming and conservation management (South Africa Government Communications 2009).

Figure 7.9 shows travel times to urban areas of various sizes, which are potential markets for agricultural products as well as sources of agricultural inputs and consumer goods for farm households. The maps do not accurately convey the state of the roads or the terrain, which can be seriously inhibiting factors when attempting to transport goods to markets. The most remote area is the central Northern Cape Province, which has a population density of about 1–2 persons per square kilometer and is the least suitable for agriculture.

FIGURE 7.9 Travel time to urban areas of various sizes in South Africa, circa 2000

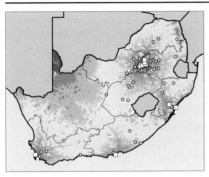

To cities of 500,000 or more people

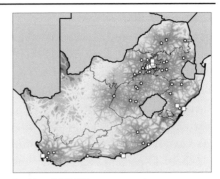

To cities of 100,000 or more people

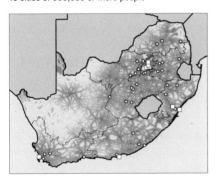

To towns and cities of 25,000 or more people

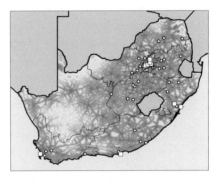

To towns and cities of 10,000 or more people

- Urban location
- < 1 hour
- 1–3 hours
- 3–5 hours
- 5–8 hours
- 8–11 hours
- 11–16 hours
- 16–26 hours
- > 26 hours

Source: Authors' calculations.

Agriculture Overview

About 8.5 million people are directly or indirectly dependent on agriculture for employment and income. The total contribution of agriculture to the economy increased from R38 billion (38 billion South African rand; US$1 ≈ R7) in 2002 to R68 billion in 2008. South Africa's dual agricultural economy comprises a well-developed commercial sector, an increasingly significant emerging sector, and a predominantly subsistence-oriented sector in the rural areas. Agricultural activities range from intensive export production (fruit, wine, and field crops) to mixed farming to cattle and sheep farming.

About 11 percent of South Africa's surface area is available for crop production, of which only 22 percent is high-potential arable land. About 1.5 percent of the country's agricultural land (1.3 million hectares) is under irrigation. About 81 percent of the total land area is farmed (crops and livestock included). However, only 70 percent of this area is suitable for grazing. Overgrazing and erosion diminish the carrying capacity of the veld and lead to land degradation. The most important limiting factor in agricultural production is water availability. Almost 50 percent of South Africa's available surface water is used for agricultural purposes.

Primary commercial agriculture contributes about 3 percent to South Africa's GDP and about 7 percent to its formal employment. However, due to strong backward and forward linkages to the economy, the agroindustrial sector is estimated to compose about 12 percent of GDP. Although the country has the potential to be self-sufficient in virtually all major agricultural products, the rate of growth in exports has been slower than that of imports.

Despite the farming industry's declining share in GDP, it remains vital to the economy and to the development and stability of the southern African region as a whole. Since 2005, agricultural exports have contributed, on average, about 7 percent of the total South African exports. Exports increased from 5 percent of agricultural production in 1988 to 38 percent in 2007; however, the growth in imports of processed goods is increasing even faster.

Major import products include wheat, rice, oil cakes, vegetable oils, and poultry meat. The estimated value of imports in 2008 was R38.4 billion, whereas exports totaled R44.3 billion in the same year. The largest export groups are wine, citrus, sugar, grapes, fruit juice, wool, and deciduous fruit, such as apples, pears, and quinces. Other important export products are avocados, pineapples, groundnuts, preserved fruit and nuts, maize (when a surplus is available), and hides and skins.

Owing to its geographic location, some parts of South Africa are prone to drought. At present, the country's agricultural sector experiences multiple stressors, including (but not limited to) variable rainfall, widespread poverty, environmental degradation, uncertainties surrounding land transfer and transformation, limited access to capital and markets, inadequate infrastructure and technology, and HIV/AIDS. Climate change—superimposed on all these other stressors—is anticipated to exacerbate these issues. With low adaptive capacity, throughout the value chain the South African agricultural sector is highly vulnerable to the effects of climate change and the anticipated increase in climate variability.

The Fourth Assessment Report of the Intergovernmental Panel on Climate Change states that "agricultural production and food security (including access to food) in many African countries and regions are likely to be severely compromised by climate change and climate variability" (Boko et al. 2007, 435). Because most of the arable land in South Africa is rainfed, increasing rainfall variability—widely projected under climate change conditions—would threaten the livelihoods of people who depend on rainfed agriculture, increasing the percentage of the population suffering from hunger and undernourishment.

Farmers, both subsistence and commercial, have developed varied strategies to cope with the current climate variability in South Africa. These strategies, however, may not be sufficient to cope with climatic changes of the future (Boko et al. 2007). In an effort to address issues related to climate change in the agricultural sector, the Department of Agriculture, Forestry, and Fisheries has developed the Climate Change Sector Plan and Climate Change Programme to ensure a sustainable, profitable agricultural sector. The plan is a response to the South African National Climate Change Response Strategy as well as the cabinet's mandated action plans calling for individual plans for all sectors. Research on the impact of climate change on agricultural production is being supported to allow policymakers to better understand and plan for likely impacts. Climate change vulnerability, mitigation, and adaptation are three key concepts that are of critical importance to agriculture.

Regardless of any impacts of climate change in South Africa, agriculture is already affected by droughts, floods, cyclones, and veld fires, among other natural hazards. The losses resulting from these impacts can be minimized by strengthening the early warning system for all natural hazards. The system communicates early warning information in the form of monthly advisories and daily extreme weather warnings to the sector for disaster risk reduction, mitigation, and preparedness in accordance with the Disaster Management Act of 2002 (Republic of South Africa, Presidency 2003). Climate variability means that drought episodes are fairly regular and are interspersed with flooding episodes, resulting in lower average agricultural yields.

TABLE 7.3 Value of production of leading agricultural commodities in South Africa, 2000 and 2008 (millions of US$)

	Production	
Commodity	2000	2008
Total	8,455	15,855
Indigenous cattle meat	1,284	1,580
Indigenous chicken meat	954	1,135
Maize	919	1,004
Grapes	674	831
Cows' milk, whole, fresh	646	771
Sugarcane	495	425
Hen eggs, in shell	241	371
Wheat	363	320
Indigenous pig meat	105	299
Oranges	205	267
Potatoes	214	264
Apples	164	221
Sunflower seed	133	205
Indigenous sheep meat	105	167
Tomatoes	95	128
Pears	87	95
Onions, dry	60	91
Wool, greasy	88	75
Groundnuts	60	74
Peaches and nectarines	79	73

Source: FAOSTAT (FAO 2010).
Note: US$ = US dollars.

Table 7.3 shows key agricultural commodities in terms of their monetary value. The value of maize as a crop is eclipsed only by the value of protein in the form of poultry. Sugarcane and wheat are the other rainfed crops that show high values; irrigated crops such as fruits and vegetables are also significant. Most livestock are fed either from natural pastures or from grains, increasing the monetary value of the latter in the food chain.

Figure 7.10 shows that the contributions of animal products, horticultural products, and field crops to the total gross value of agricultural production are

FIGURE 7.10 Value of production of agricultural commodities by type in South Africa, 2007/2008–2011/2012

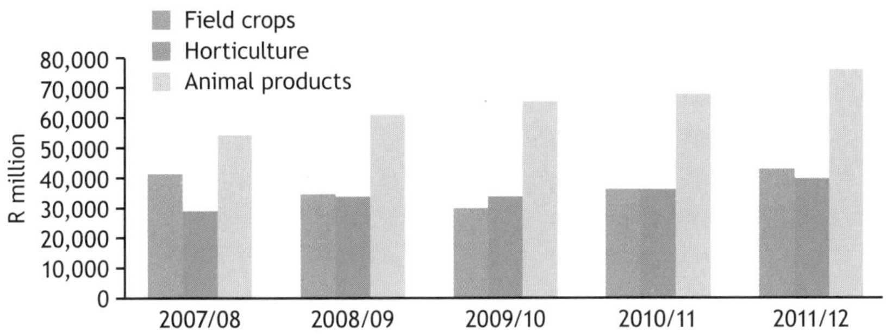

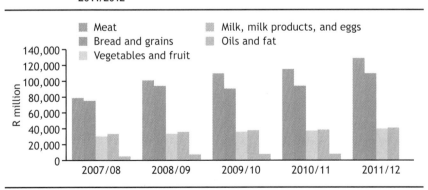

Source: Republic of South Africa, DAFF (2013).

51.3 percent, 25.7 percent, and 23.0 percent, respectively. The poultry meat industry made the largest contribution, at 18.2 percent, followed by cattle and calves slaughtered, at 11.2 percent, and maize, at 10.9 percent.

The consumption expenditure on food for the year 2009/10 shows a 2.29 percent increase over the previous year at R338,875 million compared to R331,300 million (Figure 7.11). The steadily increasing demand for meat reflects a growing trend; the increasing grain requirement for feed coincides with a recent decrease in human grain consumption.

FIGURE 7.11 Gross value of food consumption expenditure in South Africa, 2007/2008–2011/2012

Source: Republic of South Africa, DAFF (2013).
Note: R = South African rand; US$1 ≈ R7.

Maize

Maize (*Zea mays* L.) is South Africa's most important field and grain crop (Figure 7.12). Approximately 10–16 kilograms of grain are produced for every millimeter of rainfall or irrigation water "consumed" by the maize plant; in essence, this means that each maize plant will have used around 250 liters of water by maturity (Du Plessis 2003). Maize is the staple food for most of South Africa's people. The crop provides the basic household requirements of subsistence and emerging farmers, and any excess production is sold to supplement household income (Republic of South Africa, NDA 2005). The maize industry is also an important foreign exchange earner through exports of maize and maize products. The consumption requirement of maize for South Africa is around 8.7 million tons per year—around 4.8 million tons for white maize and around 3.9 million tons for yellow maize—as it is the most important feed ingredient in the beef, dairy, and poultry industries.

FIGURE 7.12 Yield (metric tons per hectare) and harvest area density (hectares) for maize in South Africa, 2005

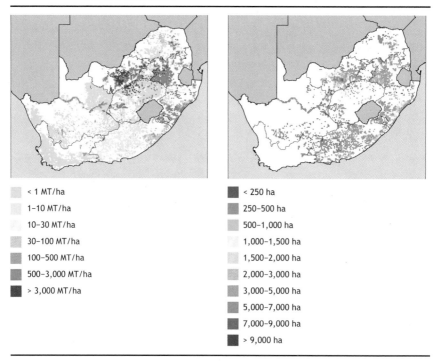

< 1 MT/ha
1–10 MT/ha
10–30 MT/ha
30–100 MT/ha
100–500 MT/ha
500–3,000 MT/ha
> 3,000 MT/ha

< 250 ha
250–500 ha
500–1,000 ha
1,000–1,500 ha
1,500–2,000 ha
2,000–3,000 ha
3,000–5,000 ha
5,000–7,000 ha
7,000–9,000 ha
> 9,000 ha

Sources: SPAM (Spatial Production Allocation Model) (You and Wood 2006; You, Wood, and Wood-Sichra 2006, 2009).
Note: ha = hectare; MT/ha = metric tons per hectare.

Maize is produced in South Africa by around 9,000 commercial farmers and provides direct employment to a workforce of around 130,000. In addition, tens of thousands of people are employed in the various industries relying on maize as a raw material (mainly for milling and stock feeds). Maize is generally planted from October to December, depending on regional and seasonal rainfall and temperature patterns—for example, as soon as sufficient soil moisture has accumulated (about 25 millimeters over a five-day period) or when soil temperatures are high enough for germination (when minimum temperatures of 10°–15°C have been maintained for a week) (Republic of South Africa, NDA 2005).

As a warm-weather crop, maize requires daily mean temperatures above 22°C: the ideal is a January mean between 19° and 24°C. The maximum temperature that can be reached before heat-related yield reductions are incurred is around 32°C. Frost can damage maize at all growth stages, and generally a frostfree period of 120–140 days is required to prevent damage. Maize can be produced in areas with a mean annual precipitation as low as 350 millimeters, but ideal rainfall levels are 450–600 millimeters during the growing season (Republic of South Africa, NDA 2005). Sustained production depends on an even distribution of rainfall throughout the growing season (Republic of South Africa, NDA 2005), but rainfall is particularly critical during the flowering season, when soil water stress reduces yields more than during other growth phases.

In the decade 1995/1996–2004/2005, maize yields in South Africa fluctuated between 2.16 and 3.04 tons per hectare, depending primarily on climatic conditions. Under irrigation, yields average 6.05 tons per hectare, with the highest yields in the Northern Cape (6.76 tons per hectare), Free State (6.31 tons per hectare), and Mpumalanga Provinces (6.27 tons per hectare) (Statistics South Africa 2002; Schulze and Walker 2007). It seems that the total area planted with maize is slowly decreasing, whereas it seems that production is increasing (Figures 7.13 and 7.14). This reflects the increasing efficiency and technological advances within the industry as well as the (mostly) favorable climatic conditions over the past decade.

Wheat

The Western Cape and Free State Provinces typically produce three quarters of the South African wheat crop (Figure 7.15). Approximately 85 percent of the total wheat area is cultivated under rainfed conditions, whereas the other 15 percent is irrigated. Most wheat produced in the Western Cape is

FIGURE 7.13 Area planted with maize, South Africa, 1993/1994–2010/2011 (thousands of hectares)

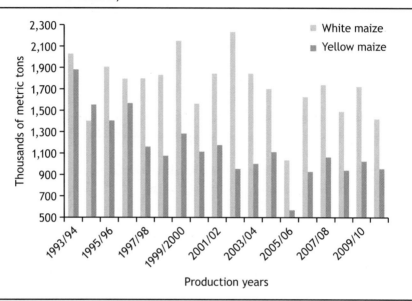

Source: Grain SA (2010).

FIGURE 7.14 Maize production, South Africa, 1993/1994–2010/2011 (thousands of metric tons)

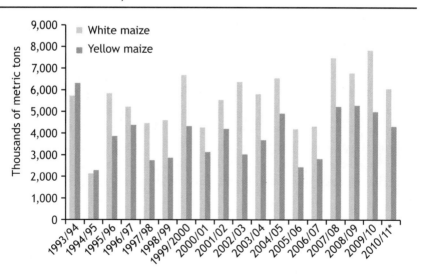

Source: Grain SA (2010).
Note: * = preliminary estimates.

FIGURE 7.15 Yield (metric tons per hectare) and harvest area density (hectares) for wheat in South Africa, 2005

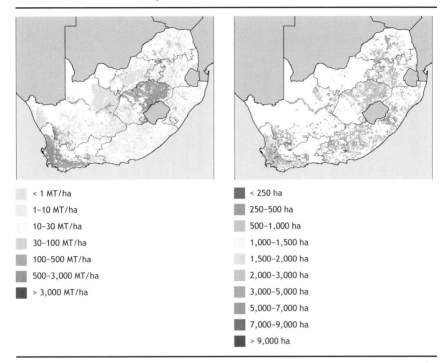

< 1 MT/ha	< 250 ha
1–10 MT/ha	250–500 ha
10–30 MT/ha	500–1,000 ha
30–100 MT/ha	1,000–1,500 ha
100–500 MT/ha	1,500–2,000 ha
500–3,000 MT/ha	2,000–3,000 ha
> 3,000 MT/ha	3,000–5,000 ha
	5,000–7,000 ha
	7,000–9,000 ha
	> 9,000 ha

Source: SPAM (Spatial Production Allocation Model) (You and Wood 2006; You, Wood, and Wood-Sichra 2006, 2009).
Note: ha = hectare; MT/ha = metric tons per hectare.

under rainfall conditions; the southwestern part of the Western Cape is one of the most important wheat-producing regions in South Africa. Unsteady and erratic rainfall in the Western Cape often produces wide variations in wheat yields and quality. Accordingly, South Africa could produce a surplus of wheat during years with very good rainfall, but it experiences shortages of wheat in most years. The planted areas are decreasing (Figure 7.16), and, though yields are increasing, total production cannot cope with increased demand.

Critics say that the shortage of wheat supply is due to a combination of factors, primarily the government's decision to open up the domestic market to global forces. But transport and infrastructure problems also make it costly for farmers to use the railways to export their product. In any case, South Africa may face a growing food crisis if a decline in domestic wheat production threatens to escalate food prices.

FIGURE 7.16 Area planted, production, and yields for wheat in South Africa, 1990/1991–2010/2011

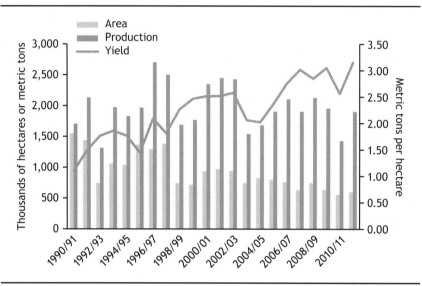

Source: Republic of South Africa, DAFF (2013).

Sugar

South Africa is ranked 13th in the world as a producer of sugarcane (*Saccharum officinarum*) (South Africa Government Communications 2005). The sugar industry in South Africa employs 85,000 people directly as well as 265,000 indirectly (for example, in the fertilizer, fuel, chemical, transport, food, and services sectors); an additional 1 million jobs are dependent on the sugar industry. The industry contributes approximately R2.0 billion annually to foreign exchange earnings. Sugarcane makes up approximately 15 percent of the gross value of the country's agricultural production and approximately 4.5 percent of its total agricultural production by tonnage. Sugarcane is grown by nearly 50,000 registered producers: 2,000 large-scale growers (average farm size 165 hectares) produce about 78 percent of the total crop; about 47,000 small-scale growers farm 2.0–2.5 hectares, on average, and produce 12 percent of the total crop. The remaining 10 percent is produced by milling companies with their own estates (Republic of South Africa, DAFF 2011). Annual production varies from 19–23 million tons, depending largely on climatic conditions. From this tonnage, around 2,500,000 tons of sugar is produced per season, of which half is marketed in South Africa and the remainder exported to markets in Africa, the Middle East, North America, and Asia. Some 95 percent of the sugarcane crop is fertilized, accounting for 18 percent

FIGURE 7.17 Yield (kilograms per hectare) and harvest area density (hectares) for sugarcane in South Africa, 2005

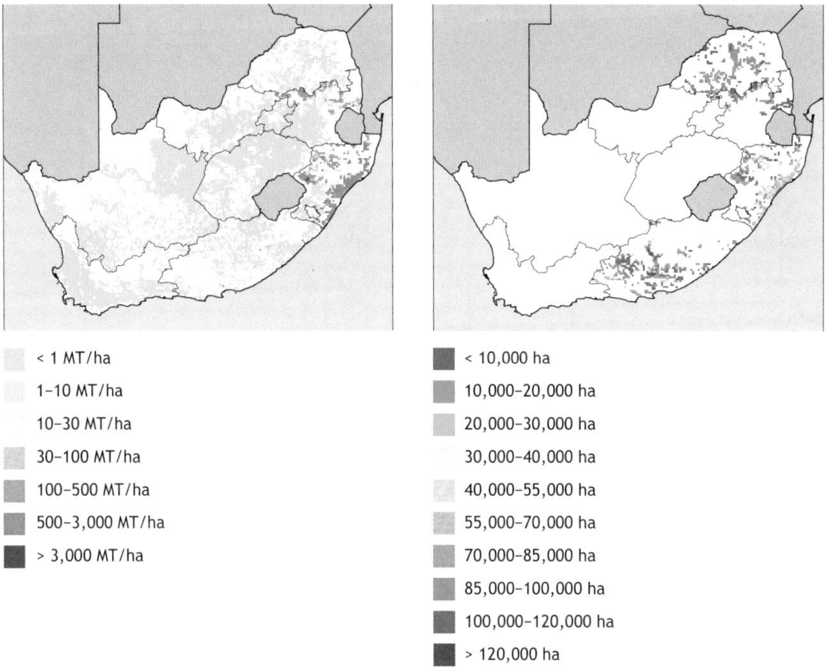

< 1 MT/ha
1–10 MT/ha
10–30 MT/ha
30–100 MT/ha
100–500 MT/ha
500–3,000 MT/ha
> 3,000 MT/ha

< 10,000 ha
10,000–20,000 ha
20,000–30,000 ha
30,000–40,000 ha
40,000–55,000 ha
55,000–70,000 ha
70,000–85,000 ha
85,000–100,000 ha
100,000–120,000 ha
> 120,000 ha

Source: SPAM (Spatial Production Allocation Model) (You and Wood 2006; You, Wood, and Wood-Sichra 2006, 2009).
Note: ha = hectare; MT/ha = metric tons per hectare.

of the fertilizer used in the country (FAO 2005a). The average cane yield is 51 tons per hectare per year (FAO 2005a). Dryland yields, however, vary considerably from year to year (Figure 7.17) (Schulze, Hull, and Maharaj 2007).

Irrigation

The total runoff per year from South African rivers is estimated at approximately 51,100 million cubic meters, but because of variable flow and high evaporation, only 30,000 million cubic meters can be economically used. The total potential groundwater delivery is estimated at 12,000 million cubic meters, of which only 5,400 million cubic meters can be readily retrieved. The estimated total water used in South Africa during 2000 was 22,500 million cubic meters.

The irrigation sector uses approximately 50 percent of the total water consumed. Groundwater irrigates 24 percent of the irrigable area, whereas surface water (from rain, rivers, dams, and canals) irrigates 76 percent. South Africa

TABLE 7.4 Distribution of irrigated area in South Africa, by province, 1999 (hectares)

Province	Commercial irrigation, permanent	Commercial irrigation, temporary	Area equipped for irrigation, total
South Africa, total	416,753	1,081,257	1,498,010
Eastern	11,070	179,995	191,065
Free State	46	68,764	68,810
Gauteng	18	16,330	16,348
KwaZulu-Natal	2,747	131,974	134,722
Mpumalanga	18,498	116,977	135,475
North West	706	114,094	114,801
Northern Cape	34,759	130,181	164,940
Northern	58,704	160,617	219,321
Western Cape	290,204	162,325	452,529

Source: FAO (2005b).

has three major rivers—the Vaal, Orange, and Limpopo—and many irrigation schemes have been developed on or near these rivers or their tributaries. Approximately 1.5 million hectares are under irrigation, amounting to 7.2 percent of the total arable land in South Africa. Table 7.4 summarizes the amount of irrigation available by province.

Scenarios for the Future

Economic and Demographic Indicators

Population
Figure 7.18 shows population projections by the UN Population Division through 2050. The low variant reflects an increase in HIV/AIDS–related deaths, as well as a reduction in the population growth rate. The medium variant is most likely, showing a small annual increase over the next 40 years.

Income
Figure 7.19 shows the three scenarios for GDP per capita used for this study. These result from combining three World Bank GDP projections with the three population projections of Figure 7.18, from United Nations population statistics: the optimistic scenario combines high GDP with low population, the baseline scenario combines the medium GDP projection with the medium

SOUTH AFRICA 195

FIGURE 7.18 Population projections for South Africa, 2010–2050

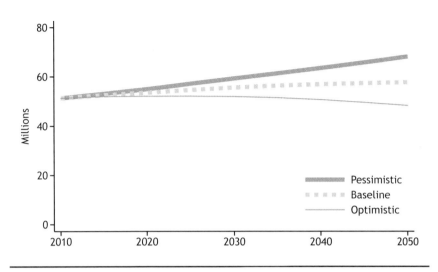

Source: UNPOP (2009).

FIGURE 7.19 Gross domestic product (GDP) per capita in South Africa, future scenarios, 2010–2050

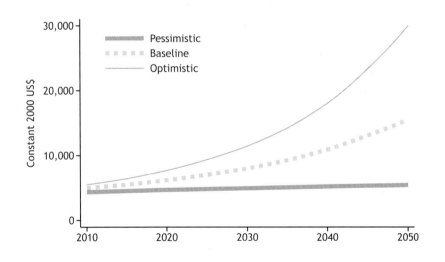

Sources: Computed from GDP data from the World Bank Economic Adaptation to Climate Change project (World Bank 2010), from the Millennium Ecosystem Assessment (2005) reports, and from population data from the United Nations (UNPOP 2009).
Note: US$ = US dollars.

population projection, and the pessimistic scenario combines the low GDP projection with the high population projection.

The worst-case scenario would assume negative trends in South Africa's political, economic, and health situations, combined with rapid population growth. The baseline scenario assumes constant growth rates for GDP and population and is more likely. The optimistic scenario reflects constant GDP growth along with a declining rate of population growth.

Biophysical Analysis

Climate Models

The latest downscaled scenarios from the Climate Systems Analysis Group were examined to find trends in the potential change in precipitation through 2050 (Republic of South Africa, DST 2011). The study uses nine general circulation models (GCMs) downscaled to a 25-kilometer grid resolution over South Africa. Note that downscaling is especially important for South Africa because the GCMs that are not downscaled fail to adequately capture the detailed spatial gradients and strong topographical forcing that are important determinants of South Africa's climate. The downscaled applications in the study focused on scenario SRES A2, a moderately high-emission scenario that envisions a heterogeneous world with an increasing global population and regionally oriented economic growth.[1]

The study shows that the models are in general agreement for September through November, with much of the country, particularly the eastern half, projected to become wetter (there is a range of wetness projected, with 10 millimeters of rain per month the approximate median). December through February is projected to be drier, particularly in the northeast, though the models are far from unanimous. Finally, the other quarterly projections show little change, on average, though they tend to lean toward predicting that it will be slightly wetter, with March through May predicted to be wetter toward the center of the country and June through August predicted to be wetter toward the east. We also note that in all quarters the lower southwestern coast of the Western Cape is projected to receive less rainfall.

Figure 7.20 shows changes in mean daily maximum temperature for the warmest month, with each map representing the outcome of a different GCM

---

[1] SRES is the Special Report on Emissions Scenarios, a report by the Intergovernmental Panel on Climate Change (IPCC 2000).

SOUTH AFRICA 197

FIGURE 7.20 Change in monthly mean maximum daily temperature in South Africa for the warmest month, 2000–2050, A1B scenario (°C)

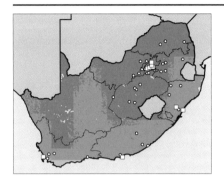

CNRM-CM3 GCM

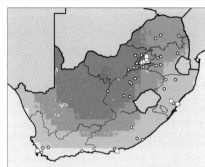

CSIRO Mark 3 GCM

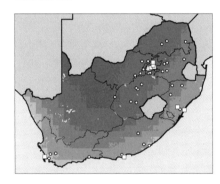

ECHAM 5 GCM

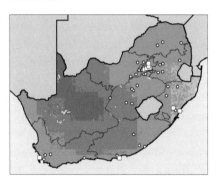

MIROC 3.2 medium-resolution GCM

- < −1
- −1 to −0.5
- −0.5 to 0
- 0 to 0.5
- 0.5 to 1
- 1 to 1.5
- 1.5 to 2
- 2 to 2.5
- 2.5 to 3
- 3 to 3.5
- > 3.5

Source: Authors' calculations based on Jones, Thornton, and Heinke (2009).
Notes: A1B = greenhouse gas emissions scenario that assumes fast economic growth, a population that peaks midcentury, and the development of new and efficient technologies, along with a balanced use of energy sources; CNRM-CM3 = National Meteorological Research Center–Climate Model 3; CSIRO = climate model developed at the Australia Commonwealth Scientific and Industrial Research Organisation; ECHAM 5 = fifth-generation climate model developed at the Max Planck Institute for Meteorology (Hamburg); GCM = general circulation model; MIROC = Model for Interdisciplinary Research on Climate, developed by the University of Tokyo Center for Climate System Research.

(four GCMs were used).[2] Of the four GCMs, the ECHAM 5 model predicts the hottest future, with the temperatures of at least half the country projected to increase by over 2.5°C and some areas well over 3°C. The coolest future is found in the projection of the CSRIO Mark 3 GCM, with most of the country seeing an increase of less than 2°C.

Climate Summary

Temperatures across the country have increased in the historical past and are seen to be increasing throughout the 21st century. The warming is expected to be greatest in the interior of the country and less along the coast. Assuming a moderate to high level of growth in greenhouse gas concentrations, by mid-century the coast is likely to warm by around 1°C and the interior by around 3°C.

Rainfall changes are expected to be regionally complex, especially in areas of strong topographical variation. In general, several trends emerge from the models. There are indications that the east coast will be wetter in summer and the west coast drier; the southwest of the country will experience drier conditions in both summer and winter. The west–east pattern of precipitation seems stable across the range of models and is physically consistent with the circulation changes; nevertheless, there is notable uncertainty about the magnitude of the response. There are indications that rainfall intensity is likely to increase—that is, even without an increase in total rainfall, the intensity is likely to be greater when it does rain. Evaporation is likely to increase due to higher temperatures. This is likely to increase the drought potential (as defined by the response of available soil moisture and available free water)—possibly even if the total rainfall of a region increases.

Crop Models

We used the Decision Support Software for Agrotechnology Transfer crop model to compute yields under current temperature and precipitation regimes and compare them to those for the year 2050 using the temperatures and precipitation levels projected. The output for key crops is mapped in Figures 7.21 (maize) and 7.22 (wheat).

---

2 CNRM-CM3 is National Meteorological Research Center–Climate Model 3. CSIRO is a climate model developed at the Australia Commonwealth Scientific and Industrial Research Organisation. ECHAM 5 is a fifth-generation climate model developed at the Max Planck Institute for Meteorology in Hamburg. MIROC is the Model for Interdisciplinary Research on Climate, developed at the University of Tokyo Center for Climate System Research.

SOUTH AFRICA 199

FIGURE 7.21 Yield change under climate change: Rainfed maize in South Africa, 2000–2050, A1B scenario

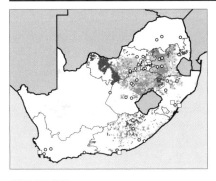

CNRM-CM3 GCM

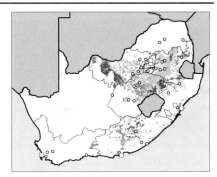

CSIRO Mark 3 GCM

ECHAM 5 GCM

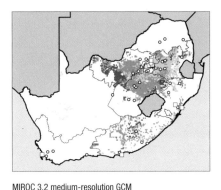

MIROC 3.2 medium-resolution GCM

- ■ Baseline area lost
- Yield loss >25% of baseline
- Yield loss 5–25% of baseline
- Yield change within 5% of baseline
- Yield gain 5–25% of baseline
- ■ Yield gain > 25% of baseline
- ■ New area gained

Source: Authors' calculations based on Jones, Thornton, and Heinke (2009).
Notes: A1B = greenhouse gas emissions scenario that assumes fast economic growth, a population that peaks midcentury, and the development of new and efficient technologies, along with a balanced use of energy sources; CNRM-CM3 = National Meteorological Research Center–Climate Model 3; CSIRO = climate model developed at the Australia Commonwealth Scientific and Industrial Research Organisation; ECHAM 5 = fifth-generation climate model developed at the Max Planck Institute for Meteorology (Hamburg); GCM = general circulation model; MIROC = Model for Interdisciplinary Research on Climate, developed by the University of Tokyo Center for Climate System Research.

FIGURE 7.22 Yield change under climate change: Rainfed wheat in South Africa (excluding Western Cape), 2000–2050, A1B scenario

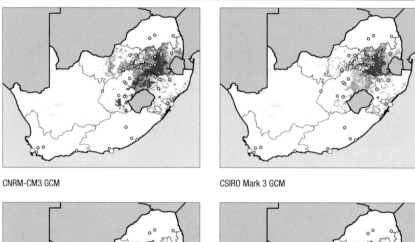

CNRM-CM3 GCM

CSIRO Mark 3 GCM

ECHAM 5 GCM

MIROC 3.2 medium-resolution GCM

- ■ Baseline area lost
- ▨ Yield loss >25% of baseline
- ▨ Yield loss 5–25% of baseline
-     Yield change within 5% of baseline
- ▨ Yield gain 5–25% of baseline
- ■ Yield gain > 25% of baseline
- ■ New area gained

Source: Authors' calculations based on Jones, Thornton, and Heinke (2009).
Notes: A1B = greenhouse gas emissions scenario that assumes fast economic growth, a population that peaks midcentury, and the development of new and efficient technologies, along with a balanced use of energy sources; CNRM-CM3 = National Meteorological Research Center–Climate Model 3; CSIRO = climate model developed at the Australia Commonwealth Scientific and Industrial Research Organisation; ECHAM 5 = fifth-generation climate model developed at the Max Planck Institute for Meteorology (Hamburg); GCM = general circulation model; MIROC = Model for Interdisciplinary Research on Climate, developed by the University of Tokyo Center for Climate System Research.

For maize, the areas that are projected to suffer significant losses are located in the medium- to high-yielding areas, and thus climate change would have fundamental impacts on food production. The possibly improved yields will be in areas that are currently marginal. Three of the four GCMs show areas of significant yield increases in North West Province. All models show significant losses in Free State. However, there are some areas in Free State that will be able to grow maize that have not been able to under the current climate. Some such areas also exist in Eastern Cape. These areas are likely ones that were previously too cold to grow maize.

Three out of four GCMs show large areas of increased wheat yield in Free State and Mpumalanga. However, in scenarios showing likely decreases in rainfall, yields are seriously under threat in the Western Cape region. Unfortunately, we failed to assess the yield changes in the Western Cape in our crop model.

Agricultural Outcomes

Figures 7.23 and 7.24 show simulation results from the International Model for Policy Analysis of Agricultural Commodities and Trade (IMPACT) associated with key agricultural crops in South Africa. Each featured crop has five graphs: production, yield, area, net exports, and world price. The simulations represent the three GDP and population scenarios—each modeled with the four GCMs—hence the box-and-whisker plots indicating the range of outcomes under different climate change models (25th percentile, median, and 75th percentile).

These maps show a continued trend of smaller planting areas for maize, with higher yields and total production at first but after 2035 succumbing to the decline in planted area. Exports are projected to increase through 2020 and afterward to decline to the point that South Africa will become a net importer of maize. It is cause for concern that total production is shown to decline—more dramatically if the scenario's increases in yield do not materialize. Thus climate change is seen to have an anticipated impact on the security of the country's maize supply, especially during years of extreme weather.

The model did not compute yield changes for wheat in the Western Cape, but it appears likely that any increases in production will be offset by expected losses in this region due to increased temperatures and lower levels of rainfall.

FIGURE 7.23 Impact of changes in GDP and population on maize in South Africa, 2010–2050

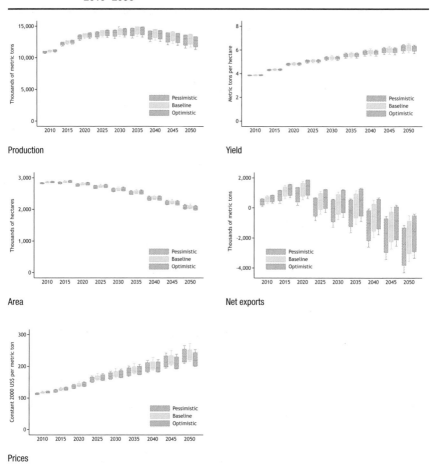

Source: Based on analysis conducted for Nelson et al. (2010).
Notes: The box and whiskers plot for each socioeconomic scenario shows the range of effects from the four future climate scenarios. GDP = gross domestic product; US$ = US dollars.

The increase in the world price will insulate South Africa against external pressure as long as domestic production is adequate to supply the demand, which seems unlikely given the divergence of production and consumption since 2001 (Figure 7.25).

Sugarcane appears to be the most resilient of the field crops to climate change (Figure 7.26). The crop's yield and area are both projected to increase, though its yield will increase by a higher percentage, and production will

FIGURE 7.24 Impact of changes in GDP and population on wheat in South Africa, 2010–2050

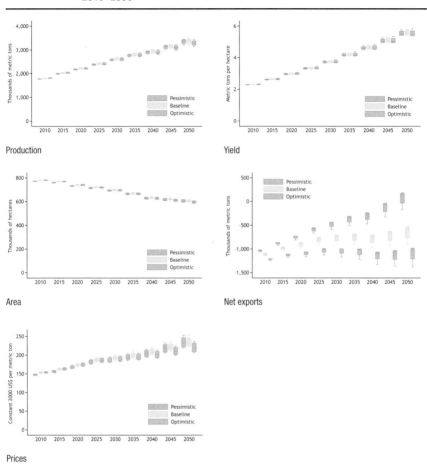

Source: Based on analysis conducted for Nelson et al. (2010).
Notes: The box and whiskers plot for each socioeconomic scenario shows the range of effects from the four future climate scenarios. GDP = gross domestic product; US$ = US dollars.

increase. Sugarcane farmers normally burn off the bagasse after harvest; this dangerous practice has to be carefully monitored to prevent runaway fires that can damage productive crops. Increases in temperature stand to increase this risk, as well the risk of veld fires, which may also damage sugar plantations. Increased rainfall, especially if more intense, may leave fields soggy or flooded. This will make it very difficult for machinery to access the crops for harvesting on both sloped and level ground.

FIGURE 7.25 Trends in wheat production and area planted in South Africa, 1970/1971–2011/2012

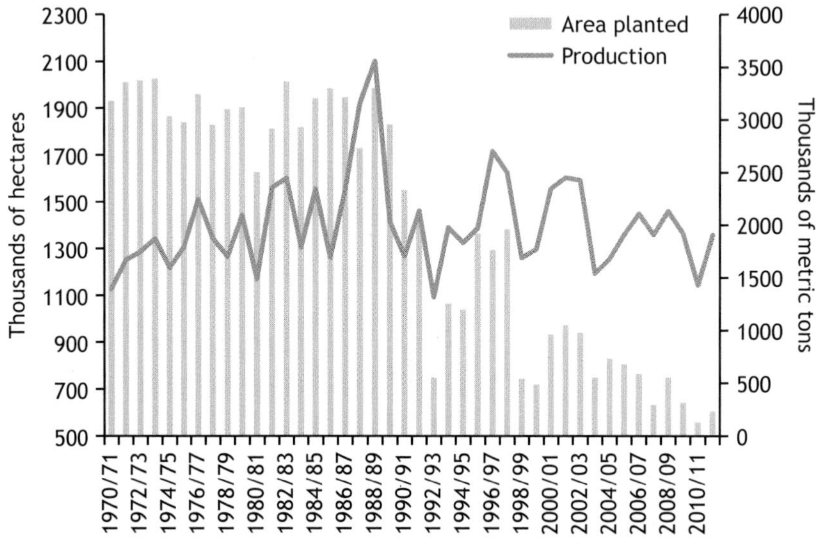

Source: USDA, FAS (2012).

Vulnerability to Climate Change

In addition to agricultural predictions, IMPACT also produces scenarios of the number of malnourished children under the age of five, as well as the available kilocalories per capita. Figure 7.27 shows the impact of future GDP and population scenarios on the number of malnourished children under age five in South Africa; Figure 7.28 shows the share of these children. After increases in the short term, levels drop off in all scenarios. In the baseline case, the number of malnourished children drops below the current level by 2035 (and by 2030 under the optimistic scenario). This outcome assumes the improved distribution and availability of food and an increase in food security (or in welfare infrastructure).

Figure 7.29 shows the kilocalories per capita available in South Africa. Because no dramatic increase in available nutrients is projected, the decline in malnourished children may be driven instead by a decline in the birth rate, resulting in fewer children under age five as a percentage of the population.

FIGURE 7.26 Impact of changes in GDP and population on sugarcane in South Africa, 2010–2050

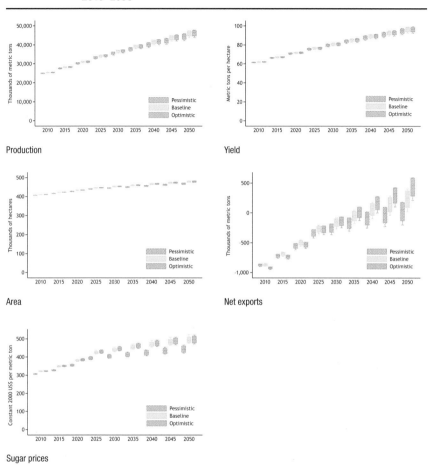

Source: Based on analysis conducted for Nelson et al. (2010).
Notes: The box and whiskers plot for each socioeconomic scenario shows the range of effects from the four future climate scenarios. GDP = gross domestic product; US$ = US dollars.

It is of concern that the pessimistic scenario shows a 20 percent reduction in kilocalories per capita. This reflects the negative impact of the large projected staple price increases, whereas GDP per capita in this scenario increases only modestly.

FIGURE 7.27 Number of malnourished children under five years of age in South Africa in multiple income and climate scenarios, 2010–2050

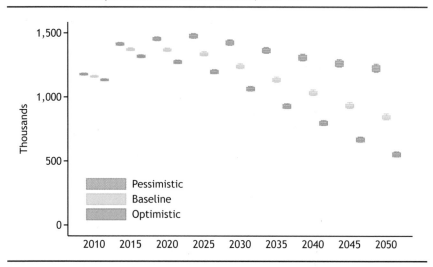

Source: Based on analysis conducted for Nelson et al. (2010).
Note: The box and whiskers plot for each socioeconomic scenario shows the range of effects from the four future climate scenarios.

FIGURE 7.28 Share of malnourished children under five years of age in South Africa in multiple income and climate scenarios, 2010–2050

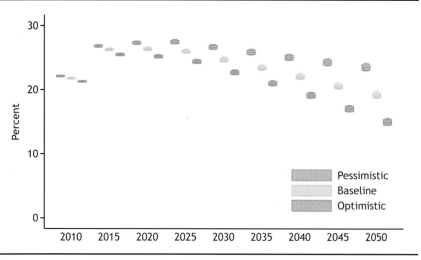

Source: Based on analysis conducted for Nelson et al. (2010).
Note: The box and whiskers plot for each socioeconomic scenario shows the range of effects from the four future climate scenarios.

FIGURE 7.29 Kilocalories per capita in South Africa in multiple income and climate scenarios, 2010–2050

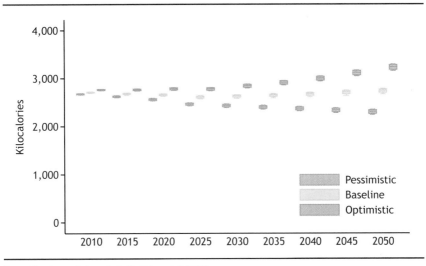

Source: Based on analysis conducted for Nelson et al. (2010).
Note: The box and whiskers plot for each socioeconomic scenario shows the range of effects from the four future climate scenarios.

Conclusions and Policy Recommendations

In this chapter we have used statistical information from various sectors and aspects of the South African society, economy, and natural environment to assess the vulnerability of the country's agriculture to the impacts of climate change. The chapter represents a useful, though not comprehensive, source of information for analysis by interested stakeholders, sector role players, and policymakers.

Agricultural production in South Africa is mainly commercial and thus less vulnerable to the impacts of climate change than other parts of the region, because adapting to climate change is, in most cases, worth the extra investment required. Resilience is a function of economic well-being; in a wealthier country, the government has more resources to assess vulnerabilities and to create and implement adaptive mechanisms, thereby increasing the resilience of even the poorest citizens.

The most vulnerable sector in agriculture was found to be that of poor farmers, including subsistence and even emerging farmers. These producers are more dependent than others on weather and climate conditions on a daily basis to provide them with basic foodstuffs and enable them to achieve a degree of food security. However, commercial agriculture provides the food for the majority of South Africans; here, too, shifts in climate patterns, growing regions, and market prices would have a major influence on plantings, yields, and total production.

It is clear that climate change will have two major impacts. First, average temperatures will increase. Higher minimum temperatures will have impacts on cold storage, frost frequency, and pest life cycles but will have the advantage of opening up areas to the cultivation of specific crops where it was previously too cold. Higher maximum temperatures will cause more extremely hot days, with increased evaporation, more days of soils drying out, and increased refrigeration requirements for fresh produce. In general, every day of 2050 is likely to be, on average, 2°C warmer than at current temperature levels. This will have severe impacts on seasonal cycles, suitable growing areas, and the way crops grow.

Second, the nature of rainfall patterns within seasons will change in both intensity and frequency. This will affect runoff, water availability, the length of dry spells, and the replenishment of groundwater. Agriculture will need to adapt to changes in the areas suitable for specific crops. Especially vulnerable will be crops that require longer-term investments in, for example, orchards, vineyards, and irrigation systems.

The future of agriculture in South Africa can be made more resilient by a more secure supply of water, specifically through irrigation. However, the country has few rivers and thus few opportunities for dam building. Increased temperatures will mean generally drier conditions. Groundwater will need to be replenished through rainfall and will be under increasing demand, not only from all forms of agriculture but also from the domestic, industrial, and mining sectors.

In the case of rangelands, the main threat to grazing will be from the effects on trees of CO_2 fertilization, which will increase the size of the root systems of trees and shrubs. On the other hand, Scheiter and Higgins (2009) point out that temperature-driven factors might favor grasses.

The implications of climate change for policy are enormous. More than ever, planning, management, and conservation of water resources will become vital for the sustainability of agriculture. The demands on fresh water for mining and industry, as well as for domestic use, must be tempered with the use of recycled, gray, or even brown water sources. Agriculture is generally already using water efficiently because it represents a significant input cost; more judicious uses can still be encouraged by wise application of tariffs and allocations.

In addition, the expansion of agriculture for previously disadvantaged groups needs to be well planned, accompanied by sufficient support to allow the growth of sustainable farming units and an increase in well-being for all farmers. Increased urbanization will leave fewer people working on the land, but this is a manageable outcome. Experience has shown that agriculture can be very efficient with a core of committed and knowledgeable farmers.

For highly vulnerable regions, such as KwaZulu-Natal, Limpopo, and the Eastern Cape, it will be essential to develop policies that do the following:

- Encourage the effective management of environmental resources (soil, water, and natural vegetation) through holistic and sustainable agricultural practices.
- Stimulate agricultural intensification and diversification. Marginal agriculture should be avoided and alternative livelihood incomes sought in degraded areas.
- In the small-scale sector, facilitate increased access to markets and participation in them.
- Develop rural infrastructure to uplift desperately poor and vulnerable communities, especially because these are the most highly vulnerable to climate change.

The main recommendation of this chapter is that decisionmakers, policymakers, and water users be educated regarding the implications of climate change on agriculture. Specific vulnerabilities need to be identified and addressed through adaptation frameworks and actions in order to build resilience and guarantee food security for the country. Because vulnerability to climate change is highly variable according to region, policymakers should tailor policies to local conditions (Gbetibouo and Ringler 2009).

References

Bartholome, E., and A. S. Belward. 2005. "GLC2000: A New Approach to Global Land Cover Mapping from Earth Observation Data." *International Journal of Remote Sensing* 26 (9): 1959–1977.

Boko, M., I. Niang, A. Nyong, C. Vogel, A. Githeko, M. Medany, B. Osman-Elasha, R. Tabo, and P. Yanda. 2007. "Climate Change 2007: Impacts, Adaptation, and Vulnerability." In *Contribution of Working Group II to the Fourth Assessment Report of the Intergovernmental Panel on Climate Change,* edited by M. L. Parry, O. F. Canziani, J. P. Palutikof, P. J. van der Linden, and C. E. Hanson. Cambridge: Cambridge University Press.

CIESIN (Center for International Earth Science Information Network), Columbia University, IFPRI (International Food Policy Research Institute), World Bank, and CIAT (Centro Internacional de Agricultura Tropical). 2004. *Global Rural–Urban Mapping Project (GRUMP), Alpha Version: Population Density Grids.* Palisades, NY, US: Socioeconomic Data and Applications Center (SEDAC), Columbia University. http://sedac.ciesin.columbia.edu/gpw.

Du Plessis, J. 2003. *Maize Production.* Potchefstroom, South Africa: Directorate Agricultural Information Services, Department of Agriculture in cooperation with ARC Grain Crops Institute.

FAO (Food and Agriculture Organization of the United Nations). 2005a. *Fertilizer Use by Crop in South Africa.* Rome. Accessed February 22, 2012. www.fao.org/docrep/008/j5509e/j5509e00.htm.

———. 2005b. AQUASTAT Country Profile: South Africa. Rome. Accessed March 17, 2012. www.fao.org/ag/agl/aglw/AQUASTAT/countries/index.stm, 28/02/2006. The particular table used is located at www.fao.org/nr/water/aquastat/irrigationmap/zaf/index.stm#table.

———. 2010. FAOSTAT. Accessed December 22, 2010. http://faostat.fao.org/site/573/default.aspx#ancor.

Gbetibouo, G. A., and C. Ringler. 2009. *Mapping South African Farming Sector Vulnerability to Climate Change and Variability: A Sub-national Assessment.* IFPRI Discussion Paper 00885. Washington, DC: International Food Policy Research Institute.

Grain SA. 2010. "NOK mielies: Oppervlakte, produksie, opbrengs per provinsie" (CEC Maize: Area Planted, Production, Yield per Province). Excel spreadsheet. Accessed November 21, 2010. www.grainsa.co.za/pages/industry-reports/production-reports.

IPCC (Intergovernmental Panel on Climate Change). 2000. "Emission Scenarios: Summary for Policy Makers." Accessed February 23, 2011. www.grida.no/climate/ipcc/spmpdf/sres-e.pdf.

Jones, P. G., P. K. Thornton, and J. Heinke. 2009. *Generating Characteristic Daily Weather Data Using Downscaled Climate Model Data from the IPCC's Fourth Assessment.* Project report. Nairobi, Kenya: International Livestock Research Institute.

Lehner, B., and P. Döll. 2004. "Development and Validation of a Global Database of Lakes, Reservoirs, and Wetlands." *Journal of Hydrology* 296 (1–4): 1–22.

Millennium Ecosystem Assessment. 2005. *Ecosystems and Human Well-being: Synthesis.* Washington, DC: Island Press. http://www.maweb.org/en/Global.aspx.

Nelson, G. C., M. W. Rosegrant, A. Palazzo, I. Gray, C. Ingersoll, R. Robertson, S. Tokgoz, et al. 2010. *Food Security, Farming, and Climate Change to 2050: Scenarios, Results, Policy Options.* Washington, DC: International Food Policy Research Institute.

Republic of South Africa, DAFF (Department of Agriculture, Forestry, and Fisheries). 2011. *A Profile of the South African Sugar Market Value Chain.* Pretoria.

———. 2013. *Abstract of Agricultural Statistics.* Pretoria.

Republic of South Africa, DST (Department of Science and Technology). 2011. *South African Risk and Vulnerability Atlas.* Accessed March 15, 2012. www.sarva.org.za/download/sarva_atlas.pdf.

Republic of South Africa, NDA (National Department of Agriculture). 2005. "Maize Profile." Pretoria. Accessed March 1, 2012. www.nda.agric.za/docs/FactSheet/maize.htm.

Republic of South Africa, Presidency. 2003. *Disaster Management Act.* No. 57 of 2002. Pretoria. Accessed May 2, 2012. www.info.gov.za/view/DownloadFileAction?id=68094.

Scheiter, Simon, and Steven I. Higgins. 2009. "Impacts of Climate Change on the Vegetation of Africa: An Adaptive Dynamic Vegetation Modelling Approach." *Global Change Biology* 15 (9): 2224–2246.

Schulze, R. E., ed. 2007. *South African Atlas of Climatology and Agrohydrology.* WRC Report 1489/1/06. Pretoria, Republic of South Africa: Water Research Commission.

Schulze, R. E., P. J. Hull, and M. Maharaj. 2007. "Sugarcane Yield Estimation." In *South African Atlas of Climatology and Agrohydrology,* edited by R. E. Schulze. WRC Report 1489/1/06, Section 16.2. Pretoria, Republic of South Africa: Water Research Commission.

Schulze, R. E., and N. J. Walker. 2007. "Maize Yield Estimation." In *South African Atlas of Climatology and Agrohydrology,* edited by R. E. Schulze. WRC Report 1489/1/06, Section 16.2. Pretoria, Republic of South Africa: Water Research Commission.

South Africa Government Communications. 2005. *South Africa Yearbook 2004/05,* 12th ed. Pretoria.

———. 2009. *South Africa Yearbook 2008/09.* Pretoria. www.gcis.gov.za/resource.

Statistics South Africa. 2002. *Report on the Survey of Large and Small Scale Agriculture.* Pretoria.

UNEP (United Nations Environment Programme) and IUCN (International Union for the Conservation of Nature). 2009. *World Database on Protected Areas (WDPA): Annual Release 2009.* Accessed 2009. www.wdpa.org/protectedplanet.aspx.

UNPOP (United Nations Department of Economic and Social Affairs–Population Division). 2009. *World Population Prospects: The 2008 Revision.* New York. http://esa.un.org/unpd/wpp/.

USDA, FAS (U.S. Department of Agriculture, Foreign Agricultural Service). 2012. "South Africa, Republic of, Grain and Feed Annual." *GAIN Report,* February 16. Accessed May 5, 2012. http://gain.fas.usda.gov/Recent%20GAIN%20Publications/Grain%20and%20Feed%20Annual_Pretoria_South%20Africa%20-%20Republic%20of_2-16-2012.pdf.

Wood, S., G. Hyman, U. Deichmann, E. Barona, R. Tenorio, Z. Guo, S. Castano, O. Rivera, E. Diaz, and J. Marin. 2010. "Sub-national Poverty Maps for the Developing World Using International Poverty Lines: Preliminary Data Release." Accessed May 6, 2010. http://povertymap.info (password protected). Some also available at http://labs.harvestchoice.org/2010/08/poverty-maps/.

World Bank. 2009. *World Development Indicators.* Washington, DC.

———. 2010. *The Costs of Agricultural Adaptation to Climate Change*. Washington, DC.

———. 2012. *World Development Indictors*. Washington, DC. http://data.worldbank.org/indicator/SI.POV.2DAY.

You, L., and S. Wood. 2006. "An Entropy Approach to Spatial Disaggregation of Agricultural Production." *Agricultural Systems* 90 (1–3): 329–347.

You, L., S. Wood, and U. Wood-Sichra. 2006. "Generating Global Crop Distribution Maps: From Census to Grid." Paper presented at the International Association of Agricultural Economists Conference in Brisbane, Australia, August 11–18.

———. 2009. "Generating Plausible Crop Distribution and Performance Maps for Sub-Saharan Africa Using a Spatially Disaggregated Data Fusion and Optimization Approach." *Agricultural Systems* 99 (2–3): 126–140.

Chapter 8

SWAZILAND

Absalom M. Manyatsi, Timothy S. Thomas, Michael T. Masarirambi,
Sepo Hachigonta, and Lindiwe Majele Sibanda

The Kingdom of Swaziland covers an area of 17,364 square kilometers bordered on the north, west, and south by the Republic of South Africa and on the east by Mozambique. The two major towns or cities are Mbabane, the capital city, and Manzini. The 2007 census put the population at 1.02 million, including 0.54 million females and 0.48 million males. About 78 percent live in rural areas and 22 percent in urban areas (Swaziland, Ministry of Economic Planning and Development 2007).

The majority of the rural people in Swaziland depend on cash income for survival strategies. Rural agriculture is insufficient to meet all food needs, but agricultural production provides a vital supplement to other food sources as well as employment opportunities during peak agricultural seasons (such as for weeding and harvesting).

Figure 8.1 shows population trends over the past 50 years, with the urban population continuing to grow but at a slower pace than it did prior to 1985. Table 8.1 shows population growth rates over the past 60 years. The total population growth rate dropped dramatically, to 0.9 percent from the 1980–1989 high of 3.6 percent, in part due to the impact of the HIV/AIDS endemic, with an adult prevalence rate currently at 26 percent. The average population density of the country is 59 persons per square kilometer, unevenly distributed across the country, with higher density in the urban areas and very sparse population in rural areas (Figure 8.2). The HIV/AIDS pandemic, affecting mainly the productive sector of the population, has had an impact on agricultural production (CANGO 2007).

About 40 percent of the 1 million inhabitants faced acute food and water shortages in 2007, with more than 60 percent of the overall population limiting or reducing their meal portions (IRIN 2007b). The drought that has persisted in southern Africa since 2002 became significantly worse in 2007. The 2007 maize harvest was the worst on record at 26,000 tons—a 60 percent reduction from 2006, causing about 50 percent of the population to be in need of food assistance until the April 2008 harvest. The drought in Swaziland

FIGURE 8.1 Population trends in Swaziland: Total population, rural population, and percent urban, 1960–2008

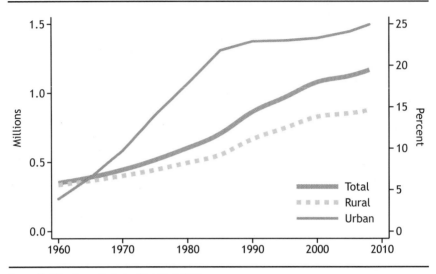

Source: *World Development Indicators* (World Bank 2009).

TABLE 8.1 Population growth rates in Swaziland, 1960–2008 (percent)

Decade	Total growth rate	Rural growth rate	Urban growth rate
1960–1969	2.4	1.8	11.6
1970–1979	3.0	2.1	9.2
1980–1989	3.6	2.9	6.4
1990–1999	2.3	2.2	2.4
2000–2008	0.9	0.7	1.7

Source: Authors' calculations based on *World Development Indicators* (World Bank 2009).

could have worsened the already severe HIV/AIDS situation, because patients on antiretroviral drugs may discontinue taking drugs in the absence of food (IRIN 2007b).

Impacts of Climate Change to Date

The shortage of water for commerce and industry has affected the Swaziland textile industry, which employed more than 30,000 people but is now

FIGURE 8.2 Population distribution in Swaziland, 2000 (persons per square kilometer)

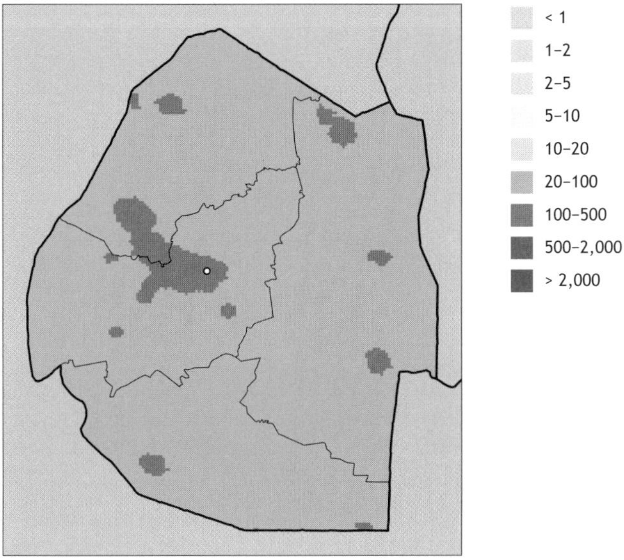

< 1
1–2
2–5
5–10
10–20
20–100
100–500
500–2,000
> 2,000

Source: CIESIN et al. (2004).

struggling to operate due to persistent drought. Investors set up the textile industry after the government promised adequate water to run their operations. The investors threatened to pull out of the country if the government did not provide adequate water supplies (IRIN 2007b).

Bush encroachment and the invasion of alien plant species may become severe due to climate change. In the lowveld and lower middleveld of Swaziland, large portions of grazing lands have been invaded by *Dichrostachyscinerea,* reducing the grazing capacity. *Chromolaena odorata,* an alien invasive plant species, continues to spread through the subtropical regions of southern Africa, a development some researchers attribute to climate change (Kriticos, Yonow, and McFadyen 2005).

The purpose of this chapter is to help policymakers and researchers better understand and anticipate the likely impacts of climate change on agriculture and on vulnerable households in Swaziland. It is anticipated that the Swaziland Climate Change Programme, together with the national focal point on climate change, will use this chapter as a resource to raise awareness among stakeholders (farmers, policymakers, nongovernmental organizations, and others) regarding issues of climate change.

Review of Current Trends

Water Resources in Swaziland

Swaziland has annual renewable water resources of $4,510 \times 10^6$ cubic meters, of which 42 percent originates in South Africa. Ten major dams store water used for irrigation, domestic, and industrial purposes, with a combined storage capacity of 588.2×10^6 cubic meters (Mwendera et al. 2002).

The five main river systems are the Lomati, the Nkomati, the Mbuluzi, the Usuthu, and the Ngwavuma. The Komati and Lomati Rivers, located in the northern part of the country, both originate in South Africa and flow out of Swaziland back into South Africa before entering Mozambique. The Mbuluzi River arises in Swaziland and flows into Mozambique. The Usuthu River, along with a number of major tributaries, originates in South Africa and flows into Mozambique, forming the border between Mozambique and South Africa. The Ngwavuma River, located in the south, arises in Swaziland and flows into South Africa before entering Mozambique. A sixth river system contributes to surface water: the Pongola River, located on the southern border with South Africa. The Jozini Dam, constructed on the South African side of the river, flooded some land in Swaziland; as compensation, an agreement was reached between the two governments to make available to Swaziland some of the water from the dam.

More than 95 percent of Swaziland's water resources are used for irrigation; only 1.2 percent are used for livestock, 2.3 percent for domestic uses, and 1.2 percent for industry (Knight Piesold 1997).

Income and Financial Indicators

Swaziland is rated as a middle-income developing country, with per capita gross domestic product (GDP) (in constant 2000 US dollars) of $1,560 in 2010. In recent years, per capita GDP has been increasing. At the same time, the share of agriculture in GDP has fallen, from an average of around 32 percent in 1970 to about 8 percent in 2008 (Figure 8.3). The real GDP growth rate was estimated at 2.0 percent in 2010, up from 1.2 percent in 2009. The agricultural sector grew by 3 percent in 2010 due to favorable weather conditions and better distribution of rainfall, which resulted in higher maize and cotton yields (Central Bank of Swaziland 2011). In 2010 the GDP composition by sector for agriculture, industry, and services was 7.4 percent, 49.2 percent, and 43.4 percent, respectively (CIA 2011).

Although the official unemployment rate is 28.2 percent (Swaziland, Ministry of Economic Planning and Development 2007), the actual figure is estimated at 40 percent, even higher in rural areas. Factors contributing to high

FIGURE 8.3 Per capita GDP in Swaziland (constant 2000 US$) and share of GDP from agriculture (percent), 1960–2008

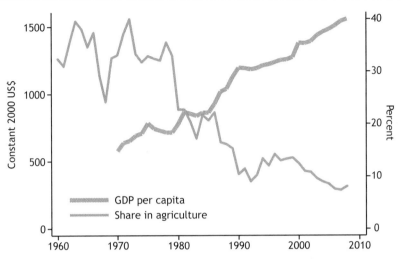

Source: *World Development Indicators* (World Bank 2009).
Note: GDP = gross domestic product; US$ = US dollars.

unemployment are the closure of major manufacturing companies in urban areas and the retrenchment from South African mines. The inflation rate slowed to about 3.5 percent in August 2010. Swaziland's currency (lilangeni) is pegged to the South African rand, subsuming Swaziland's monetary policy to South Africa's. Customs duties coming from the South African Customs Union (SACU) have contributed as much as 70 percent of government revenue over the years. However, the revenue share for Swaziland from SACU dropped by 62 percent in 2010, resulting in heavy budget cuts and growing government deficits (SACU 2010). Swaziland's economy grew by 2.4 percent in 2008 before declining to about 0.4 percent in 2009 (African Economic Outlook 2010)—far short of the government's target of 5 percent, which is required to reduce the poverty rate from 69 percent to 30 percent by 2015 in accord with the United Nations (UN) Millennium Development Goals.

Well-being Indicators in Swaziland

Maize is the staple foodcrop of Swaziland and is often used as an index of the availability of food in the country. Maize production showed a declining trend from 2004/2005 to 2006/2007, with some improvement in 2008/2009 (Figure 8.4). Swaziland is a net importer of maize, wheat, dairy products,

FIGURE 8.4 Maize production in Swaziland, 2004/2005–2009/2010

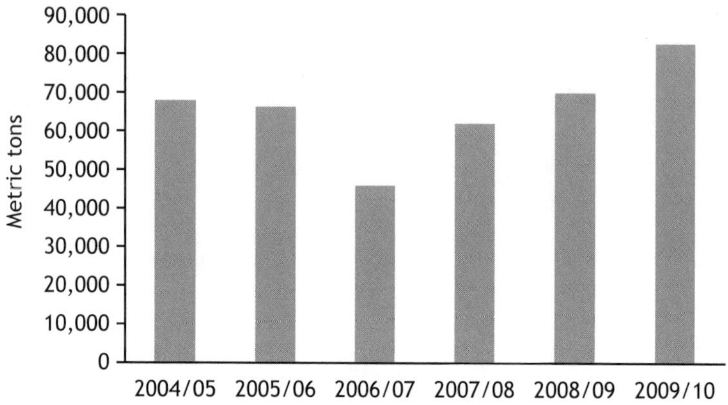

[Bar chart: Metric tons by year
- 2004/05: ~68,000
- 2005/06: ~66,000
- 2006/07: ~45,000
- 2007/08: ~62,000
- 2008/09: ~69,000
- 2009/10: ~82,000]

Source: Swaziland, Ministry of Economic Planning and Development (2010).

and other agricultural products; it normally imports some 60 percent of the food consumed in the country. With the exception of wheat, almost all food imports come from South Africa. In 2010/2011, the total maize requirement was estimated at 166,000 tons, whereas domestic production was estimated at 82,000 tons, requiring the country to import about 50 percent of its requirement (Swaziland News Stories 2010).

About 69 percent of the total population lives below the poverty line of $1 per day. The distribution of wealth and income is very skewed: the highest 10 percent of income earners garner 41 percent of the national income, and the lowest 10 percent of income earners garner just 1.6 percent of the national income (CIA 2011). About 40 percent of the population was faced with acute food and water shortages in 2007, when the prolonged drought caused the worst harvest in the country's recorded history (IRIN 2007a). About 18 percent of the population fell below the minimum dietary energy requirement of 2,100 kilocalories (IRIN 2007a). According to figures from 2000 (long before the 2007 food crisis), about 9 percent of the children under five years of age were malnourished in 2000.

Table 8.2 shows education and labor statistics for Swaziland in various periods. The government introduced free primary education in 2010 for grades 1 and 2 of the seven primary grades. Free education will be extended to an additional grade each year, covering all seven grades by 2015. This is expected to result in increased school enrollment at all levels and ultimately to improve the adult literacy rate, which stood at 81.6 percent in 2010. The government

TABLE 8.2 Education and labor statistics for Swaziland, 2000s

Indicator	Year	Value (percent)
Primary school enrollment (percent)[a]	2009	83.0
Secondary school enrollment (percent)[a]	2009	56.0
Adult literacy rate (percent)[b]	2010	81.6
Percent of active population employed in agriculture, both subsistence agriculture (informal) and commercial agriculture (formal)[c]	2008	70.0
Under-five malnutrition (percent)[a]	2010	9.1

Sources: [a]UNICEF (2011); [b]CIA (2011); [c]How We Made It in Africa (2010).

pays the school fees of orphaned and vulnerable children (OVCs) at all levels; there were about 120,000 registered OVCs in the country in 2010.

The agriculture sector (both subsistence and commercial) employs about 70 percent of the labor force. However, agriculture contributed only about 7.4 percent of GDP in 2010, because the economy has shifted to producing coal, wood pulp, sugar, drink concentrates, textiles and apparel, and other manufactured goods.

Figure 8.5 shows that life expectancy in Swaziland has declined over the past two decades, from 60 years in 1990 to just 45 years in 2005, driven

FIGURE 8.5 Well-being indicators in Swaziland, 1960–2008

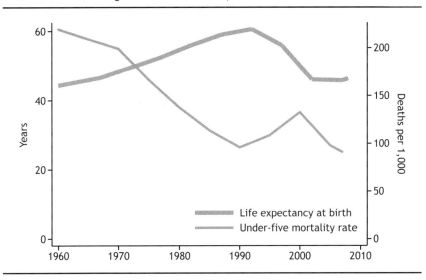

Source: World Bank (2009).

by the high prevalence of HIV/AIDS in addition to the impacts of poverty. The under-five mortality rate declined from 220 per 1,000 in 1960 to 100 per 1,000 in 1990; although the impacts of HIV/AIDS and poverty produced an increase through 2000, the rate has declined over the past decade to 90 per 1,000, due partly to the availability of antiretroviral drugs and increased awareness of HIV and AIDS, leading to changes in behavior (CANGO 2007).

Land Use Overview

Two major systems of land tenure exist in Swaziland. Title deed land (TDL) is privately owned land used mainly for ranching, forestry, or the estate production of such crops as vegetables, sugarcane, citrus, and pineapples; TDL covers 46 percent of the country. Swazi Nation land (SNL) is land held in trust by the king for the Swazi people, comprising 54 percent of the country. The average farm holding of cultivated land (SNL) is 1.5 hectares, with the maize yield averaging 1.0 tons per hectare. The average household size is eight people (Swaziland, Ministry of Works 2000).

The country is divided into four ecological zones based on elevation, landforms, geology, soil type, and vegetation. The highveld, middleveld, and lowveld each occupy about one-third of the country; the Lubombo Plateau occupies less than one-tenth. In the highveld, poor soils restrict agriculture to mainly grazing activities; only 3 percent of the region has good arable soils (Murdoch 1968). Almost 15 percent of the middleveld has arable soil of fair to good quality; about 20 percent of the lowveld has fair to good soils. Swaziland has a subtropical climate with summer rains (75 percent of all rainfall occurs in October–March). Climatic conditions range from subhumid and temperate in the highveld to semiarid in the lowveld. The long-term annual average rainfall figures for the highveld, middleveld, lowveld, and Lubombo Plateau are 950, 700, 475, and 700 millimeters, respectively (Swaziland, Ministry of Works 2000).

Swaziland has a total area of 1.7 million hectares. The main land use is extensive communal grazing, which covers about 50 percent of the area (SNL) (Figure 8.6 and Table 8.3). About 12 percent of the total area is used for small-scale crop production. Large-scale agriculture is practiced in 6 percent of the total area (TDL), with about 70,000 hectares irrigated; the dominant crop under irrigation is sugarcane. Forest plantations cover about 8 percent of the total area (see Table 8.3) and parks, reserves, and hunting areas about 4 percent. Such areas are important for the protection of fragile environments as well as for the tourism industry, which contributes about 7 percent of the GDP.

FIGURE 8.6 Land cover and land use in Swaziland, 2000

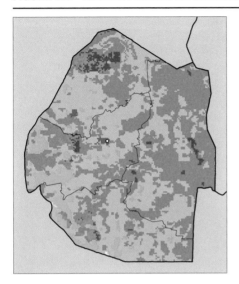

- Tree cover, broadleaved, evergreen
- Tree cover, broadleaved, deciduous, closed
- Tree cover, broadleaved, open
- Tree cover, broadleaved, needle-leaved, evergreen
- Tree cover, broadleaved, needle-leaved, deciduous
- Tree cover, broadleaved, mixed leaf type
- Tree cover, broadleaved, regularly flooded, fresh water
- Tree cover, broadleaved, regularly flooded, saline water
- Mosaic of tree cover/other natural vegetation
- Tree cover, burnt
- Shrub cover, closed-open, evergreen
- Shrub cover, closed-open, deciduous
- Herbaceous cover, closed-open
- Sparse herbaceous or sparse shrub cover
- Regularly flooded shrub or herbaceous cover
- Cultivated and managed areas
- Mosaic of cropland/tree cover/other natural vegetation
- Mosaic of cropland/shrub/grass cover
- Bare areas
- Water bodies
- Snow and ice
- Artificial surfaces and associated areas
- No data

Source: GLC2000 (Bartholome and Belward 2005).

TABLE 8.3 Land use in each ecological zone of Swaziland, 1994

	\multicolumn{2}{c	}{Highveld}	\multicolumn{2}{c	}{Middleveld}	\multicolumn{2}{c	}{Lowveld}	\multicolumn{2}{c	}{Lubombo}	\multicolumn{2}{c	}{Total}
Land use type	Hectares	Percent	Hectares	Percent	Hectares	Percent	Hectares	Percent	Hectares	Percent
Smallholder agriculture	39,100	6.9	74,000	15.4	84,200	15.1	16,100	10.9	213,400	12.3
Large-scale agriculture	1,100	0.2	18,400	3.8	81,400	16.9	2,800	1.9	103,700	6.0
Extensive communal grazing	320,600	56.7	291,000	60.3	173,800	30.5	80,400	54.3	865,800	50.0
Ranching	49,500	8.7	85,800	17.8	184,300	31.0	32,700	22.1	332,300	19.2
Forest plantations	132,300	23.4	7,500	1.6	0	0.0	0	0.0	139,800	8.1
Extraction and collection	0	0.0	0	0.0	8,400	1.3	0	0.0	8,400	0.5
Parks and reserves	20,100	3.5	900	0.4	21,100	4.7	16,100	10.9	58,200	3.4
Water reservoirs	400	0.1	0	0.0	3,800	0.7	0	0.0	4,200	0.2

Source: Rammelzwaal and Dlamini (1994).

Natural forests cover 36,556 hectares, while woodlands cover 382,261 hectares and bushlands or savannas cover 232,954 hectares (Menne and Carrere 2007). In Swaziland, plantations of black wattle (*Acacia mearnsii*) were first established in the early 1880s; the wood is used mainly for tanning in the leather industry. Wattle wood was used for mine props in local tin mines, which were thriving at that time, and later became widely used as building material and fuelwood. Subsequently, pine and eucalyptus trees were introduced as the main plantation species. The main pine species are *Pinus patula, Pinus radiate,* and *Pinus taeda,* covering about 80 percent of the planted area. The eucalyptus species are mainly *Eucalyptus saligna* and *Eucalyptus grandis,* covering 20 percent of the planted area.

About 22 percent of the population lives in urban areas, including the two main cities, Mbabane and Manzini, and 15 other towns. Manzini has a population of about 100,000 people, while Mbabane has a population of between 70,000 and 90,000. There are also some rural settlements with sizeable populations, mainly around police stations, mission schools, mining settlements, railway stations, and shopping centers.

Figure 8.7 shows the travel times to urban areas of various sizes. The first panel of Figure 8.7 shows travel times to cities of 500,000 or more people (referring to cities outside the country, because there are no cities of that size in Swaziland). These urban areas provide potential markets for agricultural products. Manzini, the largest city, is strategically located in the center of the country. The travel time between Manzini and Mbabane, as well as to several other urban areas, is less than one hour due to good road networks. For some remote areas that still have no roads, however, the travel time to the nearest town is five hours or more. Such areas tend to be underdeveloped and impoverished, with very little means of livelihood.

Agriculture Overview

Although 80 percent of the rural population is engaged in agriculture, almost all households rely on additional sources to meet their total food requirement. This chapter focuses on field and plantation crops, but livestock is also a significant sector, with 585,000 head of cattle, 276,000 goats, and 28,000 sheep counted in 2008 (FAO 2010).

About 95 percent of the water resources are used for irrigation, with sugarcane using over 84 percent of the irrigation water. Sugarcane is planted on about 53,000 hectares, exclusively under irrigation. It is grown mainly in

FIGURE 8.7 Travel time to urban areas of various sizes in Swaziland, circa 2000

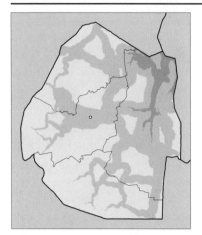

To cities of 500,000 or more people

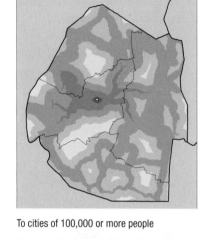

To cities of 100,000 or more people

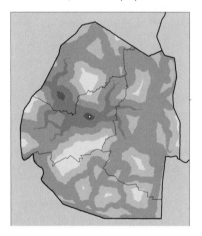

To towns and cities of 25,000 or more people

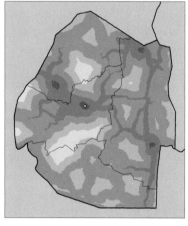

To towns and cities of 10,000 or more people

- Urban location
- < 1 hour
- 1–3 hours
- 3–5 hours
- 5–8 hours
- 8–11 hours
- 11–16 hours
- 16–26 hours
- > 26 hours

Source: Authors' calculations.

the lowveld but is increasingly being grown in the lower middleveld. The Swaziland Sugar Association (SSA), the umbrella organization for growers and millers of sugarcane, provides technical services (SSA 2010). The total cane production was 4,912,949 tons in 2008/2009 and 4,908,152 tons in 2009/2010. About 500 small-scale growers have emerged since the 1990s. The Swaziland Water and Agricultural Development Enterprise (SWADE), a government-owned company, plays a major role in promoting smallholder farmers' participation in the sugarcane industry; it facilitates the implementation of the Komati Downstream Development Project and the Lower Usuthu Smallholder Irrigation Project (LUSIP) (SWADE 2010). The sugarcane industry directly employs about 16,000 people, and another 20,000 people benefit from it indirectly (FAO 2005).

Maize is the staple food in Swaziland and is cultivated on about 47,000 hectares. The area planted in maize has been decreasing over time due to a number of factors, including the change from maize production to sugarcane production and the persistent drought and erratic rains, which have led to the abandonment of crop growing. In 1992/1993 the area under maize cultivation was 58,787 hectares (Swaziland, Ministry of Economic Planning and Development 1992). Predominantly rainfed, maize is especially vulnerable to droughts and other effects of climate change. The National Maize Corporation (NMC) is responsible for guaranteeing the availability of quality white maize all year, as well as for reducing the marketing barriers and costs to Swazi farmers and increasing the efficiency of the domestic maize market (NMC 2010). It facilitates setting a floor price for maize and controls imports and exports of the crop.

Like maize production, seed cotton production has declined over the past 15 years. In 1992/1993, about 26,600 hectares of land were under cotton production, declining to 3,000 hectares by 2008. The production of seed cotton is coordinated and promoted by the Swaziland Cotton Board, a public enterprise under the Ministry of Agriculture (Swaziland, Ministry of Agriculture 2010). Cotton is grown mainly in the lowveld because it is drought tolerant, and it is a substitute crop for maize in areas with low rainfall.

Citrus fruits (oranges and grapefruits) are grown mainly for canning and juice extraction at the Swaziland Fruit canning factory in Malkerns. Other crops, such as groundnuts, beans, and cowpeas, are grown on a small scale, often intercropped with maize. Table 8.4 shows key agricultural crops in terms of area harvested; Table 8.5 shows key agricultural commodities in terms of their consumption.

TABLE 8.4 Area and production of different crops in Swaziland, 2006–2008

Rank	Crop	Harvest area (thousands of hectares)	Total production (metric tons)
1	Sugarcane	53	5,000,000
2	Maize	47	60,765
3	Seed cotton	15	1,115
4	Roots and tubers	7	51,696
5	Oranges	7	38,960
6	Groundnuts	6	3,699
7	Grapefruits (including pomelos)	4	33,008
8	Beans	4	4,958
9	Potatoes	3	5,057
10	Pineapples	1	16,841
11	Cowpeas	2	769
12	Sweet potatoes	1	3,068

Source: FAOSTAT (FAO 2010).
Note: All values are based on a three-year average for 2006–2008.

TABLE 8.5 Consumption of leading food commodities in Swaziland, 2003–2005 (thousands of metric tons)

Rank	Crop	Percent of total	Food consumption
	Total	100.0	486
1	Maize	15.2	74
2	Fermented beverages	11.0	53
3	Roots and tubers	8.6	42
4	Wheat	8.4	41
5	Sugar	6.5	31
6	Beef	5.4	26
7	Grapefruits	4.7	23
8	Beer	4.3	21
9	Other fruits	4.2	20
10	Other vegetables	4.0	20

Source: FAOSTAT (FAO 2010).
Note: All values are based on a three-year average from 2003–2005.

Maize is the leading food commodity consumed in the country, with more than 74,000 tons consumed annually. Second are fermented beverages made from maize and sorghum, including traditional beer (see Table 8.5). Virtually all the wheat consumed in the country is imported. The total sugar produced in the country in 2006–2008 was about 631,000 tons, with 31,000 tons consumed domestically and the rest (about 95 percent) exported. Fruits and vegetables are produced mainly on a small-scale basis, and most of the country's demand is met through imports. The National Agricultural Marketing Board (NAMBOARD) is responsible for promoting the production of fruits and vegetables as well as controlling their import (NAMBOARD 2010).

Maize is produced in all the ecological zones of Swaziland but mainly in the highveld and middleveld, because yields are very low in the lowveld; the average yield in the highveld and upper middleveld is about 2 tons per hectare (Figure 8.8). Nearly all maize production is rainfed, except that of "green maize," which is produced by irrigation.

FIGURE 8.8 Yield (metric tons per hectare) and harvest area density (hectares) for rainfed maize in Swaziland, 2000

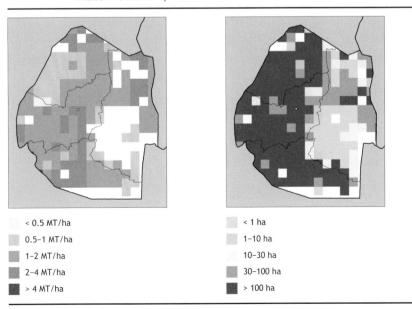

- < 0.5 MT/ha
- 0.5–1 MT/ha
- 1–2 MT/ha
- 2–4 MT/ha
- > 4 MT/ha

- < 1 ha
- 1–10 ha
- 10–30 ha
- 30–100 ha
- > 100 ha

Source: SPAM (Spatial Production Allocation Model) (You and Wood 2006; You, Wood, and Wood-Sichra 2006, 2009).
Note: ha = hectare; MT/ha = metric tons/hectare.

FIGURE 8.9 Yield (metric tons per hectare) and harvest area density (hectares) for irrigated sugarcane in Swaziland, 2000

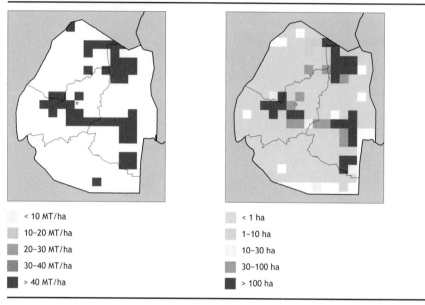

< 10 MT/ha
10–20 MT/ha
20–30 MT/ha
30–40 MT/ha
> 40 MT/ha

< 1 ha
1–10 ha
10–30 ha
30–100 ha
> 100 ha

Source: SPAM (Spatial Production Allocation Model) (You and Wood 2006; You, Wood, and Wood-Sichra 2006, 2009).
Note: ha = hectare; MT/ha = metric tons/hectare.

Sugarcane is grown predominantly under irrigation, with the average yield more than 90 tons per hectare. Rainfed sugarcane production is found in very small areas in the northern part of the country, where the average yield is more than 40 tons per hectare (Figure 8.9). Rainfed sugarcane tends to be grown by small-scale farmers on small plots of less than 1 hectare, with an average yield of less than 1 ton per hectare.

Scenarios for the Future

Economic and Demographic Indicators

Population
The high-variant population projection for Swaziland shows the population almost doubling by the year 2050. The low-variant projection shows the population increasing by about 25 percent by 2050, from 1.1 million to about 1.4 million; this variant assumes a continuing low rate of population growth (Figure 8.10). The medium variant shows the population increasing by around 50 percent by 2050. Any increase in population needs to be accompanied by

FIGURE 8.10 Population projections for Swaziland, 2010–2050

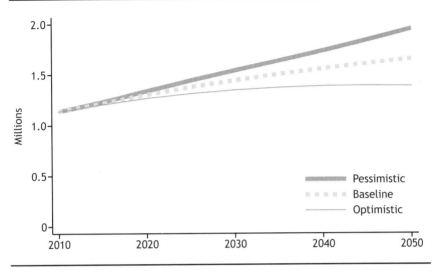

Source: UNPOP (2009).

economic gains, along with increased food production and additional facilities and infrastructure.

Income

Figure 8.11 presents three overall scenarios for GDP per capita derived by combining three GDP scenarios with the three population scenarios of Figure 8.10 (based on UN population data). The optimistic scenario combines high GDP growth with low population growth, the baseline scenario combines medium GDP growth with medium population growth, and the pessimistic scenario combines low GDP growth with high population growth. (The agricultural modeling in the next section uses these scenarios as well.)

The pessimistic scenario shows a slow increase in GDP per capita, while the optimistic scenario shows a sharp increase in GDP per capita over time. The baseline scenario, showing a moderate increase in GDP per capita, is the likely scenario for Swaziland: annual economic growth is expected to be about 2.5 percent, with a moderate population increase. Current policies aimed at improving the economy include soliciting foreign direct investment by improving the country's infrastructure, including its road networks, railway service, telecommunications, and air travel. A new international airport is under construction at Sikhuphe and should be in operation in 2013.

FIGURE 8.11 Gross domestic product (GDP) per capita in Swaziland, future scenarios, 2010–2050

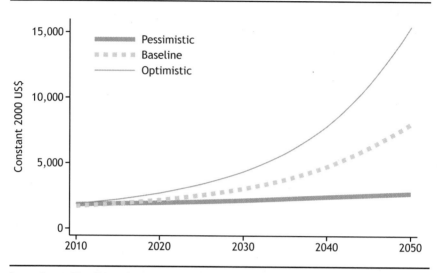

Sources: Computed from GDP data from the World Bank Economic Adaptation to Climate Change project (World Bank 2010b), from the Millennium Ecosystem Assessment (2005) reports, and from population data from the United Nations (UNPOP 2009).
Note: US$ = US dollars.

Biophysical Analysis

Climate models

Figure 8.12 shows projected precipitation changes for 2000–2050 under the four downscaled general circulation models (GCMs) we use in this monograph with the A1B scenario.[1] CNRM-CM3 shows an insignificant change in annual rainfall over the 50-year period (in the range of −50 to +50 mm). CSIRO Mark 3 shows no significant change in the northern and eastern parts of the country (mainly the highveld), and an annual increase of between 50 and 100 millimeters in the central and eastern parts of the country (the middleveld, lowveld, and Lubombo Plateau). Both ECHAM 5 and MIROC 3.2, however, show a decrease in annual precipitation over the same period (see Figure 8.12); ECHAM 5 shows an annual reduction

---

[1] The A1B scenario describes a world of very rapid economic growth, low population growth, and rapid introduction of new and more efficient technologies with moderate resource use and a balanced use of technologies. CNRM-CM3 is National Meteorological Research Center–Climate Model 3. CSIRO is a climate model developed at the Australia Commonwealth Scientific and Industrial Research Organisation. ECHAM 5 is a fifth-generation climate model developed at the Max Planck Institute for Meteorology in Hamburg. MIROC is the Model for Interdisciplinary Research on Climate, developed at the University of Tokyo Center for Climate System Research.

FIGURE 8.12 Change in mean annual precipitation in Swaziland, 2000–2050, A1B scenario (millimeters)

CNRM-CM3 GCM

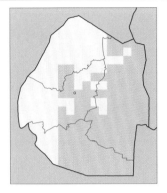

CSIRO Mark 3 GCM

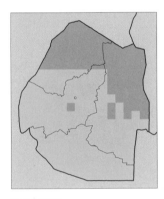

ECHAM 5 GCM

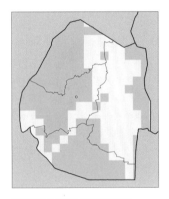

MIROC 3.2 medium-resolution GCM

- < −400
- −400 to −200
- −200 to −100
- −100 to −50
- −50 to 50
- 50 to 100
- 100 to 200
- 200 to 400
- > 400

Source: Authors' calculations based on Jones, Thornton, and Heinke (2009).

Notes: A1B = greenhouse gas emissions scenario that assumes fast economic growth, a population that peaks midcentury, and the development of new and efficient technologies, along with a balanced use of energy sources; CNRM-CM3 = National Meteorological Research Center–Climate Model 3; CSIRO = climate model developed at the Australia Commonwealth Scientific and Industrial Research Organisation; ECHAM 5 = fifth-generation climate model developed at the Max Planck Institute for Meteorology (Hamburg); GCM = general circulation model; MIROC = Model for Interdisciplinary Research on Climate, developed by the University of Tokyo Center for Climate System Research.

of as much as 200 millimeters in the northern part of the country. That level of rainfall reduction would create some hardship and would require adaptive measures.

Figure 8.13 shows the change in the mean daily maximum temperature for the warmest month with the A1B scenario, according to the GCMs. All four GCMs show an increase in the annual maximum temperatures for Swaziland (see Figure 8.13). CNRM-CM3 shows an increase of between 1.5° and 2.0°C, whereas CSIRO Mark 3 shows an increase of between 1.0° and 1.5°C. ECHAM 5 shows the greatest increase: between 2.0° and 2.5°C for the western part of the country (highveld and middleveld) and between 1.5° and 2.0°C for the eastern part (lowveld and Lubombo Plateau). MIROC 3.2 shows an increase of around 1.5°C for the entire country, with the "splotchy" pattern actually reflecting the fact that all values are very close to that figure, with some slightly higher and some slightly lower (see Figure 8.13).

The potential impacts of climate change include changes in land cover; reduced rainfall and increased temperatures would likely mean an increase in shrubs and herbaceous cover and a reduction in tall tree cover. In planning adaptation and mitigation strategies for climate change, it is advisable to keep in mind the worst-case scenario, which might include an increase in temperature of between 2.0° and 2.5°C and a reduction in precipitation of 200 millimeters. However, investment to prepare for such an outcome would be premature until there are more indications that it will be realized.

Crop Models

The Decision Support Software for Agrotechnology Transfer (DSSAT) crop modeling system was used to compare the modeled 2050 yield results (using the four GCMs) with the results for baseline yield (with unchanged climate).

The output for maize—the staple food—is mapped in Figure 8.14. The results from CNRM-CM3, CSIRO Mark 3, and MIROC 3.2 are similar, showing a reduction in maize yield of more than 25 percent of the baseline, as well as an increase in yield of more than 25 percent in some parts of the lowveld. The area shown with a yield loss of more than 25 percent is in the traditional maize-producing regions (the highveld), and it covers a larger area than that shown with a yield increase of more than 25 percent (the lowveld). This result indicates that the overall maize production would be reduced under both CNRM-CM3 and CSIRO Mark 3. ECHAM 5 also shows an overall yield loss of more than 25 percent of baseline for most of the country.

FIGURE 8.13 Change in monthly mean maximum daily temperature in Swaziland for the warmest month, 2000–2050, A1B scenario (°C)

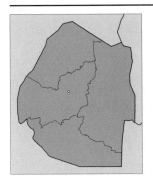

CNRM-CM3 GCM

CSIRO Mark 3 GCM

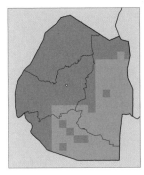

ECHAM 5 GCM

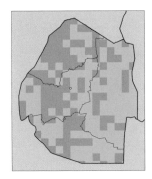

MIROC 3.2 medium-resolution GCM

- < –1
- –1 to –0.5
- –0.5 to 0
- 0 to 0.5
- 0.5 to 1
- 1 to 1.5
- 1.5 to 2
- 2 to 2.5
- 2.5 to 3
- 3 to 3.5
- > 3.5

Source: Authors' calculations based on Jones, Thornton, and Heinke (2009).
Notes: A1B = greenhouse gas emissions scenario that assumes fast economic growth, a population that peaks midcentury, and the development of new and efficient technologies, along with a balanced use of energy sources; CNRM-CM3 = National Meteorological Research Center–Climate Model 3; CSIRO = climate model developed at the Australia Commonwealth Scientific and Industrial Research Organisation; ECHAM 5 = fifth-generation climate model developed at the Max Planck Institute for Meteorology (Hamburg); GCM = general circulation model; MIROC = Model for Interdisciplinary Research on Climate, developed by the University of Tokyo Center for Climate System Research.

FIGURE 8.14 Yield change under climate change: Rainfed maize in Swaziland, 2000–2050, A1B scenario

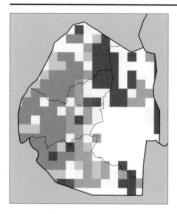

CNRM-CM3 GCM

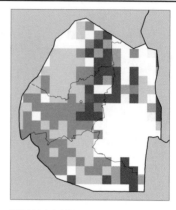

CSIRO Mark 3 GCM

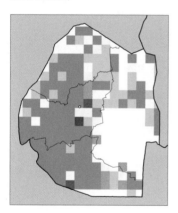

ECHAM 5 GCM

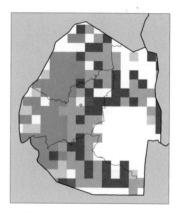

MIROC 3.2 medium-resolution GCM

- Baseline area lost
- Yield loss >25% of baseline
- Yield loss 5–25% of baseline
- Yield change within 5% of baseline
- Yield gain 5–25% of baseline
- Yield gain > 25% of baseline
- New area gained

Source: Authors' calculations.

Notes: A1B = greenhouse gas emissions scenario that assumes fast economic growth, a population that peaks midcentury, and the development of new and efficient technologies, along with a balanced use of energy sources; CNRM-CM3 = National Meteorological Research Center–Climate Model 3; CSIRO = climate model developed at the Australia Commonwealth Scientific and Industrial Research Organisation; ECHAM 5 = fifth-generation climate model developed at the Max Planck Institute for Meteorology (Hamburg); GCM = general circulation model; MIROC = Model for Interdisciplinary Research on Climate, developed by the University of Tokyo Center for Climate System Research.

Note that the baseline yield for the lowveld is lower than that for the highveld, and the latter is shown with a yield loss of 25 percent or more in *all* the models. Nevertheless, in all but the ECHAM 5 model, we see substantial areas with a yield increase of more than 25 percent under climate change. This means that there may be opportunities to adapt to climate change at the national level by shifting maize production over time to areas that are becoming more productive as a result of the change.

Agricultural Outcomes

Figures 8.15–8.17 show simulation results from the International Model for Policy Analysis of Agricultural Commodities and Trade (IMPACT) associated with maize, sugarcane, and cotton, respectively. Each featured crop has five graphs: production, yield, area, net exports, and world prices. The simulations included the three GDP and population scenarios and the GCMs.

The production of maize is expected to increase over time in all scenarios but not enough to meet domestic demand, currently at 150,000 tons per year. This is reflected in anticipated increased imports of maize (shown as a decrease in net exports in Figure 8.15). By 2050, the country is shown importing nearly 200,000 tons of maize. With the world price of maize shown to be increasing from $120 to about $250 per ton, maize could become unaffordable for the majority of the population if the pessimistic scenario comes to pass.

The harvested area of maize is shown to decrease over time in all scenarios, whereas the average yield is shown to increase, from 1.0 tons per hectare to about 1.5 tons per hectare by 2050. Hybrid maize seeds, already available in Swaziland, can produce as much as 10 tons per hectare under good management and agronomic conditions. Several programs have been initiated to increase production, including the Comprehensive African Agricultural Development Program (CAADP), the Swaziland Agricultural Development Program (SADP), and the Tractor Hire Pool program of the Ministry of Agriculture (discussed below).

Sugarcane is produced mainly for export; less than 5 percent of the sugar produced is consumed locally. Production of sugarcane is expected to increase to about 17,000 tons by 2050 (see Figure 8.16). The industry receives technical and agronomic support from organizations such as the Swaziland Sugar Association and SWADE. The area planted with sugarcane is expected to increase to more than 80,000 hectares by 2050. Only 11 percent of the area in Swaziland is considered arable, and not all of that is suitable for irrigation, which is generally required for sugarcane. Land is already being converted to

FIGURE 8.15 Impact of changes in GDP and population on maize in Swaziland, 2010–2050

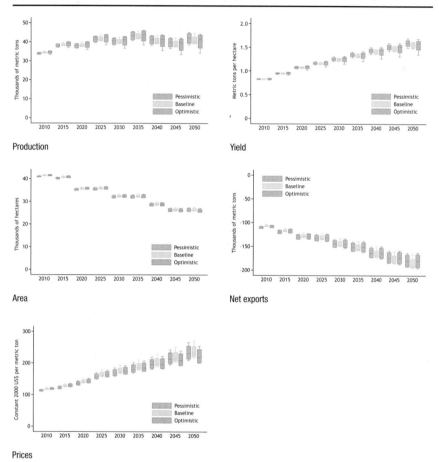

Source: Based on analysis conducted for Nelson et al. (2010).
Notes: The box and whiskers plot for each socioeconomic scenario shows the range of effects from the four future climate scenarios. GDP = gross domestic product; US$ = US dollars.

sugarcane production from other uses, such as grazing, growing cotton, and subsistence maize farming. The net export of sugar is expected to triple by 2050, bringing much-needed foreign currency (see Figure 8.16). The sugar processing industry is diversifying to producing ethanol from molasses and generating electricity from byproducts as a way to offset fluctuations in the world price of sugar. Generating electricity would also reduce the operation costs of the industry.

FIGURE 8.16 Impact of changes in GDP and population on sugarcane in Swaziland, 2010–2050

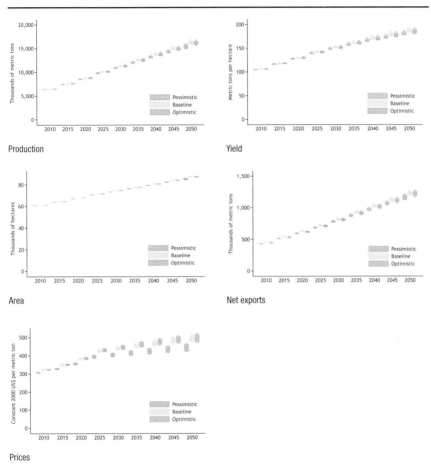

Source: Based on analysis conducted for Nelson et al. (2010).
Notes: The box and whiskers plot for each socioeconomic scenario shows the range of effects from the four future climate scenarios. GDP = gross domestic product; US$ = US dollars.

Cotton production was at its lowest in 2010 but is expected to improve with government assistance. The world price of cotton, kept artificially low by subsidies paid to farmers in developed counties, erodes profits from growing cotton. The Swaziland government has committed to provide financial assistance for the purchase of cotton inputs and to support the price of cotton in order to encourage cotton production (Fibre2fashion 2008). Moreover,

FIGURE 8.17 Impact of changes in GDP and population on cotton in Swaziland, 2010–2050

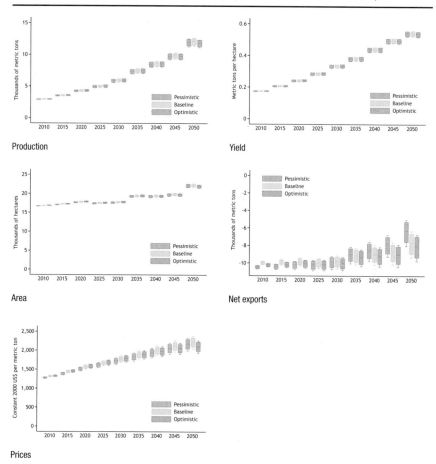

Source: Based on analysis conducted for Nelson et al. (2010).
Notes: The box and whiskers plot for each socioeconomic scenario shows the range of effects from the four future climate scenarios. GDP = gross domestic product; US$ = US dollars.

the industry is evaluating the potential of genetically modified cotton to increase yield.

Swaziland is among the countries that currently benefit from the African Growth and Opportunity Act (AGOA), which enables countries to export goods and commodities to the United States duty free and which has promoted growth in the textile and apparel industry (AGOA 2000). Local demand for cotton is likely to increase, driven by the growing textile industry (Figure 8.17). The cotton yield is projected to almost triple between 2010 and 2050, and with the area planted with cotton increasing by over 25 percent, total production might quadruple. Still, domestic demand is projected to grow, so imports of cotton are projected to decline only slightly.

Vulnerability to Climate Change

The results from running the DSSAT crop model for the entire world were input into IMPACT, which computes global agricultural commodity prices and output by country and region. IMPACT was run with four climate model and scenario combinations. In particular, the CSIRO and the MIROC models were used, with the A1B and the B1 scenarios.[2] Those four combinations were run for each of the three per capita GDP scenarios (see Figure 8.11).

In addition to agricultural predictions, IMPACT also produces scenarios of the number of malnourished children under the age of five, as well as the available kilocalories per capita. Figure 8.18 shows the impact of future GDP and population scenarios on the number of malnourished children under age five; Figure 8.19 shows the share. In the optimistic scenario, the number of malnourished children is expected to increase from about 33,000 children currently to a peak of 44,000 in 2020, and then decrease to 13,000 in 2050. The assumption is that GDP will increase after 2020, making food (kilocalories) more readily available, as reflected in Figure 8.20. In the pessimistic scenario, however, the number of malnourished children will increase to a high of 55,000 and decrease only slightly by 2050, to 45,000. All the scenarios show malnutrition growing more widespread over the next 10 years in the absence of an effective policy intervention by the government. Such a policy should be developed with the aim of reducing poverty, and it could provide assistance to the vulnerable population in the form of food aid.

Figure 8.20 shows the kilocalories consumed per capita, and that number appears to be correlated with malnutrition in children and with GDP per capita. The decline in consumption under the pessimistic scenario reflects the steep increase in staple food prices that dominates the otherwise positive effect of the modest growth in GDP per capita in that scenario.

---

2 The B1 scenario is a greenhouse gas emissions scenario that assumes a population that peaks mid-century (like the A1B), but with rapid changes toward a service and information economy, and the introduction of clean and resource-efficient technologies.

FIGURE 8.18 Number of malnourished children under five years of age in Swaziland in multiple income and climate scenarios, 2010–2050

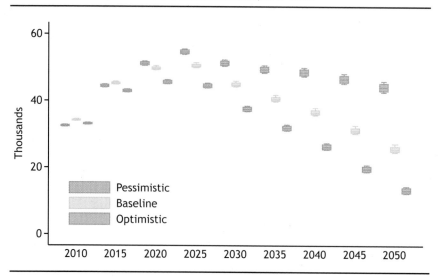

Source: Based on analysis conducted for Nelson et al. (2010).
Note: The box and whiskers plot for each socioeconomic scenario shows the range of effects from the four future climate scenarios.

FIGURE 8.19 Share of malnourished children under five years of age in Swaziland in multiple income and climate scenarios, 2010–2050

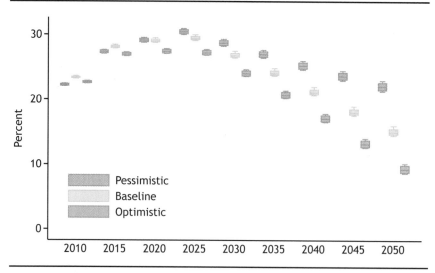

Source: Based on analysis conducted for Nelson et al. (2010).
Note: The box and whiskers plot for each socioeconomic scenario shows the range of effects from the four future climate scenarios.

FIGURE 8.20 Kilocalories per capita in Swaziland in multiple income and climate scenarios, 2010–2050

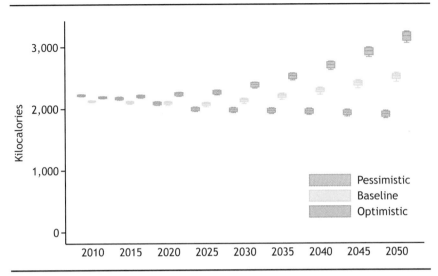

Source: Based on analysis conducted for Nelson et al. (2010).
Note: The box and whiskers plot for each socioeconomic scenario shows the range of effects from the four future climate scenarios.

Climate Change in Swaziland

Climate change and variability are evident in Swaziland in many forms, including hydrological disasters (droughts and storms), changes in rainfall regime, and extreme weather conditions (Manyatsi, Mhazo, and Masarirambi 2010). Rural communities in Swaziland do not have adequate information on climate change. Limited information is available through the local radio in the form of daily weather forecasts that provide short-term information oriented to the major towns. Seasonal forecasting is provided by the National Meteorology Department of the Ministry of Tourism and Environmental Affairs. These longer-term forecasts are often very technical and are not understood by the rural communities, including farmers. Climate change receives little coverage in the print media in Swaziland, except when it results in disasters or extreme conditions (Manyatsi 2008). A significant number of rural community members attribute climate change to nonscientifically proven causes, such as breakdowns of tradition, biblical manifestations, and supernatural powers (Manyatsi, Mhazo, and Masarirambi 2010). In formulating measures to mitigate the effects of climate change, the beliefs and concerns of the community need to be taken into consideration. The Ministry of Tourism and

Environmental Affairs has launched a competition to find a SiSwati name for climate change to help Swazis more easily understand the subject (*Swazi Observer* 2010b).

Coping with Climate Change

Several livelihood strategies have been employed by communities to cope with hydrological disasters in Swaziland (Edje 2006; Manyatsi 2006; Manyatsi, Mhazo, and Masarirambi 2010; Manyatsi et al. 2010). Indigenous knowledge is used to predict or prevent hydrological disasters, and natural resources are used as a source of food and income. Indigenous methods of predicting hydrological events include observing the behavior of animals and the appearance of specific fruits. For example, the cry of *Cuculus solitarius,* a local bird, between August and November, is taken as a signal that the wet season will begin within two weeks (Manyatsi 2006). The restless and noisy behavior of pigs, peacocks, and ducks is taken as a sign of imminent heavy storms. The abundance of wild fruits—such as emantulwa (*Vangueria infausta*) and emaganu (*Sclerocarya birrea*)—between the months of December and February is a sign of imminent famine that season (Manyatsi 2006). Indigenous vegetables and fruits are used as alternative sources of food: ligusha (*Corchorus olitorius*), emantulwa (*Vangueria infausta*), and ematapha (*Scolopia mundi*) (Edje 2006). Wetland resources serve as sources of livelihood for the rural poor in times of drought, providing materials for handicrafts as well as sources of food and traditional medicine (Manyatsi et al. 2010).

Livelihood strategies and coping mechanisms practiced by communities in Swaziland are shown in Table 8.6. They include selling livestock in order to buy food and pay school fees, pooling resources through informal societies, and receiving food aid from the government, nongovernmental organizations (NGOs), and private companies.

The traditional social support system, made up of neighbors, relatives, and members of the household, plays an important role in responding to natural disasters. In the event of famine, neighbors assist one another in providing food, either in exchange for goods or favors (*kwenanisa*) or by selling them. In the event of a natural disaster, such as lightning striking a homestead, neighbors respond by calling for help, providing shelter, and providing food if the household's food reserve has been destroyed. Neighbors and relatives help out by plowing the fields of affected families that have lost cattle due to drought or lightning, either for free or in exchange for goods or favors (*kutsatsa*). Under the traditional system called *kusisa,* an affected

TABLE 8.6 Livelihood strategies and coping mechanisms practiced by rural communities in Swaziland, 2010

Strategy	Ecological zone		
	Middleveld	Lowveld	Lubombo
Selling livestock in order to buy food and pay school fees	✓	✓	✓
Buying water from those with boreholes		✓	
Provision of food aid to community members by NGOs and private companies	✓	✓	✓
Getting food rations from other community members	✓	✓	✓
Getting water from schools that have boreholes and other reliable water sources		✓	
Harvesting water from roofs		✓	
Recycling water		✓	
Pooling resources through societies	✓	✓	✓
Giving elderly people social grants from the government	✓	✓	✓
Providing seeds and other farming inputs to needy members of the community by NGOs		✓	✓
Engaging in mixed cropping and crop diversification	✓		✓
Growing vegetables under irrigation		✓	

Source: Manyatsi, Mhazo, and Masarirambi (2010).
Note: NGOs = nongovernmental organizations.

household may be given cattle to look after in return for rights to the milk and use of the cattle for plowing.

A National Disaster Agency was set up under the deputy prime minister's office to assess the effects of natural disasters and to coordinate the response at a national level. The Coordinating Assembly of Non Governmental Organizations (CANGO), an umbrella body for all nongovernmental organizations in the country, includes about 60 organizations that are actively involved in social welfare and relief operations in the event of natural disasters, including the Swaziland Red Cross, World Vision, the Swaziland Farmers' Development Foundation, and the Women's Resource Center (CANGO 2010). The NGOs distribute food, blankets, and shelter to communities affected by natural disasters. They also play an important role in increasing food security by effecting improvements in agriculture productivity based on small-scale irrigation in the form of community gardens in which each household has a small plot.

Institutional Response to Climate Change

Swaziland ratified the United Nations Framework Convention on Climate Change (UNFCCC) in 1996 and the Kyoto Protocol in 2006. The country produced its first national communication to the UNFCCC in 2000. That report provided a national inventory of anthropogenic emissions by sources, as well as of removal of greenhouse gases by sinks. On aggregate, Swaziland's overall status was as a net carbon dioxide sink. The predominant carbon uptake was by natural forests, commercial plantations, trees in towns, and wetlands plants (Swaziland, Ministry of Works 2000). A second national communication is currently in preparation.

The Ministry of Agriculture is responsible for undertaking research on crop varieties and livestock breeds appropriate to the Swaziland environment. The ministry also disseminates information on best agricultural practices through its Department of Extension Services. The National Meteorology Department (under the Ministry of Tourism and Environmental Affairs) is the country's focal point for climate change. The ministry recently established the Swaziland Climate Change Programme (SCCP) to address climate change issues in the country, implemented by the Meteorology Department and funded by the financial mechanism of the UNFCCC (the Global Environmental Facility) through the UN Development Programme (UNDP) (SCCP 2010). A permanent office has been established under the SCCP and is headed by a climatologist. A national climate change committee was established in September 2010 to coordinate the implementation of the SCCP. The committee, composed of 21 representatives of government ministries with diverse qualifications, is chaired by the climatologist from the Ministry of Tourism and Environmental Affairs (*Swazi Observer* 2010a).

Several public, private, and nongovernmental organizations play major roles in addressing climate change issues in the country in addition to the government ministries that are represented in the National Climate Change Committee. The public sector includes the University of Swaziland (UNISWA 2010), the Swaziland Water and Agricultural Development Enterprise (SWADE 2010), the Swaziland Water Services Corporation (SWSC 2010), and the Swaziland Electricity Company (SEC 2010).

Private-sector entities that have played a major role in climate change issues in the country include the financial sector, agricultural estates (such as the Royal Swaziland Sugar Corporation, Illovo Sugar, and United Plantations), and suppliers of agriculture inputs. The majority of the NGOs in Swaziland are affiliated with the 63-member CANGO, which is engaged in coordination, capacity building, and policy advocacy, including on climate change issues. CANGO affiliates have also been engaged in developmental programs and programs to mitigate the effects of climate change; its membership includes many NGOs that work with communities at the grassroots level. CANGO was accordingly selected as the national focal point for the Food, Agriculture, and Natural Resources Policy Analysis Network (FANRPAN) (CANGO 2010).

Swaziland does not have a policy directly addressing the issues of climate change and climate change adaptation. However, several existing policies and laws are related to climate change, as shown in Table 8.7. They include the Forest Preservation Act of 1907, the National Trust Commission Act of 1972, the Water Act of 2003, and the Disaster Management Act of 2006.

Policies that address the issue of climate change and vulnerability to climate change include the Poverty Reduction Strategy and Action Plan (Swaziland, Ministry of Economic Planning and Development 2006), The Rural Settlement Policy of 2003 (Swaziland, Ministry of Agriculture and Cooperatives 2003), and the Forest Policy of 2002 (Swaziland, Ministry of Agriculture and Cooperatives 2002). The Poverty Reduction Strategy calls for fair distribution of the benefits of growth through fiscal policy and for empowerment of the poor to generate income, human capital development, and an improved quality of life. The Forest Policy aims to achieve efficient, profitable, and sustainable management and use of forest resources for the benefit of the entire society and to increase the role of forestry in environmental protection, conservation of plant and animal genetic resources, and rehabilitation of degraded land.

Projects and Programs Addressing Climate Change Issues

- National projects that have implications for agriculture include the Maguga Dam project, the Komati Downstream Development Project (KDDP), the LUSIP, the Smallholder Agricultural Development Project, and the Earth Dam Rehabilitation Project. The Maguga Dam was constructed jointly by the Swaziland government and the South African government with the aim of harnessing water for irrigation downstream of the Komati River (KOBWA 2010). In Swaziland, the dam construction culminated in the implementation of the KDDP, which was designed to benefit a population of about 22,000 people and to develop 6,000 hectares of new irrigation schemes along the Komati basin.

TABLE 8.7 Legislation in Swaziland relevant to climate change

Legislation	Relevance to climate change
Forests Preservation Act of 1907 (Swaziland, Ministry of Justice 1907)	Provides for the preservation of trees and forests growing on government and Swazi Nation land. Prohibits the cutting down, damage, removal, sale, or purchase of indigenous or government timber without the permission of the Minister of Agriculture.
Private Forests Act of 1951 (Swaziland, Ministry of Justice 1951)	Provides for the better regulation and protection of private forests in the country.
Grass Fire Act of 1955 (Swaziland, Ministry of Justice 1955)	Consolidates laws relating to grass burning and grass fires, which contribute to the accumulation of greenhouse gases. Regulates the interval and period when fire can be set on grass.
National Trust Commission Act of 1972 (Swaziland, Ministry of Justice 1972)	Provides for the establishment of national parks and monuments. The national parks are protected from anthropogenic activities, and the trees and other vegetation in them act as sinks for greenhouse gases produced elsewhere.
Waste Regulations of 2000 (Swaziland, Ministry of Justice 2000a)	Regulates the management of solid waste and liquid waste disposal on land. Prohibits persons from disposing of commercial or industrial waste or household waste produced in urban areas except at an approved waste disposal facility. Also prohibits the import of hazardous waste into the country.
Environmental Audit, Assessment and Review Regulations of 2000 (Swaziland, Ministry of Justice 2000b)	Requires any operator who undertakes development to submit an environmental assessment report and a comprehensive mitigation plan to the Swaziland Environmental Authority before permission can be given to undertake the development. The impact of the development on climate change is taken into consideration when reviewing the environmental assessment.
Environmental Management Act of 2002 (Swaziland, Ministry of Justice 2002)	Provides for and promotes the enhancement, protection, and conservation of the environment and the sustainable management of natural resources. Advocates for minimizing the generation of waste.
Water Act of 2003 (Swaziland, Ministry of Justice 2003)	Established a National Water Authority responsible for advising ministries on matters related to water use and distribution. Also established other institutions including the River Basin Authorities, the Water Apportionment Board, the Irrigation Districts, and the Water Users' Associations.
Disaster Management Act of 2006 (Swaziland, Ministry of Justice 2006)	Established a national disaster management policy in order to minimize potential losses from hazards and to provide timely and appropriate assistance to victims. Swaziland is prone to disasters associated with climate change, including droughts, famines, and occasional floods and storms.

Source: SEA and UNEP (2005).

- The LUSIP involved the construction of a dam to impound water diverted from the Usuthu River. The project is designed to develop 11,500 hectares for irrigation. Both the KDDP and the LUSIP are implemented by SWADE, a government-owned company (SWADE 2010) designed to assist the most disadvantaged agricultural producers on Swazi Nation land. Financed by the International Fund for Agricultural Development, its components include developing 185 hectares of new small-scale irrigation schemes and consolidating another 257 hectares of existing schemes to promote farmers' management of irrigation.

- The Earth Dam Rehabilitation and Construction program rehabilitated several medium-sized earth dams in the lowveld and lower middleveld, benefiting small-scale farmers.

- Government programs that address climate change issues include the CAADP, the SADP, and the Tractor Hire Pool program. The CAADP provides a strategic framework agreed to by the heads of states involved in the New Partnership for Africa's Development that is aimed at increasing national budget expenditure on agriculture to at least 10 percent and ensuring agriculture growth of at least 6 percent per year. Swaziland signed the CAADP framework in March 2010 (CAADP 2010).

- The SADP was established to improve smallholder production and marketing in order to achieve sustainable food security. Still in the formation stage, it will establish a Rural Youth Development Program, Junior Farmer Field and Life Schools, and Farmer Field Schools, among other projects. The program will also play a role in disseminating new crop varieties and livestock types approved by the Ministry of Agriculture.

- The government-subsidized tractor hire scheme, the Tractor Hire Pool, serves small-scale farmers. The government provides tractors or engages private tractor owners to plow for farmers, who pay subsidized fees for the service.

Conclusions and Policy Recommendations

Climate change may have a significant effect on agriculture in Swaziland. Scenarios show temperatures increasing by as much as 2.5°C by 2050 and annual precipitation decreasing by more than 100 millimeters in the highveld and the lowveld, the ecological zones where most of the maize is produced. Both the yield and the harvested area of maize would drop in these scenarios, resulting in an increased net import of maize with higher domestic prices of maize, making the staple food less affordable. Climate change scenarios coupled with GDP and population growth scenarios show the number of malnourished children increasing in the short run (up to the year 2025) and then decreasing.

Sugarcane, produced under irrigation, is less vulnerable to climate change as long as adequate water supplies are available. The harvested area planted in sugarcane is shown increasing as land is converted from other uses (such as livestock grazing and growing maize). Financial institutions more readily underwrite sugarcane production because of the perceived high financial returns from the sugar industry and the guaranteed demand from local sugar mills. Cotton production, too, is shown increasing but not enough to meet local demand, with most of the needed cotton being imported.

A National Committee has recently been established to spearhead the SCCP. The committee, made up of government officers, does not include representatives of public organizations, private organizations, or NGOs, and it has had little impact to date. In developing Swaziland's climate change policy, the National Committee needs to engage all relevant stakeholders to represent a broad range of interests.

There is a need to improve the understanding of local communities regarding climate change. Agricultural research needs to provide evidence to guide policy and to enable assessment of the potential impacts of climate change as well as climate change policies. Researchers need to take into account the ways farmers and communities currently adapt to weather variability and extreme events.

We make the following recommendations based on the study:

- The National Committee for the SCCP should facilitate the development of a climate change policy and climate change adaptation action plan, soliciting needed assistance (in the form of funding and expertise) from international organizations such as the UNDP, UNFCCC, and FANRPAN.

- Climate change should be addressed and streamlined in national agendas. Government agencies should facilitate dissemination of climate change information to stakeholders.

- The National Meteorology Department should produce simplified versions of seasonal weather forecast reports for farmers.

- Agricultural extension officers advising farmers on crop and livestock production should be sensitized to and trained in climate change issues, as well as in interpreting seasonal weather reports.

- The government should upscale the construction of small dams, especially in the lowveld area, which is more vulnerable to climate change. Communities could use the water captured to produce crops and vegetables to improve their livelihoods.

- If climate change reduces maize production and increases prices, people may have to replace this staple with other foods. Researchers and policymakers need to explore effective adaptations in light of this scenario.

- Farmers should grow drought-tolerant crops that can also withstand higher temperatures. National research institutions, together with seed suppliers, should ensure that drought-tolerant and heat-tolerant seeds are available and affordable, for example, through FANRPAN's Harmonized Seed Security Project.

- Vulnerable households (selected using reliable tools such as the Household Vulnerability Index) should be provided with food and agricultural inputs.

The challenge for policymakers is to respond fast enough to meet the urgency of the situation. Continuing dialogue between researchers and policymakers will provide mutual learning opportunities and ensure that the knowledge produced by researchers is both useful and used. FANRPAN has initiated national dialogues for engagement among stakeholders, and such dialogues have been useful in communicating climate change issues.

References

African Economic Outlook. 2010. "Swaziland." Accessed October 2, 2010. www.africaneconomicoutlook.org/en/countries/southern-africa/swaziland/.

AGOA (African Growth and Opportunity Act). 2000. "African Growth and Opportunity Act." Accessed October 15, 2010. www.agoa.gov/.

Bartholome, E., and A. S. Belward. 2005. "GLC2000: A New Approach to Global Land Cover Mapping from Earth Observation Data." *International Journal of Remote Sensing* 26 (9): 1959–1977.

Boko, M., I. Niang, A. Nyong, C. Vogel, A. Githeko, M. Medany, B. Osman-Elasha, R. Tabo, and P. Yanda. 2007. "Climate Change 2007: Impacts, Adaptation, and Vulnerability." In *Contribution of Working Group II to the Fourth Assessment Report of the Intergovernmental Panel on Climate Change,* edited by M. L. Parry, O. F. Canziani, J. P. Palutikof, P. J. van der Linden, and C. E. Hanson. Cambridge: Cambridge University Press.

CAADP (Comprehensive African Agriculture Development Programme). 2010. "Swaziland to Sign CAADP Compact." Accessed October 4, 2010. www.caadp.net/news/?p=558.

CANGO (Coordinating Assembly of Non Government Organizations of Swaziland). 2007. "Swaziland Human Development Report: HIV and AIDS and Culture." Accessed October 13, 2011. http://hdr.undp.org/en/reports/nationalreports/africa/swaziland/Swaziland_NHDR _2008.pdf.

———. 2010. "CANGO Request for Applications in Support of the U.S. President's Emergence Plan for AIDS Relief (PEPFAR) in Swaziland." Accessed October 3, 2010. www.cango.org.sz/.

Central Bank of Swaziland. 2011. "Annual Report 2011." Accessed October 13, 2011. www .centralbank.org.sz/index.php?option=com_docman&task=cat_view&gid=34&Itemid=236.

CIA (Central Intelligence Agency). 2011. "The World Factbook: Swaziland." Accessed October 13, 2011. www.cia.gov/library/publications/the-world-factbook/geos/wz.html.

CIESIN (Center for International Earth Science Information Network), Columbia University, IFPRI (International Food Policy Research Institute), World Bank, and CIAT (Centro Internacional de Agricultura Tropical). 2004. *Global Rural–Urban Mapping Project (GRUMP), Alpha Version: Population Density Grids.* Palisades, NY, US: Socioeconomic Data and Applications Center (SEDAC), Columbia University. http://sedac.ciesin.columbia.edu/gpw.

Edje, O. T. 2006. "Indigenous Knowledge in Nature Conservation and Utilization." In *Nature Conservation and Natural Disaster Management—Role of Indigenous Knowledge in Swaziland.* Mbabane: United Nations Environment Programme and University of Swaziland.

FAO (Food and Agriculture Organization of the United Nations). 2005. "FAO/WFP Crop and Food Supply Assessment Mission to Malawi." Accessed November 17, 2010. www.fao.org/ docrep/008/j5509e/j5509e00.htm.

———. 2010. FAOSTAT. Accessed December 22, 2010. http://faostat.fao.org/site/573/default. aspx#ancor.

Fibre2fashion. 2008. "Govt Undertakes Revival of Cotton Growing Activities." Accessed October 5, 2010. www.fibre2fashion.com/news/cotton-news/newsdetails.aspx?news_id=62464.

How We Made It in Africa. 2010. "Swaziland Economic Overview." Accessed October 13, 2011. www.howwemadeitinafrica.com/swaziland-economic-overview/2982/.

IRIN (Integrated Regional Information Networks). 2007a. "Swaziland: Facing Climate Change." July 20. Accessed August 2, 2007. www.irinnews.org/Report.aspx?ReportId=73337.

———. 2007b. "Swaziland: Coping Strategies Wear Thin in Ongoing Food Crisis." October 16. Accessed October 17, 2007. http://www.irinnews.org/report/74833/swaziland-coping-strategies-wear-thin-in-ongoing-food-crisis.

Jones, P. G., P. K. Thornton, and J. Heinke. 2009. *Generating Characteristic Daily Weather Data Using Downscaled Climate Model Data from the IPCC's Fourth Assessment.* Project report. Nairobi, Kenya: International Livestock Research Institute.

Knight Piesold Consulting Engineers. 1997. "Swaziland Water Sector Situation Report." Mbabane.

KOBWA (Komati Basin Water Authority). 2010. Website. Accessed October 15, 2010. www.kobwa.co.za.

Kriticos, D. J., T. Yonow, and R. E. McFadyen. 2005. "The Potential Distribution of *Chromolaena odorata* (Siam Weed) in Relation to Climate." *Weed Research* 45 (4): 246–254.

Manyatsi, A. M. 2006. "Indigenous Knowledge in Natural Disaster Management." In *Nature Conservation and Natural Disaster Management: Role of Indigenous Knowledge in Swaziland.* Mbabane: United Nations Environment Programme and University of Swaziland.

———. 2008. "Media Coverage of Climate Change Issues in Swaziland." In *Actions towards a Sustainable Future,* edited by M. Mlipha. Mbabane: Swaziland Environment Authority.

Manyatsi, A. M., N. Mhazo, and M. T. Masarirambi. 2010. "Climate Variability and Changes as Perceived by Rural Communities in Swaziland." *Research Journal of Environmental and Earth Sciences* 2 (3): 165–170.

Manyatsi, A. M., N. Mhazo, S. Msibi, and M. T. Masarirambi. 2010. "Utilisation of Wetland Resources for Livelihood in Swaziland: The Case of Lobamba Lomdzala Area." *Journal of Social Sciences* 2 (3): 262–268.

Menne, W., and R. Carrere. 2007. "Swaziland: The Myth of Sustainable Timber Plantations." Accessed October 14, 2011. www.wrm.org.uy/countries/Swaziland/Book_Swaziland.pdf.

Millennium Ecosystem Assessment. 2005. *Ecosystems and Human Well-being: Synthesis.* Washington, DC: Island Press. http://www.maweb.org/en/Global.aspx.

Murdoch, G. 1968. *Soils and Land Capability in Swaziland.* Mbabane: Swaziland Ministry of Agriculture.

Mwendera, E. J., A. M. Manyatsi, O. Magwenzi, and S. M. Dhlamini. 2002. *Water Demand Management Programme for Southern Africa.* Country Report for Swaziland. Pretoria, Republic of South Africa: International Union for Conservation of Nature Southern Africa Country Office.

NAMBOARD (National Marketing Board, Swaziland). 2010. Website. Accessed October 2, 2010. www.namboard.co.sz.

Nelson, G. C., M. W. Rosegrant, A. Palazzo, I. Gray, C. Ingersoll, R. Robertson, S. Tokgoz, et al. 2010. *Food Security, Farming, and Climate Change to 2050: Scenarios, Results, Policy Options*. Washington, DC: International Food Policy Research Institute.

NMC (National Maize Corporation, Swaziland). 2010. Accessed October 2, 2010. Website. http://www.nmc.co.sz.

Rammelzwaal, A., and W. S. Dlamini. 1994. *Present Land Use Map of Swaziland*. Mbabane: Swaziland Ministry of Agriculture and Cooperatives.

SACU (Southern Africa Customs Union). 2010. Website. Accessed October 10, 2010. www.sacu.int.

SCCP (Swaziland Climate Change Programme). 2010. Website. Accessed October 5, 2010. www.climatechange.org.sz/index.php?option=com_content&view=article&id=44&Itemid=125.

SEA (Swaziland Environmental Authority) and UNEP (United Nations Environment Programme). 2005. *Compendium of Environmental Laws of Swaziland*. Mbabane.

SEC (Swaziland Electricity Company). 2010. Website. Accessed October 7, 2010. www.sec.co.sz.

SSA (Swaziland Sugar Association). 2010. Website. Accessed October 6, 2010. www.ssa.co.sz.

SWADE (Swaziland Water and Agriculture Development Enterprise). 2010. Website. Accessed November 18, 2010. www.swede.co.sz.

Swaziland, Ministry of Agriculture and Cooperatives. 2002. *Forest Policy*. Mbabane.

———. 2003. *Rural Settlement Policy*. Mbabane.

Swaziland, Ministry of Economic Planning and Development. 1992. *Statistical Bulletin*. Mbabane.

———. 2006. *Poverty Reduction Strategy and Action Plan: A Summarized Version*. Mbabane.

———. 2007. *Census Results*. Mbabane.

———. 2010. *Swaziland Millennium Development Goals Report*. Mbabane.

Swaziland, Ministry of Justice. 1907. *The Forests Preservation Act of 1907*. Mbabane.

———. 1951. *The Private Forests Act of 1951*. Mbabane.

———. 1955. *The Grass Fire Act of 1955*. Mbabane.

———. 1972. *The National Trust Commission Act of 1972*. Mbabane.

———. 2000a. *The Waste Regulations of 2000*. Mbabane: Ministry of Justice.

———. 2000b. *The Environmental Audit, Assessment, and Review Regulations of 2000*. Mbabane.

———. 2002. *Environmental Management Act*. Mbabane.

———. 2003. *The Water Act of 2003*. Mbabane.

———. 2006. *The Disaster Management Act of 2006*. Mbabane.

Swaziland, Ministry of Works. 2000. *First National Communication to the UNFCCC [United Nations Framework Convention on Climate Change]*. Mbabane.

Swaziland Meteorological Services. 2010. *Location of Weather Stations*. Mbabane.

Swaziland News Stories. 2010. "Swaziland Continues to Import More Maize." Accessed October 5, 2010. www.swazilive.com/Swaziland_News/Swaziland_News_Stories.asp?News_id=1370.

Swazi Observer. 2010a. "National Climate Committee Named." September 21.

———. 2010b. "SiSwati Name for Climate Change Sought." September 14.

SWSC (Swaziland Water Services Corporation). 2010. Website. Accessed October 14, 2010. www.sec.co.sz.

UNICEF (United Nations International Children's Fund). 2011. "Swaziland Statistics." Accessed October 13, 2011. www.unicef.org/infobycountry/swaziland_statistics.html.

UNISWA (University of Swaziland). 2010. Website. Accessed October 8, 2010. www.uniswa.sz.

UNPOP (United Nations Department of Economic and Social Affairs–Population Division). 2009. *World Population Prospects: The 2008 Revision*. New York. http://esa.un.org/unpd/wpp/.

World Bank. 2009. *World Development Indicators*. Washington, DC.

———. 2010a. *World Development Indicators*. Washington, DC.

———. 2010b. *The Costs of Agricultural Adaptation to Climate Change*. Washington, DC.

You, L., and S. Wood. 2006. "An Entropy Approach to Spatial Disaggregation of Agricultural Production." *Agricultural Systems* 90 (1–3): 329–347.

You, L., S. Wood, and U. Wood-Sichra. 2006. "Generating Global Crop Distribution Maps: From Census to Grid." Paper presented at the International Association of Agricultural Economists Conference in Brisbane, Australia, August 11–18.

———. 2009. "Generating Plausible Crop Distribution and Performance Maps for Sub-Saharan Africa Using a Spatially Disaggregated Data Fusion and Optimization Approach." *Agricultural Systems* 99 (2–3): 126–140.

Chapter 9

ZAMBIA

Joseph Kanyanga, Timothy S. Thomas, Sepo Hachigonta, and Lindiwe Majele Sibanda

Agriculture in Africa south of the Sahara is becoming increasingly risky due to extreme climate variability. In recent times, scientific studies have strongly suggested that many developing countries face substantial environmental and social challenges, with food insecurity high on the list (Vogel and O'Brien 2003). With a relatively large and impoverished rural population that largely relies on rainfed agriculture, Zambia is vulnerable to the impacts of rainfall variability, which pose challenges for food security and planning. Climate change will further enhance the impacts associated with climate extremes in most parts of Zambia.

The objective of the chapter is to help policymakers, researchers, and country negotiators better understand and anticipate the likely impacts of climate change on agriculture and on vulnerable households in Zambia, given that about 70 percent of the nation's population is dependent on rainfed agriculture.

Review of Current Trends

Economic and Demographic Indicators

Population

Figure 9.1 shows the total and rural populations as well as the share of the population that is urban in Zambia. The urban population was approximately 18 percent of the total population in 1960, increased to 40 percent in 1980, and then declined between 1990 and 2000 to 35 percent, where it remained mostly steady through 2008. A number of factors contributed to this increase—in particular, the growth in Zambian industry from 1960 to 1980.

The country has urbanized at an average rate of 4.32 percent over the past five decades. The fastest decadal urban growth rate, 8.30 percent, was recorded during the decade 1960–1969 (Table 9.1). The urban growth rate could be explained by considering that there was more pronounced rural–urban

FIGURE 9.1 Population trends in Zambia: Total population, rural population, and percent urban, 1960–2008

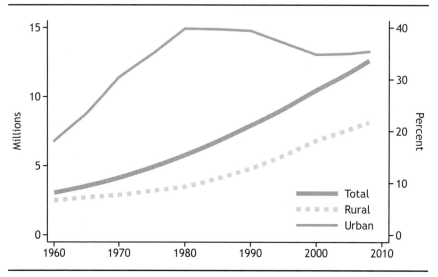

Source: *World Development Indicators* (World Bank 2009).

TABLE 9.1 Population growth rates in Zambia, 1960–2008 (percent)

Decade	Total growth rate	Rural growth rate	Urban growth rate
1960–1969	3.0	1.5	8.3
1970–1979	3.3	1.9	6.1
1980–1989	3.2	3.2	3.1
1990–1999	2.8	3.6	1.6
2000–2008	2.3	2.2	2.5

Source: Authors' calculations based on *World Development Indicators* (World Bank 2009).

migration resulting from the economic boom just after Zambia achieved independence in 1964, largely driven by the attraction of the well-performing copper mining industry in Copperbelt Province and other emerging industries along the railway line. The slowest growth rate, 1.60 percent, was experienced during 1990–1999, which was also the period of massive retrenchments of workers following the closure of many companies related to the liberalization of the Zambian economy during the third republic. The modest increase in urban growth during 2000–2008, to 2.50 percent, may be explained by renewed economic stability and growth when this environment enabled the reopening of some companies, particularly in Copperbelt Province.

The average decadal rural growth rate during 1960–2010 was 2.34 percent, though it remained low during the decades 1960–1969 (at 1.50 percent) and 1970–1979 (at 1.90 percent). The period 1980–1989 saw an increase to 3.20 percent. The highest rural growth rate was during the decade 1990–1999, when it reached 3.60 percent. This could have been related to the aforementioned retrenchment of workers returning to their rural areas. The trend in rural population growth shows a steady increase, then a decline to 2.20 percent during 2000–2008, though the rural population experienced slower growth than did the total population. The rural population in 2008 stood at about 7 million out of a total population of about 12 million.

Because most of Zambia's rural population is dependent on agriculture, the projected rural population growth will require a corresponding increase in agricultural production.

Figure 9.2 shows the geographic distribution of the population in Zambia. In 2000 the highest population distribution was mainly over parts of Copperbelt, Central, and Lusaka Provinces, with more than 100 persons per square kilometer. This was followed by most parts of Southern and Eastern Provinces and a few parts of Luapula, Copperbelt, and Northern Provinces, with a population distribution of between 20 and 100 persons per square kilometer. The

FIGURE 9.2 Population distribution in Zambia, 2000 (persons per square kilometer)

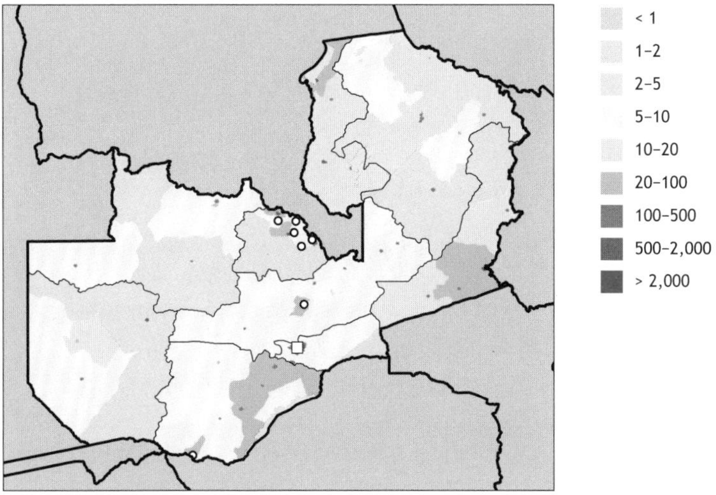

Source: CIESIN et al. (2004).

lowest population distribution, between 2 and 5 persons per square kilometer, was recorded over much of Western, North Western, and rural Copperbelt Provinces, as well as parts of Northern, Eastern, and Lusaka Provinces.

Income

Figure 9.3 shows trends in gross domestic product (GDP) per capita and the proportion of GDP from agriculture. The past five decades have shown a decreasing trend in GDP per capita: in 1960 it was about $550; in 2008, although rebounding from its low point, it was below $400. In 1964 Zambia experienced an economic boom, with GDP per capita above $600, driven by high copper prices on the international market. After 1975, however, GDP per capita started to decline rapidly, reaching its lowest point between 1995 and 2000, at around $300. Several factors contributed to this decline. First, after independence the nationalization policy brought much of the economic base under the management of the Zambian government. Second, declining copper prices on the international market meant a decrease in revenue for Zambia. Third, after 1991 there was mass closure of most mines and industries due to the economic liberalization policy.

FIGURE 9.3 Per capita GDP in Zambia (constant 2000 US$) and share of GDP from agriculture (percent), 1960–2008

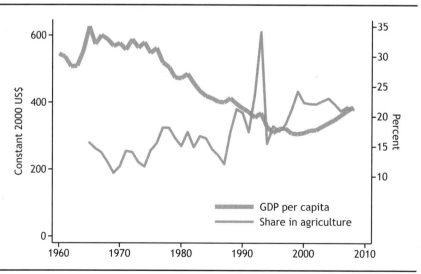

Source: *World Development Indicators* (World Bank 2009).
Note: GDP = gross domestic product; US$ = US dollars.

TABLE 9.2 Education and labor statistics for Zambia, 2000s

Indicator	Year	Value (percent)
Primary school enrollment (percent gross, three-year average)	2007	119.0
Secondary school enrollment (percent gross, three-year average)	2007	43.1
Adult literacy rate	2007	70.6
Percent employed in agriculture and related areas	2000	71.6
Percent with vulnerable employment (own farm or day labor)	2003	79.3
Under-five malnutrition (weight for age)	2002	23.3

Source: *World Development Indicators* (World Bank 2009).

Since 2000 there has been a steady increase in GDP per capita, which reached almost $400 in 2008. This is attributed to economic stability and growth driven by the reopening of some mines and improved fiscal management in the public sector. Further, Zambia has also reached the completion point for heavily indebted poor countries, which relieved the country of a significant debt burden.[1]

Overall, the contribution of the agriculture sector to national GDP shows a moderate upward trend (see Figure 9.3), from about 12 percent around 1965 to about 20 percent in 2008. Agriculture is important to the national economy and is dominated by small-scale and subsistence farmers.

Vulnerability

Table 9.2 shows, for Zambia, several indicators of a population's vulnerability and resiliency to economic and natural shocks, including level of education, literacy, and concentration of labor in poorer or less dynamic sectors. It is evident that although the level of primary school enrollment is high (119.0 percent), fewer than half of students progress to secondary school (43.1 percent). Most of the enrolled school pupils drop out after primary school. This implies that many people do not have the training and skills required for employment and income generation. Thus a less skilled workforce is entering the job market to grow the economy, including the agriculture sector. The relatively high adult literacy rate, 70.6 percent, is largely limited to primary-level literacy.

The employment rate in agriculture, 71.6 percent in 2000, shows the significance of the agriculture sector as a source of income. Because agriculture

---

[1] The Heavily Indebted Poor Countries Initiative was launched in 1996 by the International Monetary Fund and the World Bank, with the aim of ensuring that no poor country faces a debt burden it cannot manage. Since then the international financial community, including multilateral organizations and governments, has worked together to reduce to sustainable levels the external debt burdens of the most heavily indebted poor countries.

FIGURE 9.4 Well-being indicators in Zambia, 1960–2008

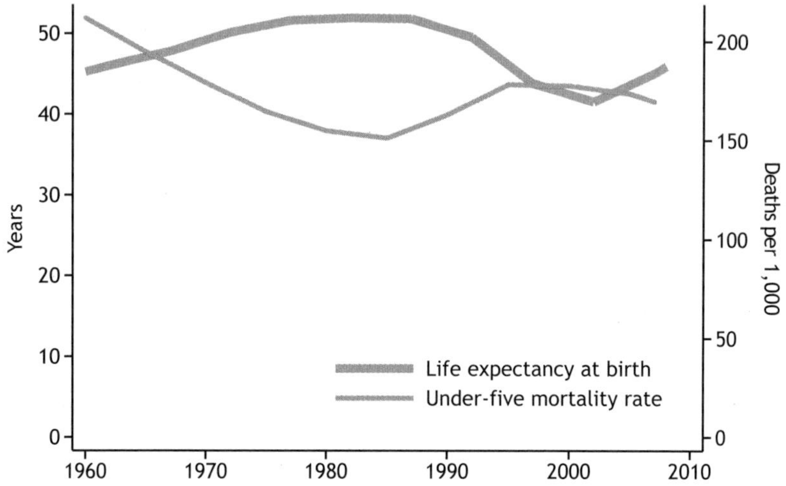

[Figure: Line chart showing Life expectancy at birth (Years, left axis, 0–50) and Under-five mortality rate (Deaths per 1,000, right axis, 0–200) from 1960 to 2010.]

Source: *World Development Indicators* (World Bank 2009).

in Zambia is mainly rainfed and thus very sensitive to variations in climatic patterns, this also means that a high percentage of the workforce in Zambia is susceptible to the vulnerability associated with the impacts of weather and climate change. Also of concern is the finding that in 2002, 23.3 percent of children under the age of five years were malnourished.

Figure 9.4 shows two noneconomic indicators of poverty: life expectancy and under-five mortality. Life expectancy in Zambia was about 45 years in 1960 and increased to 50 in the six years following independence, probably driven by enhanced employment opportunities and access to primary healthcare after independence. Life expectancy stayed level through 1995, at slightly more than 50 years, coinciding with overall good economic performance, but it declined thereafter to about 40 years in 2000. This decline may be linked to the fall in GDP per capita driven by mass closures of companies after the third republic, as well as to the emergence of the HIV/AIDS epidemic. After 2000, life expectancy started to increase, reaching 45 years in 2008.

In 1960, the under-five mortality rate was above 200 per 1,000. Thereafter it declined steadily, to slightly more than 150 per 1,000 in 1985, during the period when the Zambian economy was flourishing and access to health facilities was improving significantly. After 1985, the under-five mortality rate began to increase, reaching 170 per 1,000 in 1995, coinciding with the decline in GDP per capita after the liberalization of the economy. In addition, the

FIGURE 9.5 Poverty in Zambia, circa 2005 (percentage of population below US$2 per day)

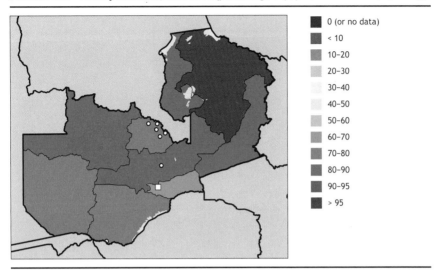

Source: Wood et al. (2010).
Note: Based on 2005 US$ (US dollars) and on purchasing power parity value.

new cost-sharing system in the public sector resulted in reduced access to antenatal primary healthcare. After 2000, the under-five mortality rate stabilized and even started to decline approaching 2008, in line with improving overall economic performance.

Figure 9.5 shows the proportion of the population currently living on less than $2 per day, which marks the poverty line. In all provinces, at least 50 percent of the population lives below this poverty line. The poorest province is Northern Province, with more than 95 percent below the poverty line. Even in Lusaka Province, a cosmopolitan administrative capital, 50–60 percent live on less than $2 per day. In Luapula, Eastern, Central, and North Western Provinces, the poverty level is 80–90 percent; in Copperbelt, Southern, and Western Provinces, it is 60–70 percent.

Land Use Overview

Figure 9.6 shows Zambia's land cover and land use as of 2000. The land cover is mainly characterized by broadleaved and deciduous trees (both open and closed) and deciduous shrub cover (both open and closed). Much of North Western Province and parts of Copperbelt, Western, Central, Luapula, and Northern Provinces have closed tree cover (broadleaved, deciduous) while much of Northern and Central Provinces has open tree cover (broadleaved,

FIGURE 9.6 Land cover and land use in Zambia, 2000

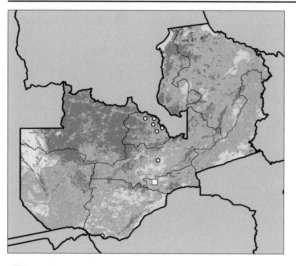

- Tree cover, broadleaved, evergreen
- Tree cover, broadleaved, deciduous, closed
- Tree cover, broadleaved, open
- Tree cover, broadleaved, needle-leaved, evergreen
- Tree cover, broadleaved, needle-leaved, deciduous
- Tree cover, broadleaved, mixed leaf type
- Tree cover, broadleaved, regularly flooded, fresh water
- Tree cover, broadleaved, regularly flooded, saline water
- Mosaic of tree cover/other natural vegetation
- Tree cover, burnt
- Shrub cover, closed-open, evergreen
- Shrub cover, closed-open, deciduous
- Herbaceous cover, closed-open
- Sparse herbaceous or sparse shrub cover
- Regularly flooded shrub or herbaceous cover
- Cultivated and managed areas
- Mosaic of cropland/tree cover/other natural vegetation
- Mosaic of cropland/shrub/grass cover
- Bare areas
- Water bodies
- Snow and ice
- Artificial surfaces and associated areas
- No data

Source: GLC2000 (Bartholome and Belward 2005).

deciduous). Eastern and Western Provinces and parts of Southern Province have shrub cover (closed-open, deciduous). Cultivated and managed areas are found mostly in Southern, Central, and Eastern Provinces.

Areas under tree cover are less vulnerable to climate variability and change because they have a higher capacity for soil moisture retention. The areas under shrub cover are more vulnerable to climate variability and change, with low rainfall and low soil moisture retention capacity as well as a higher potential for excessive evaporation. Most of Zambia's agriculture is concentrated in these more vulnerable areas, making the agriculture sector more vulnerable to climate variability and change.

The locations of protected areas, including parks and natural reserves, are shown in Figure 9.7. These locations provide important protection for fragile environments, which may also be important for the tourism industry. Much of Zambia's landmass, however, is in unprotected (customary) areas, whose use is regulated by traditional authorities. These unprotected areas are overexploited due to unsustainable land use practices, which render these vast areas susceptible to deforestation and degradation and thus more vulnerable to variations in climatic patterns.

The protected areas, despite being regulated by the government, are also under pressure due to competition for use by the tourism sector. Protected areas are normally rich in biodiversity, in both flora and fauna, and thus represent key areas for tourism. Most of Zambia's wildlife is found in these areas, including the Kafue National Park in Central Province and Luangwa National Park in Eastern Province. The opening of these areas for tourism through ventures such as safari lodges and resorts, fish camps, and game ranches compromises the environmental protection and conservation of these areas. Changing the balance of the ecosystem in protected areas can result in shifts in biodiversity regimes and even in extinction of species.

Figure 9.7 shows that the protected areas are also surrounded by rural communities. Thus they are also threatened by livelihood activities such as cultivation and settlement, illegal timber cutting and logging, uncontrolled fires, poaching, and fuelwood and charcoal burning. These activities weaken the resilience of these ecosystems to climate variability and change. They also compromise the vital function of these areas as water recharge areas and carbon sinks.

Figure 9.8 shows travel times to urban areas of various sizes, which are potential markets for agricultural products. Travel to cities of 500,000 or more people is clearly more manageable around the major towns in Lusaka, Copperbelt, and Livingstone. Farmers in these areas have shorter travel times

FIGURE 9.7 Protected areas in Zambia, 2009

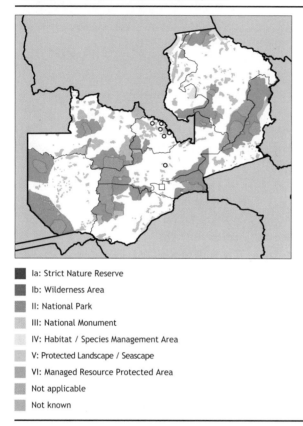

■ Ia: Strict Nature Reserve
■ Ib: Wilderness Area
■ II: National Park
■ III: National Monument
■ IV: Habitat / Species Management Area
■ V: Protected Landscape / Seascape
■ VI: Managed Resource Protected Area
■ Not applicable
■ Not known

Sources: Protected areas are from the World Database on Protected Areas (UNEP and IUCN 2009). Water bodies are from the World Wildlife Fund's Global Lakes and Wetlands Database (Lehner and Döll 2004).

of under an hour or between 1 and 3 hours owing to good road and railway networks. However, most rural farmers are outside this region. Most farmers along the rail line, in Copperbelt, Central, Lusaka, and Southern Provinces, have travel times of 3–5 hours. In these provinces, farmers have the option of using rail or road transport systems.

Elsewhere, travel times are much longer: in Solwezi (North Western Province) and the region along the Great East Road (to Eastern Province), 5–8 hours; in Western, Luapula, Northern, and much of North Western and Eastern Provinces, more than 8 hours; and in far-flung areas of Western, Luapula, and Northern Provinces, it takes more than a day (24 hours) to reach Lusaka.

Travel times to cities of more than 100,000 are between 1 and 3 hours from most urban centers and their environs. For the majority of the rural

FIGURE 9.8 Travel time to urban areas of various sizes in Zambia, circa 2000

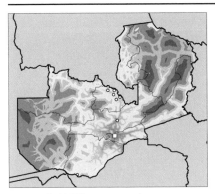

To cities of 500,000 or more people

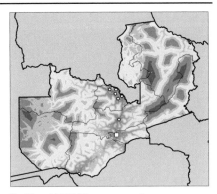

To cities of 100,000 or more people

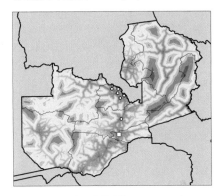

To towns and cities of 25,000 or more people

To towns and cities of 10,000 or more people

- Urban location
- < 1 hour
- 1–3 hours
- 3–5 hours
- 5–8 hours
- 8–11 hours
- 11–16 hours
- 16–26 hours
- > 26 hours

Source: Authors' calculations.

population, the travel time is more than 3 hours. For rural farmers from Western Province and parts of North Western, Eastern, and Northern Provinces the travel time is more than 8 hours. To towns of more than 25,000 or 10,000 people, the travel time is significantly shorter for most rural areas except for parts of Western, Northern, and Eastern Provinces.

Agriculture in Zambia is thus significantly limited by poor transportation infrastructure, which inhibits access to markets as well as delivery of farming implements and inputs.

Agriculture Overview

Tables 9.3 and 9.4 show key agricultural commodities ranked by area harvested and use as food for human consumption (ranked by weight). During 2006–2008, maize—the main staple foodcrop in Zambia—covered the largest harvest area, more than 700,000 hectares and almost 50 percent of the total cropped area. Most subsistence farmers grow maize.

The other significant agricultural commodities are cassava, with 187,000 hectares (13.1 percent of cropped area); seed cotton, at 118,000 hectares (8.3 percent); and groundnuts, at 91,000 hectares (6.4 percent). The other agricultural commodities each had less than 50,000 hectares of harvest area (less than 5 percent).

TABLE 9.3 Harvest area of leading agricultural commodities in Zambia, 2006–2008 (thousands of hectares)

Rank	Crop	Percent of total	Harvest area
	Total	100.0	1,426
1	Maize	49.9	711
2	Cassava	13.1	187
3	Seed cotton	8.3	118
4	Groundnuts	6.4	91
5	Other pulses	3.4	49
6	Tobacco	3.2	45
7	Millet	3.1	45
8	Other fresh vegetables	2.4	34
9	Sorghum	1.7	25
10	Sugarcane	1.7	24

Source: FAOSTAT (FAO 2010).
Note: All values are based on the three-year average for 2006–2008.

TABLE 9.4 Consumption of leading food commodities in Zambia, 2003–2005 (thousands of metric tons)

Rank	Crop	Percent of total	Food consumption
	Total	100.0	3,747
1	Maize	35.2	1,317
2	Cassava	25.1	940
3	Fermented beverages	6.7	250
4	Other vegetables	5.6	208
5	Wheat	4.8	179
6	Sugar	3.6	135
7	Other fruits	2.5	94
8	Freshwater fish	1.8	68
9	Beef	1.5	58
10	Sweet potatoes	1.5	55

Source: FAOSTAT (FAO 2010).
Note: All values are based on the three-year average for 2003–2005.

The country's high level of dependence on maize makes rural communities highly vulnerable to climate variability and change. Maize is highly vulnerable to climate fluctuations, because its phenology is sensitive to changes in moisture and temperature. Adapting to changes in climatic patterns is especially difficult because the majority of the rural farming communities have limited access to capital, technology, and extension services. The few extension officers are normally based near urban centers, which are distant from most rural farming communities.

During 2003–2005, maize was the leading food commodity consumed in Zambia, at 1,317,000 tons per year (35.2 percent of total food consumption). The second most consumed food commodity was cassava, at 940,000 tons (25.1 percent).

Figure 9.9 shows that, for the year 2000, the yield of rainfed maize was rather low, mostly in the ranges 0.5–1.0 ton per hectare and 1–2 tons per hectare. Some areas, especially in Western, Luapula, and Northern Provinces, recorded yields of less than 0.5 ton per hectare. The yield patterns indicate that there is low maize productivity over much of the country. The main areas of maize cultivation are concentrated in the eastern and southern parts of the country, while other areas (for instance, Western, North Western, Lusaka, and Central Provinces) also show large areas of maize cultivation. Although maize is the staple food, the rest of the country has minimal cultivation of maize.

FIGURE 9.9 Yield (metric tons per hectare) and harvest area density (hectares) for rainfed maize in Zambia, 2000

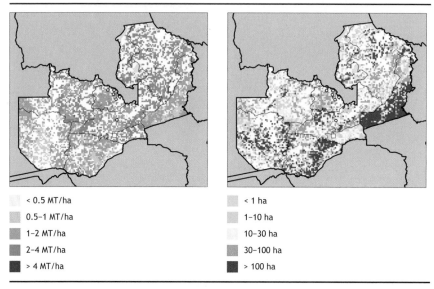

< 0.5 MT/ha	< 1 ha
0.5–1 MT/ha	1–10 ha
1–2 MT/ha	10–30 ha
2–4 MT/ha	30–100 ha
> 4 MT/ha	> 100 ha

Sources: SPAM (Spatial Production Allocation Model) (You and Wood 2006; You, Wood, and Wood-Sichra 2006, 2009).
Note: ha = hectare; MT/ha = metric tons per hectare.

Rainfed cassava is produced mainly in Luapula, Northern, and North Western Provinces (Figure 9.10), with yields mostly in the ranges 4–7 tons per hectare and 7–10 tons per hectare. The rest of the country has very low cassava yields.

Although rainfed cotton (Figure 9.11) is a cash crop, it has very low yields and limited growing area. Much of the country has yields of less than 0.5 ton per hectare and less than 1 hectare average growing area. This indicates that farming in Zambia is largely subsistence, focusing on foodcrops and only marginally growing cash crops.

Rainfed groundnuts (Figure 9.12) also have very low yields and small growing areas. In all nine provinces, yields are mostly less than 0.5 ton per hectare. Eastern Province has the largest growing area, 30–100 hectares. Parts of Southern, Central, Luapula, and Northern Provinces also have some areas under groundnut cultivation. Western and North Western Provinces have the least growing area.

The discussion of the yields and growing areas of key crops in Zambia underlines the lack of crop diversity, which renders rural agriculture highly vulnerable to climate variability and change. Changes in weather patterns can mean total crop failure for maize, the main crop of the country's rural farming community, exacerbating food insecurity and poverty.

FIGURE 9.10 Yield (metric tons per hectare) and harvest area density (hectares) for rainfed cassava in Zambia, 2000

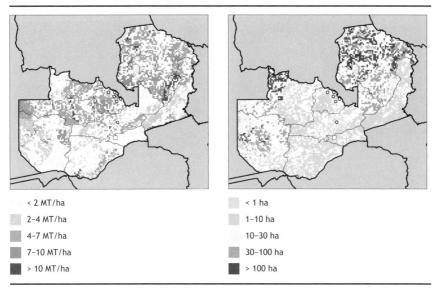

< 2 MT/ha
2–4 MT/ha
4–7 MT/ha
7–10 MT/ha
> 10 MT/ha

< 1 ha
1–10 ha
10–30 ha
30–100 ha
> 100 ha

Sources: SPAM (Spatial Production Allocation Model) (You and Wood 2006; You, Wood, and Wood-Sichra 2006, 2009).
Note: ha = hectare; MT/ha = metric tons per hectare.

FIGURE 9.11 Yield (metric tons per hectare) and harvest area density (hectares) for rainfed cotton in Zambia, 2000

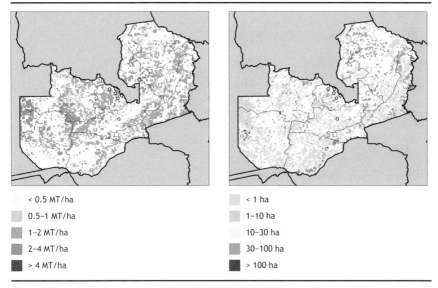

< 0.5 MT/ha
0.5–1 MT/ha
1–2 MT/ha
2–4 MT/ha
> 4 MT/ha

< 1 ha
1–10 ha
10–30 ha
30–100 ha
> 100 ha

Sources: SPAM (Spatial Production Allocation Model) (You and Wood 2006; You, Wood, and Wood-Sichra 2006, 2009).
Note: ha = hectare; MT/ha = metric tons per hectare.

FIGURE 9.12 Yield (metric tons per hectare) and harvest area density (hectares) for rainfed groundnuts in Zambia, 2000

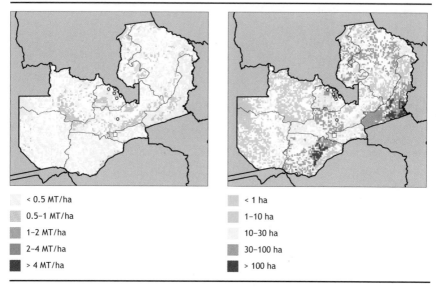

< 0.5 MT/ha
0.5–1 MT/ha
1–2 MT/ha
2–4 MT/ha
> 4 MT/ha

< 1 ha
1–10 ha
10–30 ha
30–100 ha
> 100 ha

Sources: SPAM (Spatial Production Allocation Model) (You and Wood 2006; You, Wood, and Wood-Sichra 2006, 2009).
Note: ha = hectare; MT/ha = metric tons per hectare.

Scenarios for the Future

Economic and Demographic Indicators

Population

Figure 9.13 shows population projections for Zambia for 2050. All three scenarios show a dramatic increase in population, which is projected to grow from less than 14 million in 2010 to much higher in 2050. The high variant scenario shows the population reaching more than 33 million by 2050, while the low variant scenario projects an increase to 26 million and the medium variant projects slightly fewer than 30 million.

The increase in population will imply an increased demand for food and thus food production (mainly maize and cassava), adding to the challenge of adapting Zambian agricultural production to future climate changes.

Economy

Figure 9.14 shows projections of GDP per capita for Zambia by 2050. They are derived by combining three GDP scenarios with the three population scenarios

FIGURE 9.13 Population projections for Zambia, 2010–2050

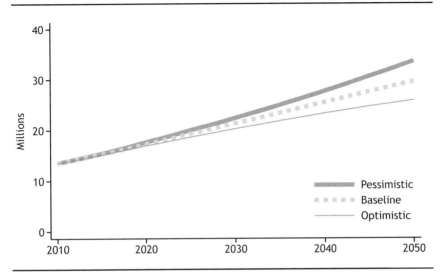

Source: UNPOP (2009).

FIGURE 9.14 Gross domestic product (GDP) per capita in Zambia, future scenarios, 2010–2050

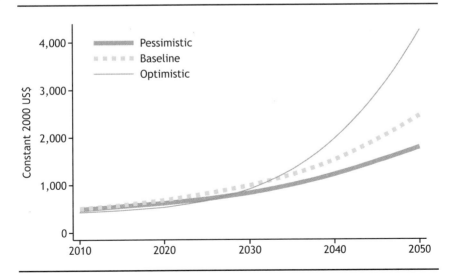

Sources: Computed from GDP data from the World Bank Economic Adaptation to Climate Change project (World Bank 2010), from the Millennium Ecosystem Assessment (2005) reports, and from population data from the United Nations (UNPOP 2009).
Note: US$ = US dollars.

of Figure 9.13 (based on United Nations population data). The optimistic scenario combines high GDP growth with low population growth, the baseline scenario combines medium GDP growth with medium population growth, and the pessimistic scenario combines low GDP growth with high population growth. (The agricultural modeling in the next section uses these scenarios as well.)

All scenarios for GDP per capita seem to show a much brighter future for Zambia. The pessimistic scenario shows that between 2010 and 2050, GDP per capita will increase by 260 percent. The baseline scenario predicts almost a fivefold increase, and the optimistic scenario predicts an almost tenfold increase.

Biophysical Analysis

Climate Models

The maps in Figure 9.15 show the precipitation changes projected by the four general circulation models (GCMs) used in this study, calculated using the A1B scenario.[2] The four GCMs that were used differ dramatically in their results. CNRM-CM3 shows an increase in mean annual precipitation in very small pockets of the western region and an increase of up to 200 mm for the extreme northern parts. CSIRO Mark 3 generally shows no significant change over most of the country but an annual increase of 50–100 millimeters for parts of Western Province (an area of currently low to medium rainfall) and a decrease of 50–100 millimeters for parts of North Western and Copperbelt Provinces (areas of high rainfall). ECHAM 5 shows a decrease in precipitation over practically the entire country. MIROC 3.2 shows an increase for most parts of the country but little change for the extreme southern region.

Figure 9.16 shows changes in the mean daily maximum temperature for the month with the highest mean daily maximum temperature. All four models show an increase in normal annual temperature, but they differ regarding the magnitude of the increase. CNRM-CM3 and ECHAM 5 show an increase of between 2.0° and 3.5°C, with temperatures in ECHAM 5 the higher of the two, while CSIRO Mark 3 and MIROC 3.2 show a smaller increase, between 1.0° and 2.0°C.

---

2 The A1B scenario describes a world of very rapid economic growth, low population growth, and rapid introduction of new and more efficient technologies with moderate resource use and a balanced use of technologies. CNRM-CM3 is National Meteorological Research Center–Climate Model 3. CSIRO is a climate model developed at the Australia Commonwealth Scientific and Industrial Research Organisation. ECHAM 5 is a fifth-generation climate model developed at the Max Planck Institute for Meteorology in Hamburg. MIROC is the Model for Interdisciplinary Research on Climate, developed at the University of Tokyo Center for Climate System Research.

FIGURE 9.15 Changes in mean annual precipitation in Zambia, 2000–2050, A1B scenario (millimeters)

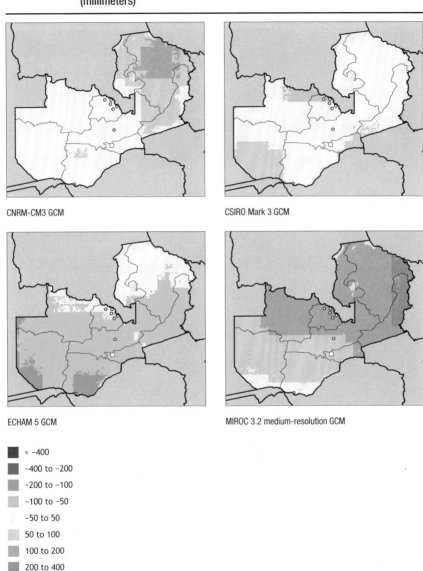

Source: Authors' calculations based on Jones, Thornton, and Heinke (2009).
Notes: A1B = greenhouse gas emissions scenario that assumes fast economic growth, a population that peaks midcentury, and the development of new and efficient technologies, along with a balanced use of energy sources; CNRM-CM3 = National Meteorological Research Center–Climate Model 3; CSIRO = climate model developed at the Australia Commonwealth Scientific and Industrial Research Organisation; ECHAM 5 = fifth-generation climate model developed at the Max Planck Institute for Meteorology (Hamburg); GCM = general circulation model; MIROC = Model for Interdisciplinary Research on Climate, developed by the University of Tokyo Center for Climate System Research.

FIGURE 9.16 Changes in monthly mean maximum daily temperature in Zambia for the warmest month, 2000–2050, A1B scenario (°C)

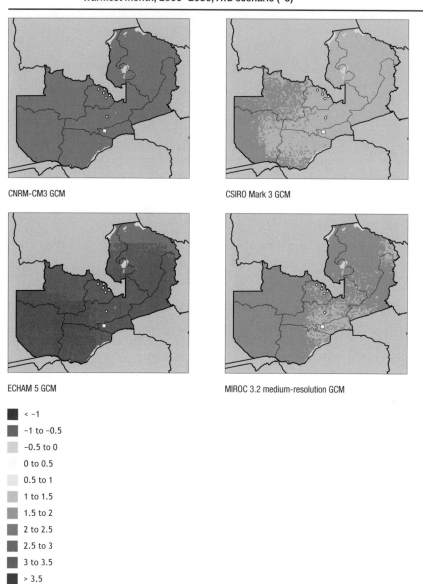

CNRM-CM3 GCM

CSIRO Mark 3 GCM

ECHAM 5 GCM

MIROC 3.2 medium-resolution GCM

- < –1
- –1 to –0.5
- –0.5 to 0
- 0 to 0.5
- 0.5 to 1
- 1 to 1.5
- 1.5 to 2
- 2 to 2.5
- 2.5 to 3
- 3 to 3.5
- > 3.5

Source: Authors' calculations based on Jones, Thornton, and Heinke (2009).
Notes: A1B = greenhouse gas emissions scenario that assumes fast economic growth, a population that peaks midcentury, and the development of new and efficient technologies, along with a balanced use of energy sources; CNRM-CM3 = National Meteorological Research Center–Climate Model 3; CSIRO = climate model developed at the Australia Commonwealth Scientific and Industrial Research Organisation; ECHAM 5 = fifth-generation climate model developed at the Max Planck Institute for Meteorology (Hamburg); GCM = general circulation model; MIROC = Model for Interdisciplinary Research on Climate, developed by the University of Tokyo Center for Climate System Research.

These differences in modeled outcomes highlight the need to consider more than one model in designing adaptation and mitigation strategies and options for climate change.

Crop Models

The Decision Support Software for Agrotechnology Transfer crop model was used to compare future yields (under four future climate scenarios) for the year 2050 with the baseline yield (with unchanged 2000 climate). For all locations, crop variety, soil, and management practices were held constant. Figure 9.17 shows the comparison between maize crop yields for 2050 with climate change and yields with an unchanged (2000) climate. All four models show similar results, though not identical. We note yield gain for maize in Western province, the eastern half of North Western Province, Copperbelt Province, and most of Northern and Luapula Provinces. Some of these yield gains exceed 25 percent. But we also note losses in Southern Province and parts of Eastern Province, and others scattered throughout the country. A small portion of the losses exceed 25 percent in each of the GCMs.

When we compare Figure 9.17 to Figure 9.9, we see that the current distribution of maize is concentrated in the areas that will likely decrease in yield with climate change rather than in the areas that are likely to increase in yield. This may suggest that policymakers should anticipate that farmers may put pressure on opening new agricultural land in areas that are not currently being used and that other areas may be abandoned or farmers may switch from planting maize to planting other crops that would do better in the future climate.

Well-being Scenarios

Figure 9.18 shows the impact of future GDP and population scenarios on the number of malnourished children under age five; Figure 9.19 shows the share of such children. Malnutrition levels among children under five years of age are shown increasing initially through 2025 and thereafter decreasing. With population increases, the increase in numbers of malnourished children through 2025 represents a decline in the proportion of children under five who are malnourished. The optimistic scenario shows a very steep decline after 2025, from about 600,000 malnourished children in 2025 to fewer than 200,000 in 2050. The pessimistic scenario indicates a slow rate of decrease, from 600,000 in 2025 to 450,000 in 2050. The baseline scenario shows an intermediate rate of decrease, from 600,000 in 2025 to 400,000 in 2025.

FIGURE 9.17 Yield change under climate change: Rainfed maize in Zambia, 2000–2050, A1B scenario

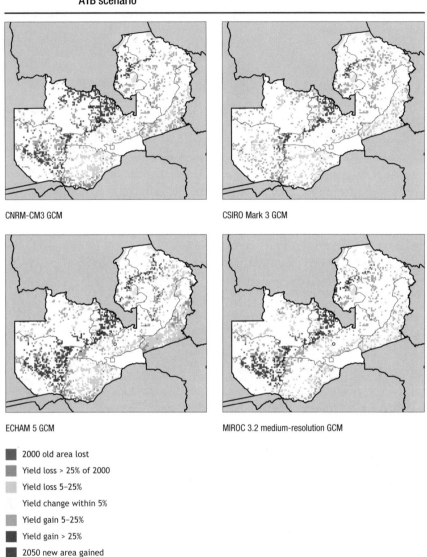

CNRM-CM3 GCM

CSIRO Mark 3 GCM

ECHAM 5 GCM

MIROC 3.2 medium-resolution GCM

- 2000 old area lost
- Yield loss > 25% of 2000
- Yield loss 5–25%
- Yield change within 5%
- Yield gain 5–25%
- Yield gain > 25%
- 2050 new area gained

Source: Authors' calculations.

Notes: A1B = greenhouse gas emissions scenario that assumes fast economic growth, a population that peaks midcentury, and the development of new and efficient technologies, along with a balanced use of energy sources; CNRM-CM3 = National Meteorological Research Center–Climate Model 3; CSIRO = climate model developed at the Australia Commonwealth Scientific and Industrial Research Organisation; ECHAM 5 = fifth-generation climate model developed at the Max Planck Institute for Meteorology (Hamburg); GCM = general circulation model; MIROC = Model for Interdisciplinary Research on Climate, developed by the University of Tokyo Center for Climate System Research.

ZAMBIA 277

FIGURE 9.18 Number of malnourished children under five years of age in Zambia in multiple income and climate scenarios, 2010–2050

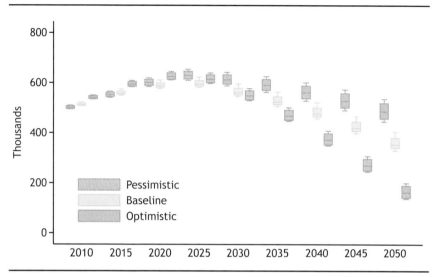

Source: Based on analysis conducted for Nelson et al. (2010).
Note: The box and whiskers plot for each socioeconomic scenario shows the range of effects from the four future climate scenarios.

FIGURE 9.19 Share of malnourished children under five years of age in Zambia in multiple income and climate scenarios, 2010–2050

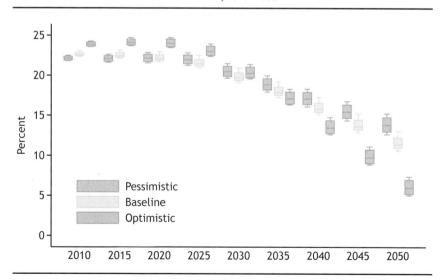

Source: Based on analysis conducted for Nelson et al. (2010).
Note: The box and whiskers plot for each socioeconomic scenario shows the range of effects from the four future climate scenarios.

FIGURE 9.20 Kilocalories per capita in Zambia in multiple income and climate scenarios, 2010–2050

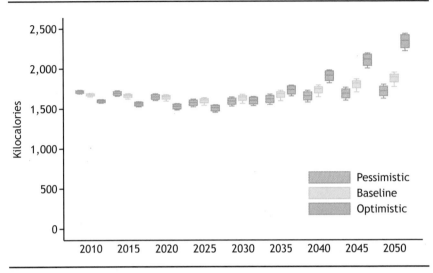

Source: Based on analysis conducted for Nelson et al. (2010).
Note: The box and whiskers plot for each socioeconomic scenario shows the range of effects from the four future climate scenarios.

Figure 9.20 shows the available kilocalories per capita. For all three scenarios, the model shows a decline in availability of kilocalories from 2010 to 2025 and a slight increase thereafter. The increase is most significant in the optimistic scenario and only moderate in the baseline and pessimistic scenarios. Overall, the increasing availability of kilocalories per capita tracks with the declining trend for levels of malnourished children under age five.

Agricultural Scenarios

Simulation results from the International Model for Policy Analysis of Agricultural Commodities and Trade (IMPACT) for key agricultural crops in Zambia are shown in Figures 9.21–9.23. Graphs of five parameters (production, yield, area, net exports, and world price) have been plotted for each featured crop.

Figure 9.21 shows the production and yield of maize increasing throughout 2010–2050. The projected increase in GDP would allow increased investment in agriculture, enabling farmers to adopt improved farming management systems and practices and improving maize yields. The projected increase in

FIGURE 9.21 Impact of changes in GDP and population on maize in Zambia, 2010–2050

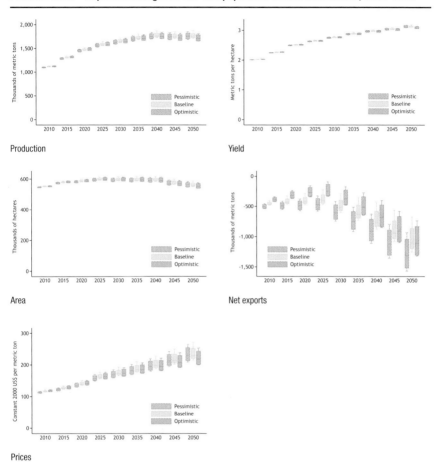

Source: Based on analysis conducted for Nelson et al. (2010).
Notes: The box and whiskers plot for each socioeconomic scenario shows the range of effects from the four future climate scenarios. GDP = gross domestic product; US$ = US dollars.

population would mean increased demand for maize as a staple food, stimulating increased production.

Nevertheless, the harvested area of maize shows a rather static trend. Net exports of maize are already negative, which means that Zambia is importing maize. IMPACT shows that imports of maize will increase, implying that the increased demand from the larger population will outstrip the increased supply. Although the doubling of maize prices can cause some concerns about the ability of people to pay for basic food, as we saw in earlier figures, income will

FIGURE 9.22 Impact of changes in GDP and population on cassava in Zambia, 2010–2050

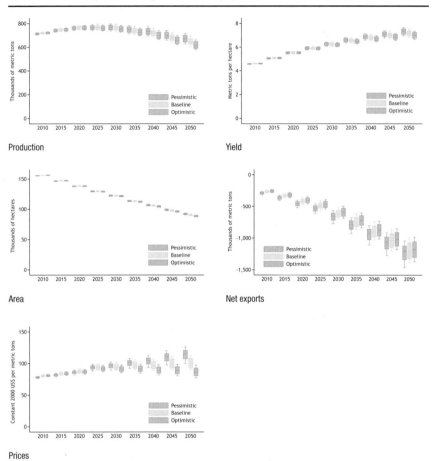

Source: Based on analysis conducted for Nelson et al. (2010).
Notes: The box and whiskers plot for each socioeconomic scenario shows the range of effects from the four future climate scenarios. GDP = gross domestic product; US$ = US dollars.

increase by much more, and the projected increase in consumption of calories indicates that people will be better off nutritionally.

For cassava, IMPACT anticipates that yields will increase by around 50 percent between 2010 and 2050; yet, due to declines of the harvested area by almost half, production will change little, showing some rise in the first half of the period and a slightly greater decline in the second half (see Figure 9.22). The net impact is that increased demand will cause cassava imports to increase by around 1 million metric tons over the period. The

FIGURE 9.23 Impact of changes in GDP and population on cotton in Zambia, 2010–2050

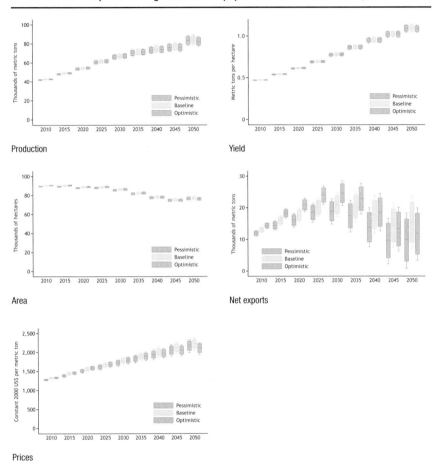

Source: Based on analysis conducted for Nelson et al. (2010).
Notes: The box and whiskers plot for each socioeconomic scenario shows the range of effects from the four future climate scenarios. GDP = gross domestic product; US$ = US dollars.

world price of cassava is projected to increase, but only very slightly, and in some cases not at all.

For cotton, the models shows an increase in production and yield in all three scenarios. The harvested area will decline slightly. Net cotton exports will clearly rise through 2030, then decline thereafter, though the models show mixed results as to whether they will fall to levels below those in 2010 (see Figure 9.23). The world price of cotton is shown to increase by around 80 percent between 2010 and 2050.

Summary

From 1960 to 2008, the population in Zambia increased by more than 100 percent, from less than 5 million in 1960 to about 10.5 million in 2008. The urban population of Zambia grew over the same period at an average decadal rate of 4.32 percent, accompanied by a steady increase in the rural population. With lagging economic growth, Zambia experienced a declining trend in GDP per capita, from about $550 in 1960 to less than $400 in 2008. These figures indicate that the economic environment has been challenging for the greater population of Zambia.

In the context of this overall challenging economic environment, several additional factors have compounded the effects on the living experience of many people considered vulnerable, particularly in sectors such as agriculture. The rural farming community, in particular, is seen to be prone to vulnerability, especially when one considers the impacts of climate change. The factors related to vulnerability are as follows:

- Low literacy levels. Although the primary school enrollment rate is high, at 119 percent, the secondary school enrollment rate is low, at 43.1 percent, implying a high primary school dropout rate.
- Low health status. Life expectancy in Zambia has remained low, around 45 years.
- High poverty levels. In all the provinces, 50 percent or more are living on less than $2 per day.
- Unsustainable land cover and land use patterns. Activities such as expanding cultivation and settlements, illegal timber cutting and logging, uncontrolled fires, poaching, and fuelwood and charcoal burning weaken the resilience of forest and shrub-cover ecosystems to climate variability and change, compromising their vital functions as water recharge areas and carbon sinks.
- A poor transport network. Most parts of Zambia have long travel times to markets for their agricultural produce, discouraging rural farming communities from expanding beyond small-scale agriculture.
- A monoculture farming system. Because most farmers are growing maize, a maize crop failure due to changes in weather patterns would essentially mean total crop failure for the country.

- The extensiveness of the farming system: Yield patterns show low productivity for maize over much of the country, so production depends on cultivating large areas. Maize production will be markedly reduced if the area cultivated is restricted by climate impacts.

The models used to develop future scenarios for agriculture show evident disagreement on several measures in terms of sign (positive or negative) as well as magnitude, as summarized below:

- Population. All three scenarios show a dramatic increase in population, which is projected to reach between 25 million and 34 million in 2050.
- Income. The scenarios for GDP per capita indicate a projected increase under all three future scenarios.
- Mean annual precipitation. The four GCMs differ dramatically in their results regarding mean annual precipitation, ranging from an overall decrease in precipitation to significant and widespread increases.
- Normal annual maximum temperature. All four models show an increase in normal annual temperature in the ranges of 1.0°–2.0°C or even 2.0°–3.5°C.
- Yield change for maize under climate change. All four models show a yield *gain* of sometimes more than 25 percent in the northern region and a yield *loss* of sometimes more than 25 percent over the rest of the country.
- Nutrition. Malnutrition among children under five years of age is shown initially increasing through 2025 and decreasing thereafter; conversely, the availability of kilocalories per capita is shown decreasing through 2025 and increasing thereafter.
- Impact of change on crops. Both the production and yield of maize are shown increasing; however, the harvest area of maize is projected to remain unchanged. Net exports of *all* modeled crops are shown declining (only cotton shows positive net exports), while all world prices are shown increasing.

Because of these differences among models, it is essential to consider a range of model outputs for policy, planning, and decisionmaking purposes and to bear in mind both the pessimistic and the optimistic scenarios when designing adaptation and mitigation measures.

Conclusions and Policy Recommendations

Agriculture in Zambia is mostly small scale and rainfed and thus is vulnerable to climate variability and change. The high dependence on maize makes rural communities especially vulnerable, because maize is highly sensitive to changes in moisture and temperature. Moreover, implementing adaptive measures is especially difficult because most rural farming communities have limited access to capital, technology, and extension services.

An integrated approach with multisectoral participation will be essential to develop effective climate change adaptation and mitigation strategies. In view of the uncertainties in the modeled future results, it is advisable to base decisionmaking on a range of scenarios.

It would be particularly helpful for more work to be done in examining the potential impacts of climate change on additional crop varieties as well as livestock. Rice and sweet potatoes are important among rural farming communities in Western Province; cattle raising is part of the farming culture in Southern and Western Provinces; and chickens, goats, and pigs are kept in almost all provinces, especially North Western Province.

We recommend that the following steps be taken in the areas indicated:

Research

- Develop and promote early-maturing and heat-resistant crop varieties.
- Invest in agricultural research on climate-resistant crop varieties, water harvesting, irrigation schemes, and water rights.
- Implement feasibility studies on the migration of farmers to agroecological regions with better potential for agriculture.

Farm Level

- Diversify from traditional maize crops to other crops that can withstand droughts, floods, and high temperatures, such as sorghum and millet.
- Rotate land use between crops and livestock to replenish soil nutrients.
- Promote conservation farming, potholing, organic fertilization through intercropping, and agroforestry.

National Level

- Introduce incentives to promote the adoption of new crop varieties in nontraditional farming areas: for example, maize in North Western, Luapula, and Northern Provinces (with high rainfall) and cassava, sorghum, and millet in the drought-prone central and southern parts of Zambia.

- Provide rural credit facilities to enable subsistence farmers to buy new varieties of seeds and other inputs.
- Disseminate information to farmers on various adaptation options through extension services.
- Remove subsidies on crops such as maize that do not perform well in a changing climate.
- Formulate appropriate policies for marketing agricultural inputs and products, geared to the needs of subsistence farmers in particular, because this most vulnerable group makes up almost 80 percent of the farming community.
- Promote efficient use of river basins for crop production.
- Invest in technological innovations such as seed banks.
- Provide opportunities for alternative rural employment in nonfarming activities—for example, beekeeping and honey making—to enable rural farmers to make a livelihood.
- Strengthen early warning systems and systems for disaster risk reduction.
- Promote communication and the dissemination of early warning information to rural farming communities, specifically on climate issues. This could be done through expanding and strengthening the radio-Internet system (RANET), which enables community radio stations to access satellite-based climate information and broadcast it in local languages.
- Promote the participation of nongovernmental organizations and the private sector in the implementation of climate change adaptation strategies for agriculture, especially in rural areas.
- Encourage and promote an integrated approach to supporting agriculture that involves allied service providers, including the Department of Water Affairs, the Meteorological Department, local authorities, roads departments, and social service providers (education and health institutions).

Although climate change will present some challenges to Zambian farmers, implementation of the above recommendations should enable them to adapt well to the potential problems identified in this chapter.

References

Bartholome, E., and A. S. Belward. 2005. "GLC2000: A New Approach to Global Land Cover Mapping from Earth Observation Data." *International Journal of Remote Sensing* 26 (9): 1959–1977.

CIESIN (Center for International Earth Science Information Network), Columbia University, IFPRI (International Food Policy Research Institute), World Bank, and CIAT (Centro Internacional de Agricultura Tropical). 2004. *Global Rural–Urban Mapping Project (GRUMP), Alpha Version: Population Density Grids*. Palisades, NY, US: Socioeconomic Data and Applications Center (SEDAC), Columbia University. http://sedac.ciesin.columbia.edu/gpw.

FAO (Food and Agriculture Organization of the United Nations). 2010. FAOSTAT. Accessed December 22, 2010. http://faostat.fao.org/site/573/default.aspx#ancor.

Jones, P. G., P. K. Thornton, and J. Heinke. 2009. *Generating Characteristic Daily Weather Data Using Downscaled Climate Model Data from the IPCC's Fourth Assessment*. Project report. Nairobi, Kenya: International Livestock Research Institute.

Lehner, B., and P. Döll. 2004. "Development and Validation of a Global Database of Lakes, Reservoirs, and Wetlands." *Journal of Hydrology* 296 (1–4): 1–22.

Millennium Ecosystem Assessment. 2005. *Ecosystems and Human Well-being: Synthesis*. Washington, DC: Island Press. http://www.maweb.org/en/Global.aspx.

Nelson, G. C., M. W. Rosegrant, A. Palazzo, I. Gray, C. Ingersoll, R. Robertson, S. Tokgoz, et al. 2010. *Food Security, Farming, and Climate Change to 2050: Scenarios, Results, Policy Options*. Washington, DC: International Food Policy Research Institute.

UNEP (United Nations Environment Programme) and IUCN (International Union for the Conservation of Nature). 2009. *World Database on Protected Areas (WDPA): Annual Release 2009*. Accessed 2009. www.wdpa.org/protectedplanet.aspx.

UNPOP (United Nations Department of Economic and Social Affairs–Population Division). 2009. *World Population Prospects: The 2008 Revision*. New York. http://esa.un.org/unpd/wpp/.

Vogel, C., and K. O'Brien. 2003. "Climate Forecasts in Southern Africa." In *Coping with Climate Variability: The Use of Seasonal Forecasts in Southern Africa*, edited by O'Brien K. and C. Vogel. Aldershot, UK: Ashgate.

Wood, S., G. Hyman, U. Deichmann, E. Barona, R. Tenorio, Z. Guo, S. Castano, O. Rivera, E. Diaz, and J. Marin. 2010. "Sub-national Poverty Maps for the Developing World Using International Poverty Lines: Preliminary Data Release." Accessed May 6, 2010. http://povertymap.info. Some also available at http://labs.harvestchoice.org/2010/08/poverty-maps/.

World Bank. 2009. *World Development Indicators*. Washington, DC.

———. 2010. *The Costs of Agricultural Adaptation to Climate Change*. Washington, DC.

You, L., and S. Wood. 2006. "An Entropy Approach to Spatial Disaggregation of Agricultural Production." *Agricultural Systems* 90 (1–3): 329–347.

You, L., S. Wood, and U. Wood-Sichra. 2006. "Generating Global Crop Distribution Maps: From Census to Grid." Paper presented at the International Association of Agricultural Economists Conference in Brisbane, Australia, August 11–18.

———. 2009. "Generating Plausible Crop Distribution and Performance Maps for Sub-Saharan Africa Using a Spatially Disaggregated Data Fusion and Optimization Approach." *Agricultural Systems* 99 (2–3): 126–140.

Chapter 10

ZIMBABWE

Francis T. Mugabe, Timothy S. Thomas, Sepo Hachigonta, and Lindiwe Majele Sibanda

Zimbabwe is a landlocked country with a total area of 390,580 square kilometers. The country shares borders with Mozambique, South Africa, Botswana, Zambia, and Namibia. Before 2000, Zimbabwe was one of the most industrialized economies in Africa south of the Sahara, with an extensive agroprocessing industry and a relatively diversified industrial sector (Carmody 1998). However, political and economic crises during the past decade negatively affected the country's agricultural sector and its economy in general.

Rainfall in Zimbabwe is erratic and ill distributed in time and space, resulting in crop failures three out of every five years. In the drier parts of the country, the coefficient of variation in annual rainfall is between 20 and 40 percent, while the variation in yields ranges between 15 and 60 percent (Lumsden and Schulze 2007). Smallholder farmers living in these marginal areas (Natural Regions IV and V) currently face food insecurity because the changing climate is causing more frequent droughts (Kalanda-Joshua et al. 2011).[1] The group most vulnerable to the effects of climate variability and change may be smallholder farmers in semiarid areas with little ability to adapt to climate variability and change (Boko et al. 2007). Semiarid conditions prevail in 75 percent of Zimbabwe.

Using a Ricardian approach, Mano and Nhemachena (2007) show that temperature and precipitation changes have significant effects on net farm revenue in Zimbabwe.[2] An increase in temperature of 2.5°C would result in a decrease in net farm revenue of $400 million for all farms in Zimbabwe—despite an increase of $300 million for farms with irrigation. A decrease in precipitation of between 7 and 14 percent would result in a decrease in farm revenue of $300 million (Mano and Nhemachena 2007). With climate change, the low-lying areas of

---

1 The natural regions are a means of classifying the agricultural potential of the country, ranging from Natural Region I, representing the high-altitude wet areas, to Natural Region V, which receives low rainfall and is dry.
2 A Ricardian approach entails econometric estimates of the impact of climatic change and other variables on land values (different from the approach used in this study).

southern Zimbabwe are likely to become non-maize-producing areas (Matarira and Makadho 1998), with growing seasons 20–25 percent shorter than at present (Auger 2008; Bwalya 2008; Scholes 2008).

Although there has been extensive research in Zimbabwe on the impacts of climate change on crop yields, there are very few data on the broader economic impacts of climate change on the agricultural sector. The objective of this chapter is to help policymakers, researchers, and country negotiators better understand and anticipate the likely impacts of climate change on agriculture and on vulnerable households in Zimbabwe, given that about 70 percent of the Zimbabwean population is dependent on rainfed agriculture and the sector contributes between 15 and 20 percent of the country's gross domestic product (GDP).

Review of Current Trends

Economic and Demographic Indicators

Population

Figure 10.1 shows the total and rural populations (left axis) as well as the share of the population that is urban (right axis). The urban population was approximately 15 percent of the total population in 1960, increasing to 23 percent in 1980 and 38 percent in 2008. A number of factors contributed to this increase—in particular, the growth in Zimbabwean industry from 1960 to 1980. The level of urbanization was low before 1980 because of prevailing regulations that did not allow women to stay with their partners in urban areas. These regulations were relaxed after Zimbabwe achieved independence in 1980, resulting in an influx of families as wives joined their husbands. Figure 10.2 shows the geographic distribution of the population in 2000.

Table 10.1 shows the total, rural, and urban population growth rates from 1960 to 2008. The rural growth rate remained constant at 2.7 percent from 1960 to 1979, while the urban growth rate decreased from a high of 6.5 percent to 5.9 percent during the same period. After increases from 1980 to 1989, both urban and rural growth rates have declined significantly since 1990. Between 2000 and 2008 the urban growth rate declined to 1.2 percent, while rural growth became negative (−0.7 percent).

The increase in the urban population has been attributed to rural–urban migration in the 1980s as people sought economic opportunities in towns

FIGURE 10.1 Population trends in Zimbabwe: Total population, rural population, and percent urban, 1960–2008

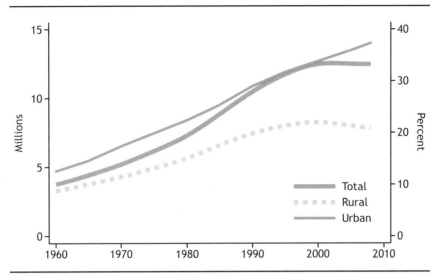

Source: *World Development Indicators* (World Bank 2009).

FIGURE 10.2 Population distribution in Zimbabwe, 2000 (persons per square kilometer)

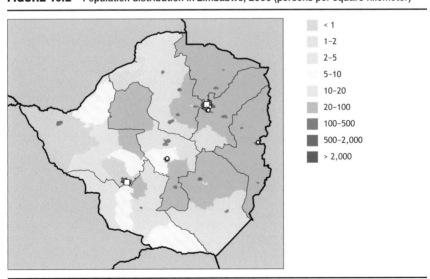

Source: CIESIN et al. (2004).

TABLE 10.1 Population growth rates in Zimbabwe, 1960–2008 (percent)

Decade	Total growth rate	Rural growth rate	Urban growth rate
1960–1969	3.3	2.7	6.5
1970–1979	3.3	2.7	5.9
1980–1989	3.7	2.9	6.3
1990–1999	1.9	1.2	3.4
2000–2008	0.0	–0.7	1.2

Source: Authors' calculations based on *World Development Indicators* (World Bank 2009).

(Potts 2010). After 2000, government policies caused significant shifts in internal migration patterns. Significant numbers moved away from jobs in the commercial farming sector because that sector was profoundly disrupted by fast-tracked land reform. Hundreds of thousands were displaced in the cities in 2005 by Operation Murambatsvina, which resulted in a reduced urban growth rate of 1.2 percent between 2000 and 2008 (see Table 10.1).[3] Many were forced into rural areas in the short term, but because they could not survive there, they subsequently returned to the towns. Figure 10.2 shows the high population densities in urban areas in 2000. The overall rural growth rate between 2000 and 2008 was negative (see Table 10.1), perhaps as a result of people's migrating to neighboring countries looking for work due to economic hardship between 2000 and 2010.

Income

Figure 10.3 shows trends in GDP per capita as well as the proportion of GDP from agriculture. Agriculture is included to indicate its importance or relative contribution to GDP and as a sector vulnerable to the impacts of climate change.

Zimbabwe's GDP increased steadily from 1968 to 1974, declined until 1979, and fluctuated between 1979 and 2000. Since 2000, per capita GDP has declined sharply, to $450 in 2008. The percentage of GDP from agriculture has fluctuated widely from the mid-1970s onward, averaging around 15–20 percent during the same period. The lowest proportion of GDP from agriculture has been associated with dry years—1991/1992, 1992/1993, and 2001/2002. This is consistent with Richardson's findings (2007) demonstrating a relationship between GDP and rainfall in Zimbabwe.

---

3 An operation of the Zimbabwean government to "clean up" the cities.

FIGURE 10.3 Per capita GDP in Zimbabwe (constant 2000 US$) and share of GDP from agriculture, 1960–2008 (percent)

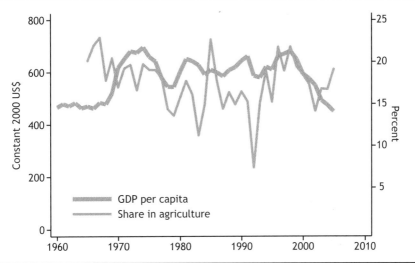

Source: *World Development Indicators* (World Bank 2009).
Note: GDP = gross domestic product; US$ = US dollars.

Clime (2007) predicted a climate change–related decline in Zimbabwean agricultural output by 2080—a decline of 29 percent with carbon fertilization and 37.9 percent without. These estimates are based on 2003 agricultural output and crops and assume an 11 percent reduction in agricultural export earnings because of reduced agricultural output.

Vulnerability to Climate Change

Table 10.2 provides some data for Zimbabwe on several indicators of a population's vulnerability and resiliency to economic and natural shocks: level of education, literacy, and concentration of labor in poorer or less dynamic sectors. The level of the country's primary school enrollment was high (101 percent) between 2005 and 2007, while its level of secondary school enrollment was low (40 percent).[4] Primary school is almost free in Zimbabwe. The other reason for high levels of primary school enrollment during this period was

---

4 The percentage of primary school enrollment may be greater than 100 because some individuals are enrolled in primary school who are older than primary school age.

TABLE 10.2 Education and labor statistics for Zimbabwe, 1900s and 2000s

Indicator	Year	Value (percent)
Primary school enrollment (percent gross, three-year average)	2006	101.2
Secondary school enrollment (percent gross, three-year average)	2006	40.0
Adult literacy rate	2007	91.2
Percent employment in agriculture	1999	60.0
Percent with vulnerable employment (own farm or day labor)	2002	61.9
Under-five malnutrition (weight for age)	2006	14.0

Source: *World Development Indicators* (World Bank 2009).

probably the feeding scheme administered by nongovernmental organizations during economic hard times in Zimbabwe, especially in the semiarid areas where drought was severe.

Figure 10.4 shows two noneconomic correlates of poverty: life expectancy and under-five mortality. Life expectancy at birth in Zimbabwe increased from 50 years to 60 years between 1960 and the late 1980s; thereafter it began to decline, reaching its lowest point—40 years—in 2004 with the onset of the HIV/AIDS pandemic. The subsequent rise in life expectancy from

FIGURE 10.4 Well-being indicators in Zimbabwe, 1960–2008

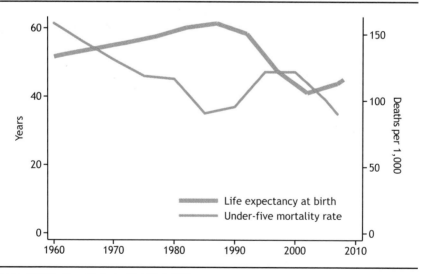

Source: *World Development Indicators* (World Bank 2009).

2004 to 2010 might have been due to the increased availability of antiretroviral medications as well as education on HIV/AIDS. Generally the percentage of people living with AIDS in Zimbabwe has decreased in recent years. The under-five mortality rate has shown a similar trend: there was a steady decrease from 1960 to 1985, followed by a sharp increase to 1998, then by another decline from 2000.

Figure 10.5 shows land cover and land use in Zimbabwe as of 2000. Agricultural activities are taking place in most parts of the country despite high variability in rainfall, except for the Gonarezhou and Whange game reserves and the Zimbabwean Great Dyke (probably because of the toxic minerals associated with the Dyke).[5]

Figure 10.6 shows the locations of protected areas, including parks and reserves. These locations provide important protection for fragile environments, which may also be important for tourism. A few of the protected areas are shown to be cultivated.

Figure 10.7 shows travel times to urban areas of various sizes, which are potential markets for agricultural products and sources of agricultural inputs and consumer goods for rural households. Except for the southeastern and northwestern regions, the travel time to cities of 100,000 people or more is 3 hours or less. The travel time to cities of more than 10,000 people is only 1–3 hours for most parts of the country. This is a result of several growth or service points established since independence as Zimbabwe decentralized its crop marketing boards to ensure that agricultural products can easily reach markets.[6]

Agriculture Overview

Tables 10.3 and 10.4 show key agricultural commodities in terms of area harvested and the provision of food for human consumption (ranked by weight). The commodities of significance to the poor, who are most vulnerable to climate change, are maize, sorghum, millet, and groundnuts. Maize is the staple food of Zimbabwe and is grown by all small-scale farmers for food security. Grain production has decreased—in part because the commercial farmers

---

5   The Dyke consists of mineral-bearing mountains that cut across Zimbabwe from northeast to southwest.
6   Most of these growth or service points are capitals of communal lands. They are generally underdeveloped and receive additional resources and incentives from the government to encourage their development into proper towns in their own right.

FIGURE 10.5 Land cover and land use in Zimbabwe, 2000

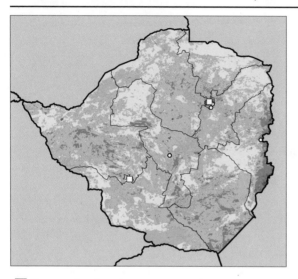

- Tree cover, broadleaved, evergreen
- Tree cover, broadleaved, deciduous, closed
- Tree cover, broadleaved, open
- Tree cover, broadleaved, needle-leaved, evergreen
- Tree cover, broadleaved, needle-leaved, deciduous
- Tree cover, broadleaved, mixed leaf type
- Tree cover, broadleaved, regularly flooded, fresh water
- Tree cover, broadleaved, regularly flooded, saline water
- Mosaic of tree cover/other natural vegetation
- Tree cover, burnt
- Shrub cover, closed-open, evergreen
- Shrub cover, closed-open, deciduous
- Herbaceous cover, closed-open
- Sparse herbaceous or sparse shrub cover
- Regularly flooded shrub or herbaceous cover
- Cultivated and managed areas
- Mosaic of cropland/tree cover/other natural vegetation
- Mosaic of cropland/shrub/grass cover
- Bare areas
- Water bodies
- Snow and ice
- Artificial surfaces and associated areas
- No data

Source: GLC2000 (Bartholome and Belward 2005).

FIGURE 10.6 Protected areas in Zimbabwe, 2009

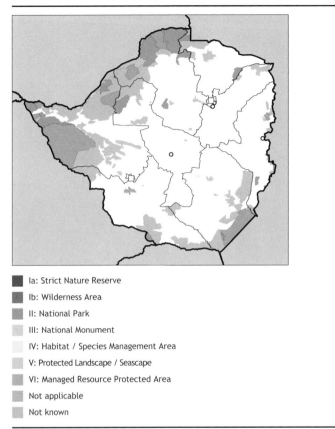

- Ia: Strict Nature Reserve
- Ib: Wilderness Area
- II: National Park
- III: National Monument
- IV: Habitat / Species Management Area
- V: Protected Landscape / Seascape
- VI: Managed Resource Protected Area
- Not applicable
- Not known

Sources: Protected areas are from the World Database on Protected Areas (UNEP and IUCN 2009). Water bodies are from the World Wildlife Fund's Global Lakes and Wetlands Database (Lehner and Döll 2004).

who once grew maize have shifted to such high-value crops as tobacco or have changed to horticulture. As a result, maize production has been left to communal farmers, whose production has gone down in the past few decades because of droughts as well as shortages of inputs (Chaguta 2010).

Figures 10.8–10.12 show the estimated yield and harvest area of key crops: maize, cotton, sorghum, millet, and groundnuts. Although millet and sorghum are drought tolerant, the area planted with them is far less than that planted with maize because Zimbabweans prefer maize as their staple food. Cotton—produced for commercial use—is grown mostly in the southeastern and northwestern regions. However, its prominence is declining because of low market prices.

FIGURE 10.7 Travel time to urban areas of various sizes in Zimbabwe, circa 2000

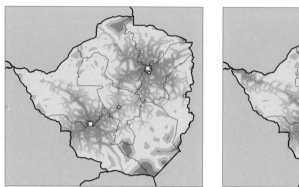
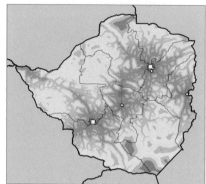

To cities of 500,000 or more people

To cities of 100,000 or more people

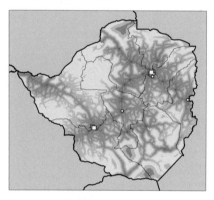
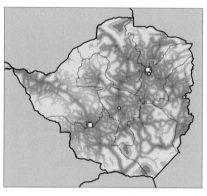

To towns and cities of 25,000 or more people

To towns and cities of 10,000 or more people

- Urban location
- < 1 hour
- 1–3 hours
- 3–5 hours
- 5–8 hours
- 8–11 hours
- 11–16 hours
- 16–26 hours
- > 26 hours

Source: Authors' calculations.

TABLE 10.3 Harvest area of leading agricultural commodities in Zimbabwe, 2006–2008 (thousands of hectares)

Rank	Crop	Percent of total	Harvest area
	Total	100.0	3,147
1	Maize	51.8	1,630
2	Seed cotton	11.3	335
3	Sorghum	8.2	260
4	Millet	6.9	216
5	Groundnuts	6.4	200
6	Soybeans	1.9	61
7	Beans	1.9	60
8	Sunflower seed	1.8	57
9	Tobacco	1.5	47
10	Cassava	1.4	44

Source: FAOSTAT (FAO 2010).
Note: All values are based on the three year average for 2006–2008.

TABLE 10.4 Consumption of leading food commodities in Zimbabwe, 2003–2005 (thousands of metric tons)

Rank	Crop	Percent of total	Food consumption
	Total	100.0	3,614
1	Maize	39.6	1,432
2	Sugar	9.5	345
3	Wheat	7.8	281
4	Cassava	4.9	178
5	Fermented beverages	4.3	156
6	Beer	3.5	126
7	Other vegetables	3.1	112
8	Beef	2.6	93
9	Alcoholic beverages	1.9	70
10	Bananas	1.9	69

Source: FAOSTAT (FAO 2010).
Note: All values are based on the three-year average for 2003–2005.

FIGURE 10.8 Yield (metric tons per hectare) and harvest area density (hectares) for rainfed maize in Zimbabwe, 2000

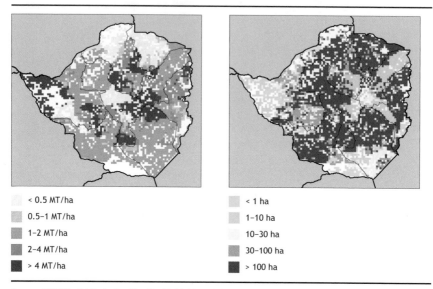

- < 0.5 MT/ha
- 0.5–1 MT/ha
- 1–2 MT/ha
- 2–4 MT/ha
- > 4 MT/ha

- < 1 ha
- 1–10 ha
- 10–30 ha
- 30–100 ha
- > 100 ha

Sources: SPAM (Spatial Production Allocation Model) (You and Wood 2006; You, Wood, and Wood-Sichra 2006, 2009).
Note: ha = hectare; MT/ha = metric tons per hectare.

FIGURE 10.9 Yield (metric tons per hectare) and harvest area density (hectares) for rainfed cotton in Zimbabwe, 2000

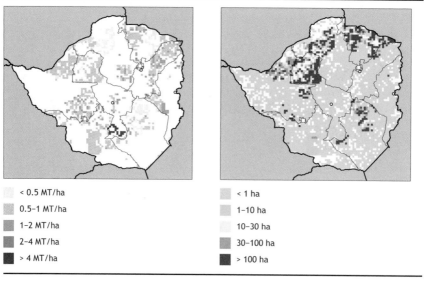

- < 0.5 MT/ha
- 0.5–1 MT/ha
- 1–2 MT/ha
- 2–4 MT/ha
- > 4 MT/ha

- < 1 ha
- 1–10 ha
- 10–30 ha
- 30–100 ha
- > 100 ha

Sources: SPAM (Spatial Production Allocation Model) (You and Wood 2006; You, Wood, and Wood-Sichra 2006, 2009).
Note: ha = hectare; MT/ha = metric tons per hectare.

FIGURE 10.10 Yield (metric tons per hectare) and harvest area density (hectares) for rainfed sorghum in Zimbabwe, 2000

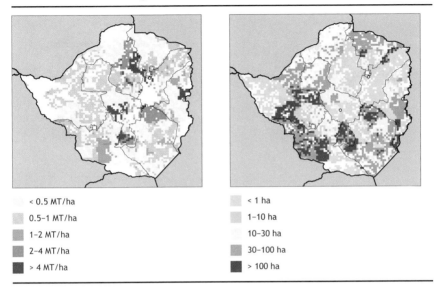

- < 0.5 MT/ha
- 0.5–1 MT/ha
- 1–2 MT/ha
- 2–4 MT/ha
- > 4 MT/ha

- < 1 ha
- 1–10 ha
- 10–30 ha
- 30–100 ha
- > 100 ha

Sources: SPAM (Spatial Production Allocation Model) (You and Wood 2006; You, Wood, and Wood-Sichra 2006, 2009).
Note: ha = hectare; MT/ha = metric tons per hectare.

FIGURE 10.11 Yield (metric tons per hectare) and harvest area density (hectares) for rainfed millet in Zimbabwe, 2000

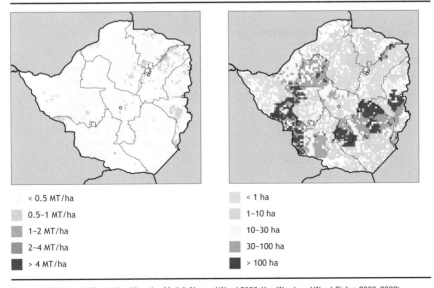

- < 0.5 MT/ha
- 0.5–1 MT/ha
- 1–2 MT/ha
- 2–4 MT/ha
- > 4 MT/ha

- < 1 ha
- 1–10 ha
- 10–30 ha
- 30–100 ha
- > 100 ha

Sources: SPAM (Spatial Production Allocation Model) (You and Wood 2006; You, Wood, and Wood-Sichra 2006, 2009).
Note: ha = hectare; MT/ha = metric tons per hectare.

FIGURE 10.12 Yield (metric tons per hectare) and harvest area density (hectares) for rainfed groundnuts in Zimbabwe, 2000

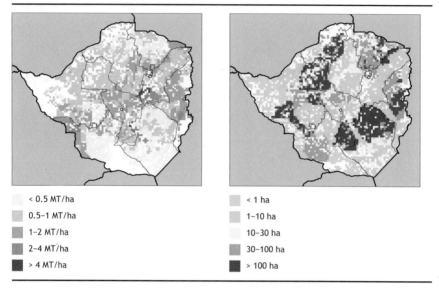

- < 0.5 MT/ha
- 0.5–1 MT/ha
- 1–2 MT/ha
- 2–4 MT/ha
- > 4 MT/ha

- < 1 ha
- 1–10 ha
- 10–30 ha
- 30–100 ha
- > 100 ha

Sources: SPAM (Spatial Production Allocation Model) (You and Wood 2006; You, Wood, and Wood-Sichra 2006, 2009).
Note: ha = hectare; MT/ha = metric tons per hectare.

Scenarios for the Future

Economic and Demographic Indicators

Population

Figure 10.13 shows population projections by the United Nations (UN) Population Division through 2050. The high variant would put enormous pressure on the social amenities of the country, given the current low economic growth rate of less than 2 percent.

Income

Figure 10.14 presents three overall scenarios for GDP per capita derived by combining three GDP scenarios with the three population scenarios of Figure 10.13 (based on UN population data). The optimistic scenario combines high GDP growth with low population growth , the baseline scenario combines medium GDP growth with medium population growth, and the pessimistic scenario combines low GDP growth with high population growth.

Although there is little difference between the baseline and pessimistic scenarios, there is a significant difference between both of these scenarios and the

FIGURE 10.13 Population projections for Zimbabwe, 2010–2050

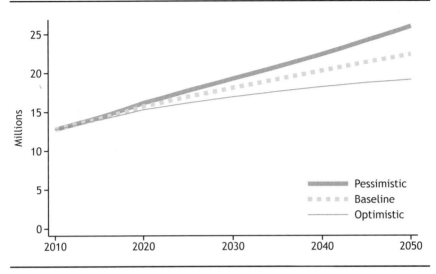

Source: UNPOP (2009).

FIGURE 10.14 Gross domestic product (GDP) per capita in Zimbabwe, future scenarios, 2010–2050

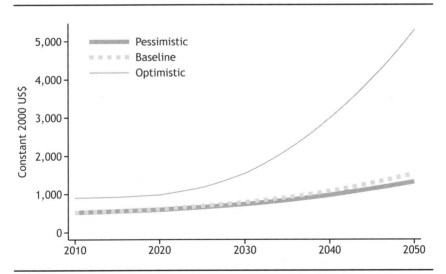

Sources: Computed from GDP data from the World Bank Economic Adaptation to Climate Change project (World Bank 2010b), from the Millennium Ecosystem Assessment (2005) reports, and from population data from the United Nations (UNPOP 2009).
Note: US$ = US dollars.

optimistic scenario. The optimistic scenario is less probable, however; given past economic performance, the baseline is more realistic.

Biophysical Analysis

Climate Models

Figure 10.15 shows projected precipitation changes in Zimbabwe in the four downscaled general circulation models (GCMs) used in this chapter with the A1B scenario.[7] ECHAM 5 shows a significant decrease in annual precipitation in the majority of the country, between −50 and −200 millimeters; CNRM-CM3 shows decreased annual precipitation only in the southern region. CSIRO Mark 3 and MIROC 3.2 show annual precipitation increasing in the extreme northern region, with little change in the rest of the country.

Figure 10.16 shows changes in the mean daily maximum temperature for the month with the highest mean daily maximum temperature. All the GCMs show an overall increase in the annual maximum temperature: CNRM-CM3 shows an increase of 2°–2.5°C for the entire country, and ECHAM 5 shows increases of 2.5°–3°C for the majority of the country and 3°–3.5°C for the eastern region, while CSIRO-Mark 3 and MIROC 3.2 show temperature increases of 1.5°–2°C for all but the northernmost regions of the country.

Crop Models

The Decision Support Software for Agrotechnology Transfer (DSSAT) crop modeling system was used to compare future yield results with the baseline yield (with unchanged climate). In addition to temperature and precipitation, we also input soil data, assumptions about fertilizer use and planting month, and additional climate data such as days of sunlight each month. The output for key crops is mapped in Figures 10.17 and 10.18.

For maize (see Figure 10.17), all the models show some yield gains (with yields in some important areas increasing by more than 25 percent of baseline) in most of the traditional maize-producing regions, as well as some areas of

---

7 The A1B scenario describes a world of very rapid economic growth, low population growth, and rapid introduction of new and more efficient technologies with moderate resource use and a balanced use of technologies. CNRM-CM3 is National Meteorological Research Center–Climate Model 3. CSIRO is a climate model developed at the Australia Commonwealth Scientific and Industrial Research Organisation. ECHAM 5 is a fifth-generation climate model developed at the Max Planck Institute for Meteorology in Hamburg. MIROC is the Model for Interdisciplinary Research on Climate, developed at the University of Tokyo Center for Climate System Research.

FIGURE 10.15 Changes in mean annual precipitation in Zimbabwe, 2000–2050, A1B scenario (millimeters)

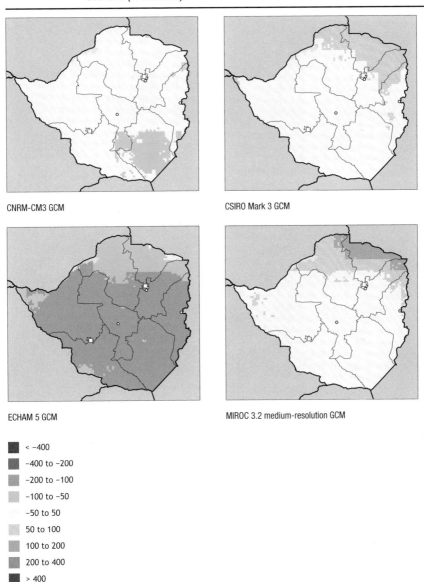

CNRM-CM3 GCM

CSIRO Mark 3 GCM

ECHAM 5 GCM

MIROC 3.2 medium-resolution GCM

- < –400
- –400 to –200
- –200 to –100
- –100 to –50
- –50 to 50
- 50 to 100
- 100 to 200
- 200 to 400
- > 400

Source: Authors' calculations based on Jones, Thornton, and Heinke (2009).

Notes: A1B = greenhouse gas emissions scenario that assumes fast economic growth, a population that peaks midcentury, and the development of new and efficient technologies, along with a balanced use of energy sources; CNRM-CM3 = National Meteorological Research Center–Climate Model 3; CSIRO = climate model developed at the Australia Commonwealth Scientific and Industrial Research Organisation; ECHAM 5 = fifth-generation climate model developed at the Max Planck Institute for Meteorology (Hamburg); GCM = general circulation model; MIROC = Model for Interdisciplinary Research on Climate, developed by the University of Tokyo Center for Climate System Research.

FIGURE 10.16 Changes in monthly mean maximum daily temperature in Zimbabwe for the warmest month, 2000–2050, A1B scenario (°C)

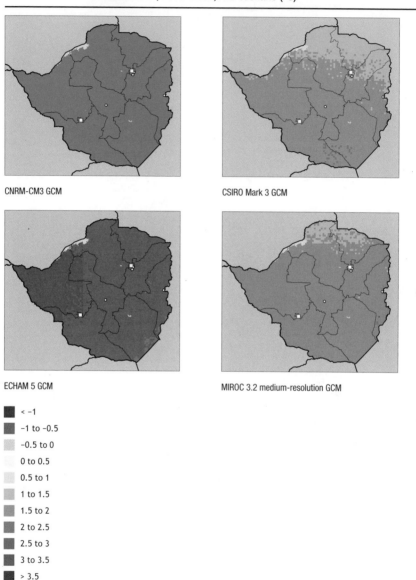

Source: Authors' calculations based on Jones, Thornton, and Heinke (2009).
Notes: A1B = greenhouse gas emissions scenario that assumes fast economic growth, a population that peaks midcentury, and the development of new and efficient technologies, along with a balanced use of energy sources; CNRM-CM3 = National Meteorological Research Center–Climate Model 3; CSIRO = climate model developed at the Australia Commonwealth Scientific and Industrial Research Organisation; ECHAM 5 = fifth-generation climate model developed at the Max Planck Institute for Meteorology (Hamburg); GCM = general circulation model; MIROC = Model for Interdisciplinary Research on Climate, developed by the University of Tokyo Center for Climate System Research.

FIGURE 10.17 Yield change under climate change: Rainfed maize in Zimbabwe, 2000–2050, A1B scenario

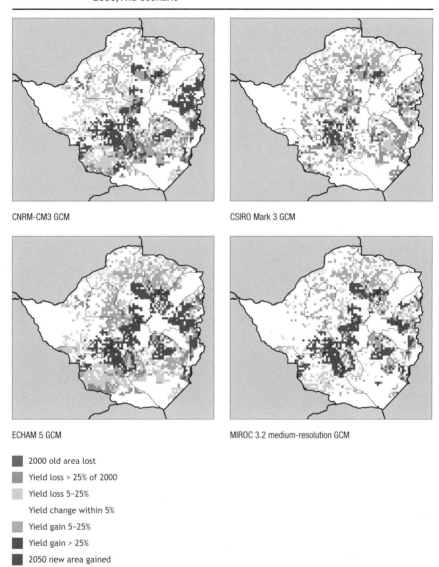

CNRM-CM3 GCM

CSIRO Mark 3 GCM

ECHAM 5 GCM

MIROC 3.2 medium-resolution GCM

- 2000 old area lost
- Yield loss > 25% of 2000
- Yield loss 5–25%
- Yield change within 5%
- Yield gain 5–25%
- Yield gain > 25%
- 2050 new area gained

Source: Authors' calculations.

Notes: A1B = greenhouse gas emissions scenario that assumes fast economic growth, a population that peaks midcentury, and the development of new and efficient technologies, along with a balanced use of energy sources; CNRM-CM3 = National Meteorological Research Center–Climate Model 3; CSIRO = climate model developed at the Australia Commonwealth Scientific and Industrial Research Organisation; ECHAM 5 = fifth-generation climate model developed at the Max Planck Institute for Meteorology (Hamburg); GCM = general circulation model; MIROC = Model for Interdisciplinary Research on Climate, developed by the University of Tokyo Center for Climate System Research.

FIGURE 10.18 Yield change under climate change: Rainfed sorghum in Zimbabwe, 2000–2050, A1B scenario

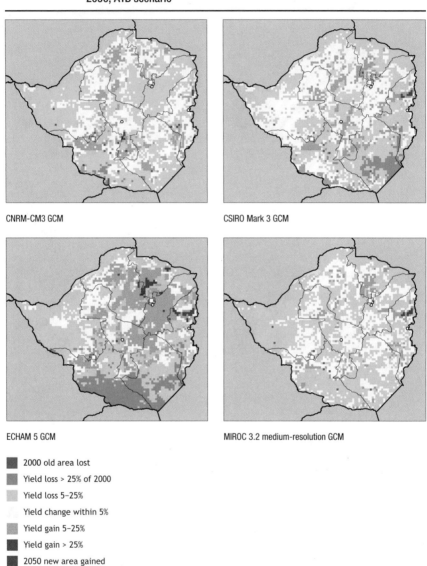

CNRM-CM3 GCM

CSIRO Mark 3 GCM

ECHAM 5 GCM

MIROC 3.2 medium-resolution GCM

- 2000 old area lost
- Yield loss > 25% of 2000
- Yield loss 5–25%
- Yield change within 5%
- Yield gain 5–25%
- Yield gain > 25%
- 2050 new area gained

Source: Authors' calculations.
Notes: A1B = greenhouse gas emissions scenario that assumes fast economic growth, a population that peaks midcentury, and the development of new and efficient technologies, along with a balanced use of energy sources; CNRM-CM3 = National Meteorological Research Center–Climate Model 3; CSIRO = climate model developed at the Australia Commonwealth Scientific and Industrial Research Organisation; ECHAM 5 = fifth-generation climate model developed at the Max Planck Institute for Meteorology (Hamburg); GCM = general circulation model; MIROC = Model for Interdisciplinary Research on Climate, developed by the University of Tokyo Center for Climate System Research.

declining yield (mostly between 5 and 25 percent) and scattered areas of baseline area lost. The CNRM, ECHAM, and CSIRO GCMs show areas with yield losses of greater than 25 percent in southern Zimbabwe.

Despite the increase in precipitation shown by CSIRO Mark 3 and MIROC 3.2, the models show yield losses in the northern parts of Zimbabwe for both maize and sorghum (see Figures 10.17 and 10.18). For sorghum, a drought-tolerant crop, CNRM-CM3, CSIRO Mark 3, and MIROC 3.2 show yield losses for almost the entire country. It would seem that the model is suggesting that maize is able to deal with higher temperatures more easily than sorghum, given the baseline climate and the precipitation anticipated.

Agricultural Outcomes

Figures 10.19–10.22 show simulation results from the International Model for Policy Analysis of Agricultural Commodities and Trade (IMPACT) (which includes climate change and the GDP and population scenarios) associated with key agricultural crops in Zimbabwe. Each featured crop has five graphs: production, yield, area, net exports, and world price.

For maize (see Figure 10.19), the harvested area is shown increasing gradually until 2025 and then decreasing after 2030 to its initial level; production is shown steadily increasing until 2050, reflecting anticipated increases in the maize yield from technological improvements. Maize is an important source of carbohydrates, providing about one-third of the daily calorie intake in Zimbabwe.

The trends for cotton (see Figure 10.20) in terms of production, yield, harvested area, net exports, and world prices are similar in all three scenarios. The production and yield of cotton increase. An increase in net export is predicted because the domestic use of cotton is limited.

The modeled trends for sorghum and millet are similar in terms of production, yield, harvested area, exports, and world prices (see Figures 10.21 and 10.22). Moreover, for all variables other than net exports, there is almost no difference among the three scenarios. It is curious that net exports fall after 2025 in the optimistic scenario, reflecting increased domestic consumption of both sorghum and millet. The area planted with sorghum and millet is shown increasing as some of the traditional maize production areas become too dry for maize production. Using Applied Simulation Technology, Dimes, Cooper, and Rao (2009) demonstrate that, in comparison to maize, groundnuts, and pigeon peas, sorghum demonstrates more resilience under climate change as a result of its heat index and water use efficiency, though this finding seems to contradict our own findings from the DSSAT crop model.

FIGURE 10.19 Impact of changes in GDP and population on maize in Zimbabwe, 2010–2050

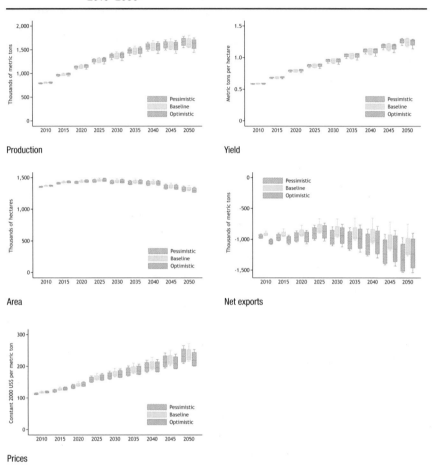

Source: Based on analysis conducted for Nelson et al. (2010).
Notes: The box and whiskers plot for each socioeconomic scenario shows the range of effects from the four future climate scenarios. GDP = gross domestic product; US$ = US dollars.

Vulnerability

Figure 10.23 shows the impact of future GDP and population scenarios on the number of malnourished children under age five in Zimbabwe; Figure 10.24 shows the share of these children. The under-five malnutrition rates are similar in the pessimistic and baseline scenarios, increasing to 600,000 by 2025, then decreasing to about the present levels in 2050. The optimistic scenario shows a similar increase in the number of malnourished children until 2025, followed by a more significant decrease by 2050. Because the population is also

FIGURE 10.20 Impact of changes in GDP and population on cotton in Zimbabwe, 2010–2050

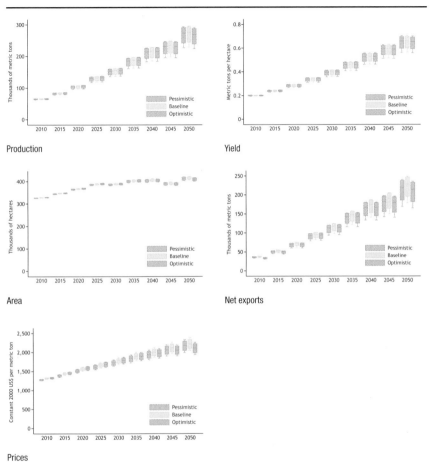

Source: Based on analysis conducted for Nelson et al. (2010).
Notes: The box and whiskers plot for each socioeconomic scenario shows the range of effects from the four future climate scenarios. GDP = gross domestic product; US$ = US dollars.

increasing, the proportion of children under age five who are malnourished falls more rapidly.

Figure 10.25 shows the kilocalories per capita available. The trends in this figure are consistent with those seen in Figure 10.23. The kilocalories per capita decrease slightly to 2025 or 2030, to around 1,400 kilocalories (2,400 is adequate) and thereafter increase again to the present levels in 2050 in both the baseline and the pessimistic scenarios. The optimistic scenario shows a slight decrease, from around 1,800 in 2010 to around 1,600 in

FIGURE 10.21 Impact of changes in GDP and population on sorghum in Zimbabwe, 2010–2050

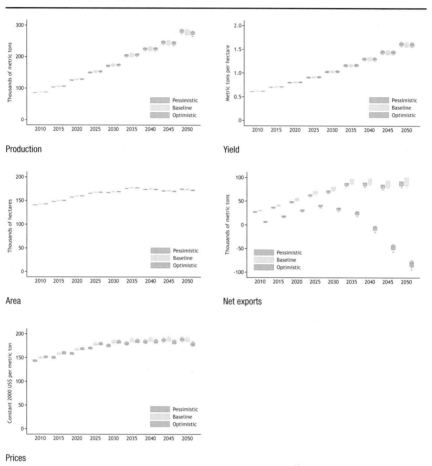

Source: Based on analysis conducted for Nelson et al. (2010).
Notes: The box and whiskers plot for each socioeconomic scenario shows the range of effects from the four future climate scenarios. GDP = gross domestic product; US$ = US dollars.

2025, and then an increase to a comfortable 2,500 by 2050—about a 60 percent increase.

Review of Policies and Programs

Zimbabwe's Initial National Communication under the United Nations Framework Convention on Climate Change (Zimbabwe, Ministry of Mines, Environment, and Tourism, 1998) describes the expected impacts of vulnerability to climate change and how that vulnerability can be reduced by adopting

FIGURE 10.22 Impact of changes in GDP and population on millet in Zimbabwe, 2010–2050

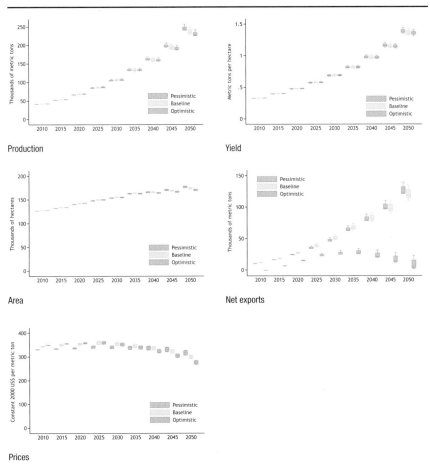

Source: Based on analysis conducted for Nelson et al. (2010).
Notes: The box and whiskers plot for each socioeconomic scenario shows the range of effects from the four future climate scenarios. GDP = gross domestic product; US$ = US dollars.

adaptation strategies. The Communication does not spell out the country's policy on adaptation strategies to reduce the vulnerability of the agricultural sector to climate change. However, there are a number of programs and projects that have been and are being implemented throughout the country. Some are subregional, implemented in more than one of the Southern Africa Development Community (SADC) countries; others are being implemented in Zimbabwe only (Chishakwe 2010) or only in particular wards of a single district. Of these, most are nongovernmental; only one is implemented by the government.

FIGURE 10.23 Number of malnourished children under five years of age in Zimbabwe in multiple income and climate scenarios, 2010–2050

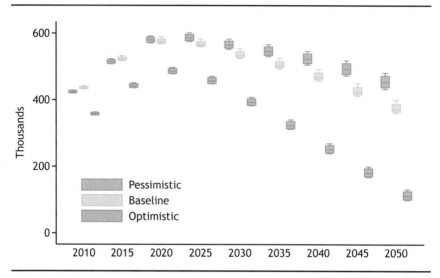

Source: Based on analysis conducted for Nelson et al. (2010).
Note: The box and whiskers plot for each socioeconomic scenario shows the range of effects from the four future climate scenarios.

FIGURE 10.24 Share of malnourished children under five years of age in Zimbabwe in multiple income and climate scenarios, 2010–2050

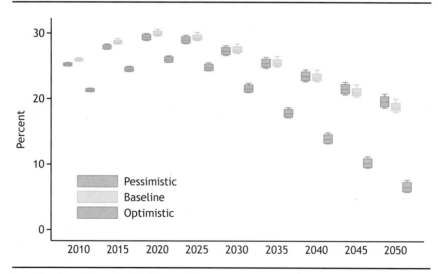

Source: Based on analysis conducted for Nelson et al. (2010).
Note: The box and whiskers plot for each socioeconomic scenario shows the range of effects from the four future climate scenarios.

FIGURE 10.25 Kilocalories per capita in Zimbabwe in multiple income and climate scenarios, 2010–2050

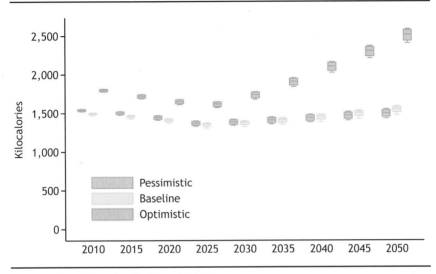

Source: Based on analysis conducted for Nelson et al. (2010).
Note: The box and whiskers plot for each socioeconomic scenario shows the range of effects from the four future climate scenarios.

Southern African programs are the Southern African Regional Climate Outlook Forum, the SADC Task Force for Monitoring Weather Conditions, the SADC Regional Early Warning System, the SADC Regional Remote Sensing Unit, and the SADC Regional Remote Sensing Unit (Chishakwe 2010).

Zimbabwe has one government-sponsored project—Coping with Drought and Climate Change—which was implemented in Chiredzi District between 2006 and 2011 and funded by the UN Development Programme. The project seeks to develop and pilot a range of long-term adaptation measures in the agriculture sector to reduce the vulnerability of smallholder farmers and pastoralists in rural Zimbabwe to current and future climate change–related shocks. It also seeks to develop long-term policy-oriented approaches for adaptation to climate change among rural men and women in agriculture. The goal is to develop and implement a range of viable pilot demonstration measures in response to identified climate risks and to develop local capacity to use climate early warning systems to strengthen adaptation and livelihood strategies.

Civil society initiatives are listed in Table 10.5. Project activities are limited to strengthening the adaptive capacity of communities and farmers with regard to their existing livelihoods. Some were designed to alleviate food shortages caused by droughts by providing emergency humanitarian support.

TABLE 10.5 Civil society–driven projects and initiatives in Zimbabwe for strengthening the adaptive capacity of farmers and areas of implementation, 2000s and ongoing

Project	Project description	Implementing agency or organization	Location of implementation	Project period
Assessment of Impacts and Adaptation to Climate Change (AIACC) project	Focused on linkages between AIACC regional studies and national communications to the United Nations Framework Convention on Climate Change (UNFCCC). Activities included vulnerability and adaptation assessments and strategy development regarding food security, water, land use vulnerabilities and adaptation, benefit–cost analysis of adaptation responses, and climate change and adaptive responses.	AIACC (UNFCCC)	Zambezi River basin	2001–2007
Oxfam Australia Programme on Climate Adaptation	Focuses on providing emergency humanitarian support, helping poor communities adapt to climate change, and campaigning for strong global action on climate change.	Oxfam	Entire country	2010–2015
Disaster Management (DM) Programme	Focuses on building strong early warning and early action capacities at national and provincial levels and on helping national and provincial authorities to improve the national disaster management framework (national disaster response plan) and to address issues related to climate change and climate adaptation. Community-based disaster risk reduction initiatives will include community preparedness for floods, volcanic eruptions, and epidemics (e.g., cholera). Building community resilience to food insecurity will also be a large component of the program.	International Federation of Red Cross and Red Crescent Societies (IFRC)	Entire country	Ongoing
Integration of climate risk management tools into existing programs and activities of the IFRC	Focused on natural disaster mitigation through developing communication and raising awareness of the consequences of climate change and opportunities for risk reduction, building climate change knowledge and capacity, serving as an advocate for Red Cross concerns and proposals in international climate policy, supporting people with high-quality analyses of knowledge about and experiences with climate risk management, and using and promoting the use of climate information to reduce vulnerability to extreme weather events and climate variability.	IFRC Climate Change Centre	Mzingwane catchment	2006–2009
Southern African Research and Documentation Centre (SARDC) Programme on Environment, Water, Research Documentation	Focuses on improving access to information on the impacts of climate change on the water resources of the Zambezi River basin.	SARDC	Zambezi River basin	Ongoing

Project	Description	Partner/Funder	Location	Period
Regional Climate Change Programme for Southern Africa Project	Sought to mitigate the impact of climate change on vulnerable livelihoods in southern Africa through adaptation. Also emphasized the importance of capacity building and the value of using a risk-based assessment approach to climate modeling and scenario building.	Department for International Development (DFID) and One World	Southwestern Zimbabwe	2007–2012
Southern Africa Regional Plan	Currently focuses on relief; preparedness, adaptation, and mitigation; developing policy and strategies for managing climate variation; and incorporating climate change data into food security and water policies.	DFID	Entire country	Ongoing
Limpopo Basin Development Challenge (LBDC)	A scientific and development challenge that seeks to increase the productivity of rainfed agriculture, increase the resilience of small-scale farmers, and reduce the risks associated with an unpredictable climate. The geographic area of the LBDC falls within the borders of Botswana, Mozambique, South Africa, and Zimbabwe.	Food Agriculture and Natural Resources Policy Analysis Network (FANRPAN)	Limpopo Basin	Ongoing
Resilience and the African Smallholder: Enhancing the Capacity of Communities to Adapt to Climate Change	Focused on enhancing the ability of households, communities, and relevant institutions to respond to changing circumstances with a view to reducing future threats to food security and environmental integrity. It did so by working with farmers to identify improved farming technologies and translating the results into action plans at the appropriate institutional level (local or national).	UZ, Climate Change Adaptation in Africa (CCAA), and International Development Research Centre (IDRC)	Makoni and Whedza Districts	2007–2010
Building Adaptive Capacity to Cope with Increasing Vulnerability Due to Climate Change	Focused on building adaptive capacity to cope with increasing vulnerability due to climate change in southern Zambia and southwestern Zimbabwe. The strategy built on participatory approaches and capitalized on farmers', scientists', and development workers' knowledge, products, and experiences in agricultural research for development to bring out positive changes in people's livelihoods.	Michigan State University, CCAA, and IDRC	Lower Gweru and Lupane Districts	2007–2010
Community-Based Adaptation to Climate Change in Africa (funded by CCAA and IDRC)	Focused on helping communities adapt to climate change and share lessons learned from project activities with key stakeholders at local, national, regional, and international levels to elicit their support for climate change adaptation.	ZERO, CCAA, and IDRC	Goromonzi District	2008–2010

Source: Chishakwe (2010).

Note: ZERO = ZERO Regional Environment Organization. See www.climatenetwork.org/profile/member/zero-regional-environment-organization.

Most are short-term projects limited to particular wards in a single district. For example, Building Adaptive Capacity was a three-year project (2007–10) implemented in Lower Gweru and Lupane Districts.

There are still a number of gaps to be addressed to reduce smallholder farmers' vulnerability to climate change. Civil society organizations engaged in climate change adaptation are not coordinated through a national policy. Any adaptation strategies should be supported by enabling government policies in order to ensure effective implementation and sustainability. Several gaps in the country's policies on climate change adaptation have been identified, and we make the following recommendations on how to fill them:

- Government should institute clear policies and programs to facilitate adaptation, taking account of the full range of adaptation strategies (market responses, institutional changes, and technological development), given the need for an enabling policy environment to facilitate adaptation.
- The meteorological department needs new, state-of-the-art equipment to collect quality data for climate forecasting.
- Most farmers in Zimbabwe do not have access to the seasonal climate forecast (SCF) issued by the Southern Africa Development Community's subregional intergovernmental programs on climate change or they cannot interpret it because of poor literacy. Meteorological departments do not assist the local farmers in making decisions on cropping patterns; they only make long-range predictions for the whole nation (Mugabe et al. 2010). Engaging farmers with the SCF can bring together meteorological officials, researchers, and extension agents in working with the farmers (Mugabe et al. in preparation).
- Civil society–driven projects have limited impact. Geographically, they serve only a few wards in a district, and they are limited to three to five years, with no upscaling of results to other areas. Projects' activities could go further toward addressing diversification in order to reduce dependence on climate-sensitive sectors such as agriculture (Chishakwe 2010). There is a need to develop a platform to share project results.
- Courses on climate change and adaptation should be introduced in universities and colleges of agriculture to prepare for better implementation of climate change and adaptation initiatives of the agricultural extension services (Twomlow et al. 2008).

- Agritex should introduce short courses on climate change and adaptation to mainstream these issues into the extension services. Extension workers need to be trained to further train the farmers they work with.

- Beyond the microlevel adaptation options, such as diversification and intensification of crop production, there is a need for broader adaptation approaches, including market responses, institutional changes, and technological developments.

- Development and promotion of new crop varieties and hybrids is crucial, especially to anticipate shorter growing seasons in southern Zimbabwe. Government support for the Department of Research for Development has declined, however, hindering their breeding program.

- Subsistence farmers are limited by shortages of inputs. The government and nongovernmental organizations should ensure more access to credit and aid facilities and should also assist farmers to acquire livestock and important farm assets.

Conclusions and Policy Recommendations

According to the climate models used in this study, the rainfall predictions for the northern parts of Zimbabwe over the next 40 years are mixed, with two models showing more rainfall, one showing no change, and one showing a decline in rainfall. The results for the southern parts of the country are also mixed, with two models showing no change and two showing a decline. Temperatures are shown increasing throughout the country relatively uniformly, but one model shows the western part with greater warming. That model shows a dramatic temperature increase for Zimbabwe, averaging just under 3°C.

The consequent shift in agroecological zones shows Natural Region V increasing while Regions III and IV—more favorable to agriculture—decline in size. The semiarid areas already have high variability in rainfall; this variability and extreme events are expected to increase with climate change. Some of the maize-producing areas will likely shift to drought-tolerant crops such as sorghum and millet, though our analysis also pointed to the fact that sorghum might be more severely stressed by the heat than maize. Certainly there will be a need to develop new varieties of both maize and sorghum that can handle higher temperatures and dry conditions at the same time.

In terms of vulnerability to climate change, according to IMPACT, the number of malnourished children under age five will increase, then start decreasing after 2025 in all three scenarios (optimistic, baseline, and pessimistic). In the optimistic scenario, fewer children under the age of five are malnourished: the number of children undernourished decreases to 120,000 as compared to 400,000 and 480,000, respectively, in the baseline and the pessimistic scenarios. The model shows similar available calories per capita between 2000 and 2050 in both the baseline and the pessimistic scenarios; in the optimistic scenario, the available calories increase by just under 40 percent.

The models show yields and production increasing for the three crops that are important to the poor—maize, sorghum, and millet—and harvest area increasing for sorghum and millet. However, achieving these results will require access to needed inputs. Currently, maize yields are very low in the smallholder farming sector because inputs such as fertilizers and seeds are unavailable or unaffordable.

Net exports of cotton increase in all the scenarios, while exports of sorghum and millets decrease in the optimistic scenario. Net exports of maize are projected to be largely unchanged, though the variance in the predictions is great. World prices increase in all the scenarios for maize and cotton but remain almost constant for millet and sorghum.

The Government of Zimbabwe does not have a clear policy on climate change adaptation in the agriculture sector. The current programs are largely sponsored by regional and nongovernmental organizations:

- SADC initiatives on climate forecasting and early warning systems provide periodic country updates on climate and drought monitoring that are broadcast on radio and television and provided in print media. However, farmers lack access to this information or cannot interpret it so as to make crop management decisions that respond to seasonal climate forecasts.

- Civil society organizations have implemented projects in various parts of the country: building institutional and farmers' capacity for climate change adaptation; enhancing the ability of households and communities to respond to changing circumstances; addressing food security needs in a changing environment.

The government needs to develop an enabling policy on adaptation to climate change with a view to reducing vulnerability, especially for the rural poor, who will most suffer from the impacts of climate change. Most of the adaptation strategies have to be implemented at the micro level. A holistic approach is needed that includes all four types of adaptation: microlevel strategies, market responses, institutional changes, and technological developments.

References

Auger, M. 2008. "Perspectives on Food Security in Sub-Saharan Africa." *Proceedings of the 3rd SADC–EU International Scientific Symposium—Towards Meeting the Challenges of Climate Change: Institutional Structures and Best Practices in Land and Water Management in Southern Africa,* Lusaka, Zambia, May 27–30.

Bartholome, E., and A. S. Belward. 2005. "GLC2000: A New Approach to Global Land Cover Mapping from Earth Observation Data." *International Journal of Remote Sensing* 26 (9): 1959–1977.

Boko, M., I. Niang, A. Nyong, C. Vogel, A. Githeko, M. Medany, B. Osman-Elasha, R. Tabo, and P. Yanda. 2007. "Climate Change 2007: Impacts, Adaptation, and Vulnerability." In *Contribution of Working Group II to the Fourth Assessment Report of the Intergovernmental Panel on Climate Change,* edited by M. L. Parry, O. F. Canziani, J. P. Palutikof, P. J. van der Linden, and C. E. Hanson.

Bwalya, M. 2008. "Towards Meeting the Challenges of Climate Change: Institutional Structures and Best Practices in Land and Water Management in Southern Africa." In *Proceedings of the 3rd SADC–EU International Scientific Symposium—Towards Meeting the Challenges of Climate Change: Institutional Structures and Best Practices in Land and Water Management in Southern Africa,* Lusaka, Zambia, May 27–30.

Carmody, P. 1998. "Neoclassical Practice and the Collapse of Industry in Zimbabwe: The Cases of Textiles, Clothing, and Footwear." *Economic Geography* 74 (4): 319–343.

Chaguta, T. 2010. *Climate Change Vulnerability and Preparedness in Southern Africa; Zimbabwe Country Report.* Cape Town, South Africa: Heinrich Böll Stiftung Southern Africa. Accessed May 22, 2011. http://www.za.boell.org/downloads/hbf_web_zim_21_2.pdf.

Chishakwe, N. E. 2010. "Southern Africa Sub-regional Framework on Climate Change Report." Accessed February 26, 2011. www.unep.org/roa/amcen/docs/AMCEN_Events/climate-change/southAfrica/SADC_Report.pdf.

CIESIN (Center for International Earth Science Information Network), Columbia University, IFPRI (International Food Policy Research Institute), World Bank, and CIAT (Centro Internacional de Agricultura Tropical). 2004. *Global Rural–Urban Mapping Project (GRUMP), Alpha Version: Population Density Grids.* Palisades, NY, US: Socioeconomic Data and Applications Center (SEDAC), Columbia University. http://sedac.ciesin.columbia.edu/gpw.

Clime, W. 2007. *Global Warming and Agriculture: Impact Estimates by Country.* Washington, DC: Center for Global Development.

Dimes, J., P. Cooper, and K.P.C. Rao. 2009. "Climate Change Impact on Crop Productivity in the Semi-arid Tropics of Zimbabwe in the 21st Century." Paper presented at the CGIAR Challenge Program on Water and Food Workshop on Increasing the Productivity and Sustainability of Rainfed Cropping Systems of Poor, Smallholder Farmers, Tamale, Ghana, September 2008. Colombo, Sri Lanka: CGIAR Challenge Program on Water and Food.

FAO (Food and Agriculture Organization of the United Nations). 2010. FAOSTAT. Accessed December 22, 2010. http://faostat.fao.org/site/573/default.aspx#ancor.

Jones, P. G., P. K. Thornton, and J. Heinke. 2009. *Generating Characteristic Daily Weather Data Using Downscaled Climate Model Data from the IPCC's Fourth Assessment.* Project report. Nairobi, Kenya: International Livestock Research Institute.

Kalanda-Joshua, M., C. Ngongondo, L. Chipeta, and F. Mpembeka. 2011. "Integrating Indigenous Knowledge with Conventional Science: Enhancing Localised Climate and Weather Forecasts in Nessa, Mulanje, Malawi." *Journal of Physics and Chemistry of the Earth.* doi: 10.1016/j.pce.2011.08.001.

Lehner, B., and P. Döll. 2004. "Development and Validation of a Global Database of Lakes, Reservoirs, and Wetlands." *Journal of Hydrology* 296 (1–4): 1–22.

Lumsden, T. G., and R. E. Schulze. 2007. "Application of Seasonal Climate Forecasts to Predict Regional Scale Crop Yields in South Africa." In *Climate Prediction and Agriculture: Advances and Challenges,* edited by J. W. Hansen and M. V. K. Sivakumar. Berlin: Springer.

Mano, R., and C. Nhemachena. 2007. *Assessment of the Economic Impacts of Climate Change on Agriculture in Zimbabwe: A Ricardian Approach.* Policy and Research Working Paper 4292. Washington, DC: World Bank.

Matarira, C. H., and J. M. Makadho. 1998. "Zimbabwe: Climate Change Impact on Maize Production and Adaptive Measures for the Agricultural Sector." *SACCAR [Southern African Centre for Cooperation in Agricultural Research and Training] Newsletter,* June 1998.

Millennium Ecosystem Assessment. 2005. *Ecosystems and Human Well-being: Synthesis.* Washington, DC: Island Press. http://www.maweb.org/en/Global.aspx.

Mugabe, F. T., C. P. Mubaya, D. H. Nanja, P. Gondwe, A. Munodawafa, E. Mutswangwa, I. Chagonda, P. Masere, J. Dimes, and C. Murewi. 2010. "Using Indigenous Knowledge for Climate Forecasting and Adaptation in Southern Zambia and Southwestern Zimbabwe." *Zimbabwe Journal of Technological Sciences* 1: 19–28.

Mugabe, F. T., J. Dimes, D. Nanja, P. Gondwe, I. Chagonda, P. Masere, M. Murewi, and M. Mundawafa. In preparation. "Engaging Smallholder Farmers with Seasonal Climate Forecasts to Design Crop Management Options: Experience from Southern Zambia and Northwestern Zimbabwe." Submitted to *Climate and Development.*

Nelson, G. C., M. W. Rosegrant, A. Palazzo, I. Gray, C. Ingersoll, R. Robertson, S. Tokgoz, et al. 2010. *Food Security, Farming, and Climate Change to 2050: Scenarios, Results, Policy Options.* Washington, DC: International Food Policy Research Institute.

Potts, D. 2010. "Internal Migration in Zimbabwe: The Impact of Livelihood Destruction in Rural and Urban Areas." In *Zimbabwe's Exodus: Crisis, Migration, Survival,* edited by J. Crush and D. Tevera. Ottawa, Canada: International Development Research Centre.

Richardson, C. J. 2007. "How Much Did Droughts Matter? Linking Rainfall and GDP Growth in Zimbabwe." *African Affairs* 106 (424): 463–478.

Scholes, B. 2008. "IPCC Findings and Implications for Southern Africa." *Proceedings of the 3rd SADC–EU International Scientific Symposium, Towards Meeting the Challenges of Climate Change: Institutional Structures and Best Practices in Land and Water Management in Southern Africa,* Lusaka, Zambia, May 27–30.

Twomlow, S., F. T. Mugabe, M. Mwale, R. Delve, D. Nanja, P. Carberry, and M. Howden. 2008. "Building Adaptive Capacity to Cope with Increasing Vulnerability Due to Climate Change—A New Approach." *Journal of Chemistry and Physics of the Earth* 33: 780–787.

UNEP (United Nations Environment Programme) and IUCN (International Union for the Conservation of Nature). 2009. *World Database on Protected Areas (WDPA): Annual Release 2009.* Accessed 2009. www.wdpa.org/protectedplanet.aspx.

UNPOP (United Nations Department of Economic and Social Affairs–Population Division). 2009. *World Population Prospects: The 2008 Revision.* New York. http://esa.un.org/unpd/wpp/.

World Bank. 2009. *World Development Indicators.* Washington, DC.

———. 2010a. *World Development Indicators.* Washington, DC.

———. 2010b. *The Costs of Agricultural Adsaptation to Climate Change.* Washington, DC.

You, L., and S. Wood. 2006. "An Entropy Approach to Spatial Disaggregation of Agricultural Production." *Agricultural Systems* 90 (1–3): 329–347.

You, L., S. Wood, and U. Wood-Sichra. 2006. "Generating Global Crop Distribution Maps: From Census to Grid." Paper presented at the International Association of Agricultural Economists Conference in Brisbane, Australia, August 11–18.

———. 2009. "Generating Plausible Crop Distribution and Performance Maps for Sub-Saharan Africa Using a Spatially Disaggregated Data Fusion and Optimization Approach." *Agricultural Systems* 99 (2–3): 126–140.

Zimbabwe, Ministry of Mines, Environment, and Tourism. 1998. *Zimbabwe's Initial National Communication on Climate Change.* Harare.

Contributors

Patrick Gwimbi (pgwimbi@yahoo.com), Lecturer, Department of Environmental Health, National University of Lesotho

Sepo Hachigonta (SHachigonta@fanrpan.org), Programme Manager, Climate Change, Food, Agriculture, and Natural Resources Policy Analysis Network (FANRPAN)

Peter Johnston (peter@csag.uct.ac.za), Climate Impacts Researcher, Climate Systems Analysis Group, University of Cape Town

Joseph Kanyanga (jk_kanyanga@yahoo.com), Chief Meteorologist, Zambia Meteorological Department

Absalom M. Manyatsi (manyatsi@agric.uniswa.sz), Senior Lecturer and Head, Agricultural and Biosystems Engineering Department, University of Swaziland

Michael T. Masarirambi (mike@uniswa.sz), Senior Lecturer, Horticulture Department, Faculty of Agriculture, University of Swaziland

Daniel Mason-d'Croz (d.mason-dcroz@cgiar.org), Research Analyst, International Food Policy Research Institute (IFPRI)

Genito A. Maure (genito.maure@gmail.com), Lecturer and Researcher, Eduardo Mondlane University, Mozambique

Francis T. Mugabe (ftmugabe@yahoo.co.uk), Dean and Researcher, School of Agricultural Sciences and Technology, Chinhoyi University of Technology, Zimbabwe

Gerald C. Nelson (g.nelson@cgiar.org), Professor Emeritus, University of Illinois, Urbana-Champaign, and former Senior Research Fellow, International Food Policy Research Institute (IFPRI)

Amanda Palazzo (palazzo@iiasa.ac.at), Research Scholar, International Institute for Applied Systems Analysis, Austria. While working on this monograph, she was a Research Analyst at the International Food Policy Research Institute (IFPRI).

Gorata Ramokgotlwane (ramokgotlhwaneg@yahoo.com), Consultant, Energy and Environment and Climate Change Research Center (EECG), Botswana

Richard Robertson (r.robertson@cgiar.org), Research Fellow, International Food Policy Research Institute (IFPRI)

John D. K. Saka (jsaka@chanco.unima.mw), Coordinator, Natural Resources and Environment Centre, Faculty of Science, Chancellor College, University of Malawi

Pickford Sibale (psibale@worldbank.org), Research Specialist, The World Bank

Lindiwe Majele Sibanda (lmsibanda@fanrpan.org), CEO, Food, Agriculture, and Natural Resources Policy Analysis Network (FANRPAN)

Tichakunda Simbini (simbinits@gmail.com), Consultant, Energy and Environment and Climate Change Research Center (EECG), Botswana

Timothy S. Thomas (t.s.thomas@cgiar.org), Research Fellow, International Food Policy Research Institute (IFPRI)

Peter P. Zhou (pzhou@global.bw), Director, Energy and Environment and Climate Change Research Center (EECG), Botswana

Index

Page numbers for entries occurring in figures are followed by an *f*; those for entries in notes, by an *n*; and those for entries in tables, by a *t*.

A1B scenario: assumptions of, 12n2, 27n1; precipitation predictions of, 12, 13f; temperature predictions of, 14f; use of, 27. *See also* Climate model projections

A2 scenario, 27, 27n1, 196

ActionAID International, 142

African Growth and Opportunity Act (AGOA), 238

Agricultural mapping models. *See* Decision Support Software for Agrotechnology Transfer

Agriculture: climate change impacts on, 1; climate change vulnerability of, 35–36. *See also* Cash crops; Foodcrops; Yields; *individual countries*

Agriculture as share of GDP: in Botswana, 50, 66; in Lesotho, 71, 76, 77f; in Malawi, 5, 116, 117f; in Mozambique, 150, 151f; in South Africa, 176–78, 177f, 184; in southern Africa, 5, 9, 9t; in Swaziland, 216, 217f; in Zambia, 5, 258f, 259; in Zimbabwe, 5, 290, 292, 293f

Agroecological zones: in Botswana, 44; in Lesotho, 71, 73, 77, 80; in Malawi, 111; in Mozambique, 152–53; in South Africa, 175; in southern Africa, 3–5, 4f; in Swaziland, 220, 222t; in Zimbabwe, 289, 319

Angola: agriculture in, 6; population trends in, 6

AR4. *See* Fourth Assessment Report of the IPCC

Australia Commonwealth Scientific and Industrial Research Organization. *See* CSIRO Mark 3 model

B1 scenario, 27, 27n1

Baseline scenario (median variant): assumptions of, 11–12, 12t, 33t; for Botswana, 54–56; incomes and population growth in, 11–12, 11t, 12t, 32–35, 33t, 34t; for Lesotho, 88–89, 98; for Malawi, 128–29, 134–36; for Mozambique, 159–60, 166, 168–69; for South Africa, 194–96, 204; for Swaziland, 228–29, 239; for Zambia, 270–72, 275, 278; for Zimbabwe, 302–4, 310–12

Batisani, N., 52

Batjes, N., 29

Beans, in Malawi, 125, 127f

Biodiversity, 180. *See also* Protected areas

Botswana: agricultural development policies of, 41–44, 48–50, 67; agriculture in, 45, 48–54, 52t, 59, 62, 64–67; climate change policies of, 15; climate change vulnerability of, 62; climate of, 44–45; climate scenarios in, 56, 59; exports of, 52, 53, 53t; food imports of, 52, 62, 67; food security in, 62; geography of, 44; income distribution in, 47, 47t; incomes in, 46–47, 47t, 54–56, 55f; land use in, 48, 49f; migration to, 6; *National Development Plans* of, 42, 67; policy recommendations for, 68–69; population distribution in, 45–46; population projections for, 54, 55f; population trends in, 45–46, 45f, 46t; poverty in, 46–47, 47f, 47t; protected areas in, 48, 50f; scenarios for future, 56, 59, 62; travel times in, 48, 51f; urban areas in, 45–46, 46t, 48, 51f; vegetation in, 48, 49f; *Vision 2016*, 41–42, 46, 67; well-being indicators in, 62–63, 65f, 66f

Box-and-whisker graphs, 37–38, 38f

CAADP. *See* Comprehensive African Agriculture Development Programme

Calorie availability: in Botswana, 62, 63, 66f, 67; data on, 35, 36; determinants of, 35; in Lesotho, 98–100, 99f; in Malawi, 134–36, 135f; in Mozambique, 166–68, 168f; in South Africa, 204–5, 207f; in Swaziland, 239, 241f; in Zambia, 278, 278f; in Zimbabwe, 311–12, 315f. *See also* Child malnutrition; Food consumption

CANGO. *See* Coordinating Assembly of Non Governmental Organizations

Carbon dioxide fertilization, 30

Cash crops: in Malawi, 119, 123, 136–37; in Mozambique, 157, 157t, 158t; in South Africa, 184, 186t; in Swaziland, 223–25, 226t; in Zambia, 266, 266t, 268. *See also* Commodity prices; Exports; Yields

Cassava: in Malawi, 123, 125, 126f, 136, 138f; in Mozambique, 157–59, 160f, 168–69, 170f; in Zambia, 266, 267, 269f, 280–81, 280f

Cattle. *See* Livestock production

CCNUCC. *See* United Nations Framework Convention on Climate Change

Cereal production: major crops, 5, 5t; in southern Africa, 5–6, 7f. *See also* Maize; Millet; Rice; Sorghum

Child malnutrition: in Botswana, 62, 65f; determinants of, 35, 36t; in Lesotho, 78, 79t, 98, 98f, 99f; in Malawi, 117t, 134, 134f, 135f; in Mozambique, 151, 151t, 166, 167f; in South Africa, 178t, 204, 206f; in Swaziland, 218, 219t, 239, 240f; in Zambia, 259t, 260, 275, 277f; in Zimbabwe, 294t, 310–11, 314f

Child mortality, under-five: in Lesotho, 78, 78f; in Malawi, 116–17, 118f; in Mozambique, 151–52, 152f; in South Africa, 179f; in southern Africa, 9, 10t; in Swaziland, 219f, 220; in Zambia, 260–61, 260f; in Zimbabwe, 294f, 295

Climate, definition of, 1. *See also* Precipitation; Temperatures

Climate change impacts: adaptation recommendations, 20–21; agricultural vulnerability to, 35–36; on agriculture, 1, 112, 208; current, 72, 87, 214–15; direct and indirect, 1, 72, 112; on food prices, 15, 72; on food security, 1, 36; on health, 72; of increased temperatures, 1; on livelihoods, 36, 242, 243t; on livestock production, 87; on poverty, 35n3; vulnerability outcomes, 35–36; weather-related, 1. *See also* Scenarios; Yield changes under climate change; *individual countries*

Climate model projections: for Botswana, 56, 57f, 58f, 59, 67–68; for Lesotho, 89–91, 90f, 92f; for Malawi, 129–31, 130f, 132f, 142–43; for Mozambique, 162, 163f, 164f, 165, 171; for South Africa, 196–98, 197f, 201–3; for

Swaziland, 230–32, 231f, 233f; for Zambia, 272, 273f, 274f, 275, 283; for Zimbabwe, 304, 305f, 306f. *See also* Scenarios; Yield changes under climate change

Climate models, 25. *See also* General circulation models; Scenarios

Clime, W., 293

CNRM-CM3 (National Meteorological Research Center – Climate Model 3), 2, 2n1, 13, 27. *See also* Climate model projections; General circulation models

Commodity prices, 15, 31

Commonwealth Scientific and Industrial Research Organization. *See* CSIRO Mark 3 model

Comprehensive African Agriculture Development Programme (CAADP), 235, 247

Cooper, P., 309

Coordinating Assembly of Non Governmental Organizations (CANGO), 244–45

Cotton: in Malawi, 125, 126f, 136–37, 139f; in Swaziland, 225, 237–39, 238f; in Zambia, 266, 268, 269f, 281, 281f; in Zimbabwe, 297, 300f, 309, 311f. *See also* Cash crops

Crop models. *See* Decision Support Software for Agrotechnology Transfer; Spatial Production Allocation Model

Crops. *See* Agriculture; Cash crops; Foodcrops; Yields

CSIRO Mark 3 model, 2, 2n1, 27. *See also* A1B scenario; Climate model projections; General circulation models

Death rates. *See* Child mortality

Decision Support Software for Agrotechnology Transfer (DSSAT): description of, 29; use of, 25, 29–30, 32. *See also* Yield changes under climate change

Deforestation: in Malawi, 119–20; in Zambia, 263

Dimes, J., 29, 309

Droughts: in Botswana, 45, 52, 67; in Lesotho, 76, 97; in Malawi, 112, 120; in Mozambique, 147; in South Africa, 184; in Swaziland, 213–15; in Zimbabwe, 289. *See also* Precipitation

DSSAT. *See* Decision Support Software for Agrotechnology Transfer

ECHAM 5 model, 2, 2n1, 13, 27. *See also* Climate model projections; General circulation models

Economic growth, in scenarios for future, 11–12, 12t, 32, 33–35, 33t. *See also* Incomes per capita

Education: female access to secondary, 35; in Lesotho, 79, 79t; in Malawi, 115, 116, 117t; in Mozambique, 150–51, 151t; in South Africa, 178, 178t; in Swaziland, 218–19, 219t; in Zambia, 259, 259t; in Zimbabwe, 293–94, 294t

Employment: in Lesotho, 79, 79t; in Malawi, 116, 117t; migrant labor, 74–75, 79; in South Africa, 178t; in Swaziland, 216–17, 219, 219t; in Zambia, 259–60, 259t; in Zimbabwe, 294t

Exports: of Botswana, 52, 53, 53t; of Malawi, 136–37; of Mozambique, 168; of South Africa, 52, 184, 188, 192, 218; of Swaziland, 235, 236, 238; of Zambia, 281; of Zimbabwe, 309, 320. *See also* Cash crops

FAO. *See* Food and Agriculture Organization of the United Nations

Fertility rates, in Lesotho, 74

Fertilizers, 30

Food and Agriculture Organization of the United Nations (FAO), 67, 86

Food consumption: in Lesotho, 84–86, 85t; in Malawi, 123, 124t; in Mozambique, 157, 158t, 169; in South Africa, 186, 187f; in Swaziland, 226t, 227; in Zambia, 267, 267t; in Zimbabwe, 295–97, 299t. *See also* Calorie availability; Child malnutrition

Foodcrops: in Lesotho, 84–86, 85t; in Malawi, 123–25, 123t, 124t; in Mozambique, 157–59, 157t, 158t, 168–69; in South Africa, 186t, 188–91; in southern Africa, 5–6, 5t; in Swaziland, 217, 226t, 227; in Zambia, 266–67, 266t; in Zimbabwe, 295–97, 299t. *See also* Cereal production; Legumes; Root crops; Yields

Food imports: of Botswana, 52, 62, 67; of Malawi, 136; of Mozambique, 157; of South Africa, 184; of Swaziland, 217–18, 227; of Zambia, 280

Food price elasticities, 36–37, 37t, 98–100

Food prices: in Botswana, 62; calorie availability and, 36–37; climate change impacts on, 15, 72. *See also* Commodity prices

Food production units (FPUs), 31, 31f

Food security: in Botswana, 62; climate change impacts on, 1, 36; in Lesotho, 86; in Malawi, 115; in Swaziland, 213–14, 218. *See also* Calorie availability; Child malnutrition

Fourth Assessment Report of the IPCC (AR4), 1, 25, 26–27, 113, 185

FPUs. *See* Food production units

GCMs. *See* General circulation models

GDP (Gross Domestic Product). *See* Agriculture as share of GDP; Incomes per capita

General circulation models (GCMs), 2; CNRM-CM3, 2, 13, 27; CSIRO Mark 3, 2, 27; development of, 25; ECHAM 5, 2, 13, 27; estimates of, 26, 27, 28t; global average changes in, 27, 28t; MIROC, 2, 27. *See also* Climate model projections; Yield changes under climate change

Greenhouse gas emissions, 26–27. *See also* Climate change impacts

Gross Domestic Product (GDP). *See* Agriculture as share of GDP; Incomes per capita

Groundnuts: in Malawi, 125, 127f; in Zambia, 266, 268, 270f; in Zimbabwe, 297, 302f

Haddad, L, 35

Harmonized World Soil Database (HWSD), 29

Health, 72. *See also* HIV/AIDS; Well-being indicators

Heavily Indebted Poor Countries Initiative, 259n

Heinke, J., 26

Higgins, Steven I., 208

High-variant scenario. *See* Optimistic scenario

Highways. *See* Roads

HIV/AIDS: in Botswana, 54; in South Africa, 178, 179f; in Swaziland, 213–14, 219–20; in Zambia, 260; in Zimbabwe, 294–95

HWSD. *See* Harmonized World Soil Database

IAMs. *See* Integrated assessment models

IFPRI. *See* International Food Policy Research Institute

IMPACT. *See* International Model for Policy Analysis of Agricultural Commodities and Trade

Imports. *See* Food imports

Incomes per capita: in Botswana, 46–47, 47t, 54–56, 55f; in Lesotho, 75–76, 77f, 88–89, 89f; in Malawi, 115–16, 117f, 129, 129f; in Mozambique, 150, 151f, 160, 161f; in scenarios for future, 11–12, 12t, 32, 33–35, 33t, 34t; in

South Africa, 176–78, 177f, 194–96, 195f; in southern Africa, 8–9, 9t; in Swaziland, 216, 217, 217f, 229, 230f; in Zambia, 258–59, 258f, 270–72, 271f; in Zimbabwe, 292, 293f, 302, 303f. *See also* Poverty

Integrated assessment models (IAMs), 2

Integrated Support Program for Arable Agriculture Development (Botswana), 42–43

Intergovernmental Panel on Climate Change (IPCC), Fourth Assessment Report of, 1, 25, 26–27, 113, 185

International Food Policy Research Institute (IFPRI), 25, 31

International Model for Policy Analysis of Agricultural Commodities and Trade (IMPACT): calorie availability data in, 35; description of, 31–32; food production units, 31, 31f; framework of, 25, 26f; inputs of, 27, 32

International Model for Policy Analysis of Agricultural Commodities and Trade (IMPACT) simulations: for Botswana, 62; for Lesotho, 100; for maize, 15, 16t; for Malawi, 136–39; for millet, 15, 17t; for Mozambique, 166–69; for sorghum, 15, 18t; for South Africa, 201–3; for Swaziland, 235–39; for Zambia, 278–81; for Zimbabwe, 309

IPCC. *See* Intergovernmental Panel on Climate Change

Irrigation: in Botswana, 42; in South Africa, 184, 193–94, 194t; in Swaziland, 216, 223, 228

Jones, P. G., 26

Koo, J., 29
Kyoto Protocol, 100, 244

Labor. *See* Employment

Land use: in Botswana, 48, 49f; in Lesotho, 80, 81f; in Malawi, 119–20, 119f; in Mozambique, 152–53, 154f; in South Africa, 180, 181f; in southern Africa, 3–5, 4f; in Swaziland, 220, 221f, 222t, 223; in Zambia, 261–63, 262f; in Zimbabwe, 295, 296f. *See also* Protected areas

Legumes: in Lesotho, 84; in Malawi, 123. *See also* Beans; Groundnuts

Lesotho: agriculture in, 71, 76, 84–87, 85t, 97, 100; climate change policies of, 19, 72, 100–104; climate change vulnerability of, 71–72, 77–80, 84, 86–87, 97–100, 104–5; climate of, 71, 72, 73; climate scenarios in, 89–91; education in, 79, 79t; employment in, 79, 79t; food security in, 86; geography of, 71, 73; incomes in, 75–76, 77f, 88–89, 89f; land use in, 80, 81f; policy recommendations for, 19, 105–6; population distribution in, 74–75, 76f; population projections for, 88, 88f; population trends in, 73–75, 74t, 75f; poverty in, 79–80; protected areas in, 80, 82f; roads in, 80; scenarios for future, 88–93, 100; travel times in, 80–82, 83f; urban areas in, 74–75, 80–82; vegetation in, 73, 81f; water resources of, 75, 82–84, 97; well-being indicators in, 78–80, 78f, 98–100

Lesotho Highlands Water Project (LHWP), 75, 82–84

Life expectancies: female, 35; in Lesotho, 78, 78f; in Malawi, 116, 118f; in Mozambique, 152, 152f; in South Africa, 178, 179f; in southern Africa, 9, 10t; in Swaziland, 219–20, 219f; in Zambia, 260, 260f; in Zimbabwe, 293–94, 294f

Livestock Management and Infrastructure Development Program (LIMID), Phase 1 (Botswana), 43

Livestock production: in Botswana, 47, 47t, 52–54, 53t, 62, 66–67; in Lesotho, 87, 97–98; in South Africa, 186–87, 187f, 208; in Swaziland, 223; in Zambia, 284

Low-variant scenario. *See* Pessimistic scenario

Maize: in Botswana, 52, 53t, 59, 60f, 62, 63f, 67; in Lesotho, 84, 86, 91–93, 94f, 100, 101f; in Malawi, 123, 125, 125f, 131, 133f, 136, 137f; in Mozambique, 157, 159f, 162, 165f, 168, 169f; prices of, 15; in South Africa, 5, 6, 188–89, 188f, 190f, 199f, 201, 202f; in southern Africa, 5, 5t, 6, 7f; in Swaziland, 217, 218, 218f, 225, 227, 227f, 232, 234f, 235, 236f; yields of, 7f, 15, 16t; in Zambia, 266, 267, 268f, 275, 276f, 278–80, 279f; in Zimbabwe, 295–97, 300f, 304, 307f, 309, 310f. *See also* Cereal production

Malawi: agriculture in, 5, 112, 120, 123–25, 123t, 124t, 136–39; climate change policies of, 140–42, 143; climate change vulnerability of, 112, 134–36, 142; climate of, 112; climate scenarios in, 129–31, 142–43; education in, 115, 116, 117t; employment in, 116, 117t; exports of, 136–37; food imports of, 136; food security in, 115; geography of, 111; incomes in, 115–16, 117f, 129, 129f; land tenure system of, 120; land use in, 119–20, 119f; policy recommendations for, 143–44; population distribution in, 115, 116f; population projections for, 128, 128f; population trends in, 113–15, 114f, 114t; poverty in, 117, 118f, 142; protected areas in, 120, 121f; scenarios for future, 128–31, 142–43; stakeholder consultations in, 139, 140–41t; travel times in, 121, 122f; urban areas in, 114, 114t, 115, 129; vegetation in, 119–20, 119f; well-being indicators in, 115–17, 118f, 134–36

Malnutrition. *See* Child malnutrition

Mano, R., 289

Matarira, C. H., 72

Max Planck Institute for Meteorology. *See* ECHAM 5 model

MDGs. *See* Millennium Development Goals

Median-variant scenario. *See* Baseline scenario

Migration: in Botswana, 46; of labor to South Africa, 74–75, 79; in Lesotho, 74–75; of refugees, 114–15; rural-urban, 8, 46, 74–75, 256, 290–92; from Zimbabwe, 6

Millennium Development Goals (MDGs), 8, 115, 217

Millet: in Botswana, 53t; in southern Africa, 5t, 6, 7f; yields of, 7f, 15, 17t; in Zimbabwe, 297, 301f, 309, 313f. *See also* Cereal production

Ministry of Agriculture (MoA), Botswana, 41, 42–44, 45, 53, 54, 67

Ministry of Agriculture, Swaziland, 235, 244

Model for Interdisciplinary Research on Climate (MIROC), 2, 2n1, 27. *See also* A1B scenario; Climate model projections; General circulation models

Models. *See* Climate models; General circulation models

Mortality. *See* Child mortality

Mozambique: agriculture in, 147–48, 157–59, 157t, 158t, 162, 166–69; civil war in, 148; climate change policies of, 20, 147–48; climate change vulnerability of, 150–52, 171; climate scenarios in, 162, 166–69, 171; education in, 150–51, 151t; exports of, 168; floods in, 147; food imports of, 157; future research in, 170–71; incomes in, 150, 151f, 160, 161f; land use in, 152–53, 154f; policy recommendations for, 20, 171–72; population distribution in, 149, 150f; population projections for, 159, 161f; population trends in, 147, 148–49, 149f, 149t; poverty in, 152, 153f; protected areas in, 153, 155f; roads in, 155; scenarios for future, 159–60, 166–69; stakeholder consultation in, 170–71; travel times in, 155, 156f; urban areas in, 148–49, 149t; vegetation in, 153, 154f; well-being indicators in, 151–52, 152f, 166–68

Muller, C., 93

NAMBOARD. *See* National Agricultural Marketing Board

NAMPAADD. *See* National Master Plan for Arable Agriculture and Dairy Development

National Adaptation Programmes of Action (NAPAs): of Lesotho, 72, 77, 79, 82, 100–101, 103–4; of Malawi, 143; of Mozambique, 20, 147–48

National Agricultural Marketing Board (NAMBOARD), Swaziland, 227

National Disasters Management Institute, 165

National Maize Corporation (NMC), Swaziland, 225

National Master Plan for Arable Agriculture and Dairy Development (NAMPAADD), Botswana, 42, 43–44

National Meteorological Research Center (CNR). *See* CNRM-CM3

National parks. *See* Protected areas

Natural resources protection. *See* Protected areas

Nelson, G. C., 1, 30

NGOs. *See* Nongovernmental organizations

Nhemachena, C., 289

NMC. *See* National Maize Corporation

Nongovernmental organizations (NGOs): in Swaziland, 243, 244–45; in Zimbabwe, 315, 316–17t, 318

Nuts. *See* Groundnuts

Oilseeds. *See* Groundnuts

Optimistic scenario (high variant): assumptions of, 11–12, 12t, 33t; for Botswana, 54–56; incomes and population growth in, 11–12, 12t, 32–35, 33t, 34t; for Lesotho, 88–89, 98; for Malawi, 128–29, 134–36; for Mozambique, 159–60, 166, 168–69; for South Africa, 194–96, 204; for Swaziland, 228–29, 239; for Zambia, 270–72, 275, 278; for Zimbabwe, 302–4, 310–12

Owusu-Ampomah, K., 88

Parks and reserves. *See* Protected areas

Partial equilibrium model. *See* International Model for Policy Analysis of Agricultural Commodities and Trade

Pessimistic scenario (low variant): assumptions of, 11–12, 12t, 33t; for Botswana, 54–56; incomes and population growth in, 11–12, 12t, 32–35, 33t, 34t; for Lesotho, 88–89, 98; for Malawi, 128–29, 134–36; for Mozambique, 159–60, 166, 168–69; for South Africa, 194–96, 204–5; for Swaziland, 228–29, 239; for Zambia, 270–72, 275, 278; for Zimbabwe, 302–4, 310–12

Policy recommendations: for Botswana, 68–69; for Lesotho, 19, 105–6; for Malawi, 143–44; for Mozambique, 20, 171–72; for South Africa, 20, 208–9; for southern Africa, 20–21; for Swaziland, 19, 248–49; for Zambia, 19, 284–85; for Zimbabwe, 318–19, 320

Population distribution: in Botswana, 45–46; in Lesotho, 74–75, 76f; in Malawi, 115, 116f; in Mozambique, 149, 150f; in South Africa, 176, 177f; in Swaziland, 213, 215f; in Zambia, 257–58, 257f; in Zimbabwe, 290, 291f. *See also* Urban areas

Population growth. *See* Population trends

Population projections: for Botswana, 54, 55f; for Lesotho, 88, 88f; for Malawi, 128, 128f; for Mozambique, 159, 161f; in scenarios for future, 11, 11t, 32–33, 33t, 34t; for South Africa, 194, 195f; for southern Africa, 11, 11t; for Swaziland, 228–29, 229f; for Zambia, 270, 271f; for Zimbabwe, 302, 303f

Population trends: in Botswana, 45–46, 45f, 46t; in Lesotho, 73–75, 74t, 75f; in Malawi, 113–15, 114f, 114t;

Population trends (*continued*)
in Mozambique, 147, 148–49, 149f, 149t; in South Africa, 175–76, 176f, 176t; in southern Africa, 6–8, 8t; in Swaziland, 213, 214f, 214t; in Zambia, 255–57, 256f, 256t; in Zimbabwe, 6, 290–92, 291f, 292t. *See also* Migration

Potatoes, in Lesotho, 84

Poverty: in Botswana, 46–47, 47f, 47t; climate change impacts on, 35n3; in Lesotho, 79–80; in Malawi, 117, 118f, 142; in Mozambique, 152, 153f; in South Africa, 178, 180f; in Swaziland, 218; in Zambia, 261, 261f. *See also* Incomes per capita

Precipitation: in Botswana, 44, 56, 57f, 59, 67; changes in southern Africa, 27; in Lesotho, 71, 73, 86, 89–91, 90f, 97; in Malawi, 112, 129–31, 130f; in Mozambique, 162, 163f; in scenarios for future, 12, 13–14, 13f, 26, 28t; in South Africa, 185, 196, 198, 208; in southern Africa, 3, 3f; in Swaziland, 220, 230–32, 231f; in Zambia, 272, 273f; in Zimbabwe, 289, 304, 305f, 319. *See also* Climate model projections; Droughts

Prices. *See* Commodity prices; Food prices

Productivity. *See* Yields

Protected areas: in Botswana, 48, 50f; in Lesotho, 80, 82f; in Malawi, 120, 121f; in Mozambique, 153, 155f; in South Africa, 180–82, 182f; in Zambia, 263, 264f; in Zimbabwe, 295, 297f

Pulses. *See* Legumes

Rainfall. *See* Droughts; Precipitation

Rao, K. P. C., 309

Refugees, in Malawi, 114–15

Reserves. *See* Protected areas

Rice: in Malawi, 123; in Mozambique, 171; in southern Africa, 5t, 7f; yields of, 7f. *See also* Cereal production

Richardson, C. J., 292

Roads: in Lesotho, 80; in Mozambique, 155; in Swaziland, 223. *See also* Transportation; Travel times

Root crops: in Lesotho, 84; in Malawi, 123. *See also* Cassava

SACU. *See* South African Customs Union

SADC. *See* Southern Africa Development Community

SCCP. *See* Swaziland Climate Change Program

Scenarios: A2, 27, 27n1, 196; B1, 27, 27n1; GDP growth rates in, 11–12, 12t, 32, 33–35, 33t; interpreting results of, 34–35; population growth in, 11, 11t, 32–33, 33t; precipitation changes in, 12, 13–14, 13f; temperature changes in, 12–13, 14, 14f. *See also* A1B scenario; Baseline scenario; Climate model projections; Optimistic scenario; Pessimistic scenario

Scheiter, Simon, 208

Sechaba Consultants, 79

Smith, L., 35

Sorghum: in Botswana, 52, 53t, 59, 61f, 62, 64f, 67; in Lesotho, 84, 86, 93, 95f, 100, 102f; in southern Africa, 5t, 6, 7f; in Swaziland, 227; yields of, 7f, 15, 18t; in Zimbabwe, 297, 301f, 308f, 309, 312f. *See also* Cereal production

South Africa: agriculture in, 5, 20, 175–76, 184–94, 186t, 187f, 198, 201–3, 207–9; climate change policies of, 185; climate change vulnerability of, 178, 184–85, 204–5, 207–8; climate of, 175, 184–85; climate scenarios in, 196–98, 201, 204–5; education in, 178, 178t; employment in, 178t; exports of, 52, 184, 188, 192, 218; food consumption in, 187, 187f; food imports of, 184; incomes in, 176–78, 177f, 194–96, 195f; land use in, 180, 181f; migrant labor in, 74–75,

79; policy recommendations for, 20, 208–9; population distribution in, 176, 177f; population projections for, 194, 195f; population trends in, 175–76, 176f, 176t; poverty in, 178, 180f; protected areas in, 180–82, 182f; scenarios for future, 194–96, 204–5; tourism in, 180; travel times in, 182, 183f; urban areas in, 175, 176t; vegetation in, 180, 181f; water imports of, 75, 82–83; water resources of, 184, 193, 208, 216; well-being indicators in, 178, 179f, 204–5

South African Customs Union (SACU), 217

Southern Africa: agriculture in, 3–6, 15; climate change programs of, 315; climate of, 3; countries comprising, 2; incomes in, 8–9, 9t; land use in, 3–5, 4f; population of, 6–8, 8t; population projections for, 11, 11t. *See also* Agroecological zones; *individual countries*

Southern Africa Development Community (SADC), 5, 314, 315

Spatial Production Allocation Model (SPAM), 27–29, 32

Special Report on Emissions Scenarios (SRES), 28t, 196. *See also* A1B scenario

SSA. *See* Swaziland Sugar Association

Staple crops. *See* Foodcrops

Sugarcane: in South Africa, 192–93, 193f, 202–3, 205f; in Swaziland, 223–25, 227, 228, 228f, 235–36, 237f, 248

SWADE. *See* Swaziland Water and Agricultural Development Enterprise

Swaziland: agriculture in, 213, 216, 217, 223–28, 226t, 232, 235–39; climate change awareness in, 241–42; climate change policies of, 215, 242–47, 246t; climate change vulnerability of, 217–20, 239; climate of, 220; climate scenarios in, 230–32, 239; disaster response in, 242–43; education in, 218–19, 219t; employment in, 216–17, 219, 219t; exports of, 235, 236, 238; food imports of, 217–18, 227; food security in, 213–14, 218; geography of, 213; incomes in, 216, 217, 217f, 229, 230f; land tenure systems of, 220; land use in, 220, 221f, 222t, 223; policy recommendations for, 19, 248–49; population distribution in, 213, 215f; population projections for, 228–29, 229f; population trends in, 213, 214f, 214t; poverty in, 218; roads in, 223; scenarios for future, 228–32, 239; textile industry of, 214–15, 238; travel times in, 223, 224f; urban areas in, 213, 214t, 223; vegetation in, 215, 220, 221f, 223; water resources of, 214–15, 216; well-being indicators in, 218–20, 219f, 239

Swaziland Climate Change Programme (SCCP), 215, 244, 248

Swaziland Cotton Board, 225

Swaziland Sugar Association (SSA), 225, 235

Swaziland Water and Agricultural Development Enterprise (SWADE), 225, 235, 244, 247

Temperatures: in Botswana, 44–45, 56, 58f, 59, 67–68; in Lesotho, 71, 73, 91, 92f; in Malawi, 112, 131, 132f; in Mozambique, 162, 164f; in scenarios for future, 12–13, 14, 14f, 26, 28t; in South Africa, 196–98, 197f, 208; in southern Africa, 27; in Swaziland, 232, 233f; in Zambia, 272, 274f; in Zimbabwe, 304, 306f. *See also* Climate model projections

Thornton, P. K., 26

Tobacco, 119, 157. *See also* Cash crops

Tourism: in South Africa, 180; in Zambia, 263. *See also* Protected areas

Transportation: of agricultural products, 80; in Lesotho, 80–82; in southern Africa, 9–10. *See also* Roads; Travel times

Travel times: in Botswana, 48, 51f; in Lesotho, 80–82, 83f; in Malawi, 121, 122f; maps of, 37; in Mozambique, 155, 156f; in South Africa, 182, 183f; in southern Africa, 10, 10f; in Swaziland, 223, 224f; in Zambia, 263–66, 265f; in Zimbabwe, 295, 298f

Under-five mortality. *See* Child mortality
UNDP. *See* United Nations Development Programme
Unemployment. *See* Employment
United Nations Department of Economic and Social Affairs–Population Division (UNPOP), 11, 128, 159, 194, 302
United Nations Development Programme (UNDP), 143, 244, 315
United Nations Educational, Scientific, and Cultural Organization, 120
United Nations Framework Convention on Climate Change (CCNUCC), 100, 244, 312
United Nations Population Fund, 88
University of Tokyo Center for Climate System Research. *See* Model for Interdisciplinary Research on Climate
UNPOP. *See* United Nations Department of Economic and Social Affairs–Population Division
Urban areas: in Botswana, 45–46, 46t, 48, 51f; in Lesotho, 74–75, 80–82; in Malawi, 114, 114t, 115, 129; migration to, 8, 46, 74–75, 256, 290–92; in Mozambique, 148–49, 149t; in South Africa, 175, 176t; in southern Africa, 6–8, 8t; in Swaziland, 213, 214t, 223; in Zambia, 255–57, 256t, 263–64; in Zimbabwe, 290–92, 292t. *See also* Travel times

Vision 2016, Botswana, 41–42, 46, 67

Water resources: of Lesotho, 75, 82–84, 97; of South Africa, 75, 82–83, 184, 193, 208, 216; of Swaziland, 214–15, 216. *See also* Irrigation; Precipitation
Well-being indicators: in Botswana, 62–63, 65f, 66f; determinants of, 35; in Lesotho, 78–80, 78f, 98–100; in Malawi, 115–17, 118f, 134–36; in Mozambique, 151–52, 152f, 166–68; in South Africa, 178, 179f, 204–5; in southern Africa, 9–10, 10t; in Swaziland, 218–20, 219f, 239; in Zambia, 259–61, 260f, 275, 278; in Zimbabwe, 293–95, 294f. *See also* Calorie availability; Child malnutrition; Child mortality; Life expectancies
Wheat: in Lesotho, 84, 93, 96f, 103f; Mozambican imports of, 157; in South Africa, 6, 189–91, 191f, 192f, 200f, 201–2, 203f, 204f; in southern Africa, 5t, 6, 7f; yields of, 7f. *See also* Cereal production
Women: education of, 35; life expectancies of, 35
World Bank, 178, 194
World Food Program, 86
World prices. *See* Commodity prices

Yield changes under climate change: in Botswana, 59, 60f, 61f, 68; in Lesotho, 91–93, 94f, 95f, 96f, 100; of maize, 15, 16t; in Malawi, 131, 133f, 136–39; of millet, 15, 17t; in Mozambique, 162, 165, 165f; of sorghum, 15, 18t; in South Africa, 198, 199f, 200f, 201; in Swaziland, 232, 234f, 235; in Zambia, 275, 276f; in Zimbabwe, 304, 307f, 308f. *See also* Decision Support Software for Agrotechnology Transfer
Yields: of cereals, 7f; in Lesotho, 86; in Malawi, 125, 125f, 126f, 127f; in Mozambique, 157–59, 159f, 160f; in South Africa, 188f, 189, 191, 191f, 193, 193f; in southern Africa, 6, 7f; in Swaziland, 227–28, 227f, 228f; in Zambia, 267–68, 268f, 269f, 270f; in Zimbabwe, 297, 300f, 301f, 302f

Zambia: agriculture in, 5, 263, 266–68, 266t, 275, 278–81, 282–83, 284–85; climate change vulnerability of, 255, 259–61, 268, 282–83; climate scenarios in, 272, 275, 283; education in, 259, 259t; employment in, 259–60, 259t; exports of, 281; food imports of, 280; incomes in, 258–59, 258f, 270–72, 271f; land use in, 261–63, 262f; policy recommendations for, 19, 284–85; population distribution in, 257–58, 257f; population projections for, 270, 271f; population trends in, 255–57, 256f, 256t; poverty in, 261, 261f; protected areas in, 263, 264f; scenarios for future, 270–72, 275, 278; tourism in, 263; travel times in, 263–66, 265f; urban areas in, 255–57, 256t, 263–64; vegetation in, 261–63, 262f; well-being indicators in, 259–61, 260f, 275, 278

Zimbabwe: agriculture in, 5, 289, 292–93, 295–97, 299t, 304, 309, 318–21; civil society initiatives in, 315, 316–17t, 318; climate change policies of, 312–15, 318; climate change vulnerability of, 19–20, 289–90, 293–95, 310–12, 318–21; climate of, 289; climate scenarios in, 304, 309; education in, 293–94, 294t; employment in, 294t; exports of, 309, 320; geography of, 289; incomes in, 292, 293f, 302, 303f; land use in, 295, 296f; migration from, 6; policy recommendations for, 318–19, 320; population distribution in, 290, 291f; population projections for, 302, 303f; population trends in, 6, 290–92, 291f, 292t; protected areas in, 295, 297f; scenarios for future, 302–4; travel times in, 295, 298f; urban areas in, 290–92, 292t; vegetation in, 295, 296f; well-being indicators in, 293–95, 294f